MATHEMATICAL LOGIC
IN ASIA

Proceedings of the 9th Asian Logic Conference

MATHEMATICAL LOGIC
IN ASIA

Novosibirsk, Russia   16 – 19 August 2005

Editors

S S Goncharov Russian Academy of Sciences, Russia

R Downey Victoria University of Wellington, New Zealand

H Ono Japan Advanced Institute of Science and Technology, Japan

World Scientific

NEW JERSEY • LONDON • SINGAPORE • BEIJING • SHANGHAI • HONG KONG • TAIPEI • CHENNAI

Published by

World Scientific Publishing Co. Pte. Ltd.

5 Toh Tuck Link, Singapore 596224

USA office: 27 Warren Street, Suite 401-402, Hackensack, NJ 07601

UK office: 57 Shelton Street, Covent Garden, London WC2H 9HE

British Library Cataloguing-in-Publication Data
A catalogue record for this book is available from the British Library.

MATHEMATICAL LOGIC IN ASIA

ISBN 981-270-045-5

Printed in Singapore by World Scientific Printers (S) Pte Ltd

PREFACE

The Asian Logic Conference has occurred every three years since its inception in Singapore in 1981. It rotates among countries in the Asia Pacific region with interests in the broad area of logic including theoretical computer science. It is now considered a major conference in this field and is regularly sponsored by the Association for Symbolic Logic. This volume contains papers, many of them surveys by leading experts, of the 9th meeting in Novosibirsk, Russia.

We were very pleased to find that *World Scientific* were enthusiastic to support this venture. Authors were invited to submit articles to the present volume, based around talks given at either meeting. The editors were very concerned to make sure that the planned volume was of high quality. We think the resulting volume is fairly representative of the thriving logic groups in the Asia-Pacific region, and also fairly representative of the meetings themselves.

The Ninth Asian Logic Conference was organised by Sobolev Institute of Mathematics of the Siberian Branch of the Russian Academy of Sciences and Novosibirsk State University under the sponsorship of Russian Foundation for Basic Research, Association for Symbolic Logic, Department of Mechanics and Mathematics of Novosibirsk State University, Siberian Foundation for Algebra and Logic, Novosibirsk Center of Information Technologies UniPro Co., Ltd., LLC Alekta, and Transtext Co. Ltd.

The conference took place in Novosibirsk, Akademgorodok, Russia, from August 16 to August 19, 2005.

The programme consisted of plenary lectures delivered by invited speakers and contributions in four sections.

Plenary speakers were Pavel Alaev (Russia), Lev Beklemishev (Russia, Netherlands), Su Gao (USA), Yurii Ershov (Russia), Sanjay Jain (Singapore), Vladimir Kanovei (Russia), Bakhadyr Khoussainov (New Zealand), Andrei Mantsivoda (Russia), Joe Miller (USA), Hiroakira Ono (Japan), Vladimir Rybakov (Russia, Great Britain), Masahiko Sato (Japan), Moshe Vardi (USA), Andrei Voronkov (Great Britain), Xishun Zhao (China). The total number of plenary lectures was 15. Contributed lectures on recursion theory, set theory, proof theory, model theory and universal algebra, non-classical logic, and logic in computer science were presented in the following sections: Computability theory, Model theory and Set theory, Non-classical logics, Proof theory, and universal algebra, and Applications of logic in computer science. The total number of contributed talks was 58.

The geography of the event included Russia, China, Japan, Singapore, USA, New Zealand, Great Britain, Korea, Canada, Germany, Greece, Kazakhstan. The number of participants was about 100 scientists.

We are grateful to Ekaterina Fokina for the great work with authors and referees while preparing the Proceedings and to Vladimir Vlasov for making the camera-ready manuscript.

Preparing of the Proceedings was supported by the grant of President of the Russian Federation for Leading Scientific Schools 4413.2006.1.

Sincerely yours, the editors:
Rod Downey, Sergey Goncharov, and Hiroakira Ono.

CONTENTS

ANOTHER CHARACTERIZATION OF THE DEDUCTION-DETACHMENT THEOREM

SERGEI V. BABYONYSHEV

Krasnoyarsk State University
Krasnoyarsk, Russia
E-mail: bsv70@yahoo.com

In Abstract Algebraic Logic, a Hilbert-style deductive system is identified with the set of its theories. This set of theories must be algebraic and must be closed under arbitrary intersections and inverse substitutions. Similarly, a Gentzen-style deductive system can be defined by providing a set of theories with similar properties, but now each theory must be a set of sequents, not just formulas. There are various kinds of Gentzen-style structures that naturally arise in connection with Hilbert systems, but in generally they fall short of being Gentzen systems. One of such structures is a family of axiomatic closure relations. Each of axiomatic closure relations is defined as a set of consequences that can be derived in the Hilbert system by modulo of some its theory, taken as the set of additional axioms. The main result of this work is the proof that a Hilbert system $\mathcal{S}$ admits the Deduction-Detachment Theorem if and only if the set of all axiomatic closure relations for $\mathcal{S}$ forms a Gentzen system.

1. Introduction

In Algebraic Logic, an abstract Hilbert-style deductive system $\mathcal{S}$ is identified with a family of sets, called $\mathcal{S}$-theories, of formulas of a given propositional language, such that this family, usually denoted by $\operatorname{Th}\mathcal{S}$, is

1) closed under arbitrary intersections, i.e., $\operatorname{Th}\mathcal{S}$ is a *closure system*,

2) closed under unions of upward-directed families, i.e., $\operatorname{Th}\mathcal{S}$ is *algebraic*,

3) closed under inverse substitutions, i.e., a preimage of any $\mathcal{S}$-theory under an arbitrary substitution is an $\mathcal{S}$-theory again.

We call $\mathcal{S}$ "abstract" because its definition does not refer to any particular axiomatization. It is easy to see that the closure operator

$$(.)^{\mathcal{S}} : X \mapsto \bigcap\{T \in \operatorname{Th}\mathcal{S} \mid X \subset T\},$$

associated with such closure system $\operatorname{Th}\mathcal{S}$, defines a finitary, substitutional consequence relation as follows

$$X \vdash_{\mathcal{S}} \alpha \iff \alpha \in X^{\mathcal{S}}.$$

It was suggested by researchers of Barcelona group to treat the Gentzen case similarly, and identify an abstract Gentzen-style deductive system with the set of its theories with two distinctive features

    1) a theory is a set of *sequents*, i.e., sequences of formulas,

    2) a substitution acts on sequents componentwise.

Out of numerous and intricate connections between Hilbert- and Gentzen-style deductive systems we will consider in this paper just one: *the deduction-detachment theorem*, discovered independently by Tarski and Herbrand. We define it in a slightly more general form.

A Hilbert-style deductive system admits the multiterm deduction-detachment theorem if there is a finite set of formulas $\Delta = \{\delta_i(x,y)\}_{i\in I}$ *such that for all formulas* α, β *and every set of formulas* Γ

$$\Gamma, \alpha \vdash_S \beta \iff (\forall \delta \in \Delta)\ \Gamma \vdash_S \delta(\alpha,\beta).$$

Even though the deduction-detachment theorem can and usually is formulated by the Gentzen rules, it was not known what abstract Gentzen-style deductive system corresponds to this axiomatization. It turns out that the key to this correspondence is axiomatic closure relations:

Let S *be an abstract Hilbert-style deductive system and* $T \in \mathrm{Th}\,S$. *Then*

$$\{\langle \alpha_1, \ldots, \alpha_n, \beta\rangle \mid T, \alpha_1, \ldots, \alpha_n \vdash_S \beta\}$$

is called an axiomatic closure relation *for* S.

In other words, an axiomatic closure relation list all consequences that are possible in S if we add all formulas from some S-theory as axioms (not axiom schemes). In this paper we will show that

An abstract Hilbert-style deductive system S *admits the multiterm deduction-detachment theorem if and only if the set of axiomatic closure relations for* S *forms an abstract Gentzen-style deductive system, i.e., it is*

    1) closed under arbitrary intersections,

    2) closed under unions of upward-directed families,

    3) closed under inverse substitutions.

In the following, abstract Hilbert-style deductive systems will be referred to as simply Hilbert systems, and similarly for abstract Gentzen-style deductive systems.

2. Definitions and Preliminaries

Sometimes the contraction "iff" or the symbol " $\iff$ " will be used for the phrase "if and only if", "$\square$" for "the end of proof" or "the end of definition",

"$\stackrel{def}{=}$" for "equals by definition", "$\forall$" stands for "for all", " $\implies$ " for "implies".

Suppose A is a set. Then $\mathcal{P}(A) := \{X \mid X \subseteq A\}$ is the *power-set* of A. We write $X \subseteq_\omega A$ if X is a finite subset of A, furthermore $\mathcal{P}_\omega(A) := \{X \mid X \subseteq_\omega A\}$. For a family of sets $\mathcal{C} \subseteq \mathcal{P}(A)$, we define $\bigcup \mathcal{C} := \bigcup_{X \in \mathcal{C}} X$, $\bigcap \mathcal{C} := \bigcap_{X \in \mathcal{C}} X$. The *n-th cartesian power* of a non-empty set A is the set $A^n := \prod_{i \in n} A$ of all vectors of length n with elements from A. A^+ denotes $\bigcup_{n=1}^{\infty} A^n$, the set of all non-empty finite sequences of elements of A. An arbitrary element of A^+ we write as $\bar{a}$. If $\bar{a} = \langle a_1, \ldots, a_n \rangle$, we also write $\{\bar{a}\}$ for $\{a_1, \ldots, a_n\}$. A function $f : A^n \to A$ is called an *n*-ary *operation on A*. Instead of $f\langle \bar{a} \rangle$ or $f(\langle \bar{a} \rangle)$ we will often write $f(\bar{a})$. A *unary operation* $f : A \to A$ is also called a *mapping* on A.

A binary relation $R \subseteq A \times A$ is *reflexive* if for all $a \in A$, aRa; *symmetric* if for all $a, b \in A$, aRb implies bRa; *transitive* if for all $a, b, c \in A$, from aRb and bRc it follows that aRc; *antisymmetric* if for all $a, b \in A$, aRb and bRa implies that $a = b$. We call $R \subseteq A \times A$ a *partial order on A* if R is reflexive, transitive and antisymmetric.

If $\leqslant$ is a partial order on A and $X \subseteq A$, an element $a \in A$ such that for all $x \in X$, $x \leqslant a$ is called an *upper boundary of X*; dually, an element $a \in A$ such that for all $x \in X$, $x \geqslant a$ is called a *lower boundary of X*; inf X is the largest (if it exists) element of A among the lower boundaries of X; similarly, sup X is the smallest (if it exists) element of A among the upper boundaries of X. If inf (sup) exists for any two-element subset of A, A is called a *lower (upper) semi-lattice*. In that case, $\inf\{a, b\}$ is usually denoted by $a \wedge b$, and $\sup\{a, b\}$ as $a \vee b$, and interpreted as binary operations on A. If both $\wedge$ and $\vee$ defined for any pair of elements of A, A is called a *lattice*. If inf and sup exists for any non-empty subset of A, A is called a *complete lattice*.

For a mapping $h : A \to A$ the operator-style notation ha will be routinely used instead of the function-style notation $h(a)$. Also any mapping h defined on A can be uniquely extended to a mapping on A^+ by the following definition:

$$h\langle a_1, \ldots, a_n \rangle = \langle ha_1, \ldots, ha_n \rangle, \quad \text{for all } \langle a_1, \ldots, a_n \rangle \in A^+.$$

The latter defines a *complex* (defined on sets of elements) mapping on $\mathcal{P}(A^+)$ as follows,

$$hX = \{h\langle \bar{a} \rangle \mid \langle \bar{a} \rangle \in X\}, \quad \text{for all } X \subseteq A^+.$$

Note that the same symbol h will be used routinely for all these mappings.

A *propositional language type* is any non-empty set $\mathcal{L}$. The elements of $\mathcal{L}$ are called *functional symbols* in an algebraic context or *logical connectives* in a logical context. With $\mathcal{L}$ is associated an *arity* function $\rho : \mathcal{L} \to \omega$ such that ρf is the *arity* or *rank* of the functional symbol $f \in \mathcal{L}$. For each $n \in \omega$: $\mathcal{L}_n := \{f \in \mathcal{L} \mid \rho f = n\}$. An *algebra* $\mathbf{A}$ of type $\mathcal{L}$ is a pair $\langle A, \mathcal{L}^{\mathbf{A}} \rangle$, where A is a non-empty set called *universe* of $\mathbf{A}$ and $\mathcal{L}^{\mathbf{A}} = \{f^{\mathbf{A}} \mid f \in \mathcal{L}\}$ is a list of operations over the set A such that for every $f \in \mathcal{L}_n$, $f^{\mathbf{A}} : A^n \to A$. Members of $\mathcal{L}^{\mathbf{A}}$ are called *basic operations* of $\mathbf{A}$. If $\mathbf{A}$, $\mathbf{B}$ are algebras of the same type, then a mapping $h : A \to B$ is called a homomorphism of $\mathbf{A}$ into $\mathbf{B}$ (written $h : \mathbf{A} \to \mathbf{B}$), if for every $f \in \mathcal{L}_n$ and every $\langle \bar{a} \rangle \in A^n$, $h f^{\mathbf{A}} \langle \bar{a} \rangle = f^{\mathbf{B}} h \langle \bar{a} \rangle$. A homomorphism $h : \mathbf{A} \to \mathbf{A}$ is called an *endomorphism of* $\mathbf{A}$; if h is also surjective and injective, then h is an *automorphism of* $\mathbf{A}$.

Let $X = \{x_i\}_{i \in I}$ be a non-empty set. The set $\mathrm{Fm}_{\mathcal{L}} X$ of *formulas (or terms) of type $\mathcal{L}$ over the set of generators X* is defined recursively as follows

1. $X \subseteq \mathrm{Fm}_{\mathcal{L}} X$,

2. if $f \in \mathcal{L}_n$ and $\alpha_1, \ldots, \alpha_n \in \mathrm{Fm}_{\mathcal{L}} X$, then $\langle f, \alpha_1, \ldots, \alpha_n \rangle \in \mathrm{Fm}_{\mathcal{L}} X$.
Traditionally a formula $\langle f, \alpha_1, \ldots, \alpha_n \rangle$ is written as $f(\alpha_1, \ldots, \alpha_n)$. Formulas will be denoted usually by small Greek letters. We write $\alpha(p_1, \ldots, p_n)$ or $\mathrm{Var}(\alpha) \subseteq \{p_1, \ldots, p_n\}$, if $\alpha \in \mathrm{Fm}_{\mathcal{L}}\{p_1, \ldots, p_n\}$. A vector $\langle \alpha_1, \ldots, \alpha_k \rangle$ of $\mathrm{Fm}_{\mathcal{L}}^{+}$ is called a *sequent* and will be written usually in the form $\alpha_1, \ldots, \alpha_{k-1} \triangleright \alpha_k$.

We can induce the structure of an algebra on $\mathrm{Fm}_{\mathcal{L}} X$ by associating with each $f \in \mathcal{L}_n$ a n-ary operation $f^{\mathbf{Fm}_{\mathcal{L}} X}$ on the set $\mathrm{Fm}_{\mathcal{L}} X$ defined by $f^{\mathbf{Fm}_{\mathcal{L}} X} \langle \bar{\alpha} \rangle = f(\bar{\alpha})$. The superscript in this case is usually omitted. This algebra $\mathbf{Fm}_{\mathcal{L}} X$ is called the *algebra of formulas (terms) of type $\mathcal{L}$ over the set of variables X*. We fix a countable set $\mathrm{Var} = \{x_0, x_1, x_2, \ldots\}$ of *propositional variables*. Then $\mathbf{Fm}_{\mathcal{L}} \mathrm{Var}$ is called *the formula algebra over the language of type $\mathcal{L}$* and will be denoted $\mathbf{Fm}_{\mathcal{L}}$. The universe of $\mathbf{Fm}_{\mathcal{L}}$ is denoted as $\mathrm{Fm}_{\mathcal{L}}$.

An algebra $\mathbf{Fm}_{\mathcal{L}} X$ is an *absolutely free algebra over the set X* in the class of all algebras of type $\mathcal{L}$. This means that, for every algebra $\mathbf{A}$ of type $\mathcal{L}$, an arbitrary mapping $h : X \to A$ can be uniquely extended to a homomorphism $h : \mathbf{Fm}_{\mathcal{L}} X \to A$. In particular any homomorphism $h : \mathbf{Fm}_{\mathcal{L}} X \to \mathbf{A}$ is determined by the mapping $h : X \to A$. A homomorphism $h : \mathbf{Fm}_{\mathcal{L}} \to \mathbf{A}$ is called an *evaluation*; a homomorphism $h : \mathbf{Fm}_{\mathcal{L}} \to \mathbf{Fm}_{\mathcal{L}}$ is called a *substitution*.

A family $\mathcal{C} \subseteq \mathcal{P}(A)$ is *upward-directed* if for every pair $X, Y \in \mathcal{C}$ there is $Z \in \mathcal{C}$ such that $X, Y \subseteq Z$. A subset $\mathcal{C} \subseteq \mathcal{P}(A)$ is *algebraic* if $\bigcup \mathcal{D} \in \mathcal{C}$ for every upward-directed subfamily $\mathcal{D} \subseteq \mathcal{C}$. A family $\mathcal{C} \subseteq \mathcal{P}(A)$ is called a *closure system on A* if $A \in \mathcal{C}$ and $\bigcap \mathcal{D} \in \mathcal{C}$ for every non-empty subfamily $\mathcal{D} \subseteq \mathcal{C}$. A closure system $\mathcal{C}$ on $\mathbf{Fm}_{\mathcal{L}}$ is *(surjectively) invariant* if for any (surjective) substitution σ and any $T \in \mathcal{C}$, $\sigma^{-1}T := \{\alpha \mid \sigma\alpha \in T\} \in \mathcal{C}$, or, in other words, if $\sigma^{-1}\mathcal{C} \subseteq \mathcal{C}$ for all (surjective) $\sigma : \mathbf{Fm}_{\mathcal{L}} \to \mathbf{Fm}_{\mathcal{L}}$. Similarly, a closure system $\mathcal{C}$ on $\mathbf{Fm}_{\mathcal{L}}^{+}$ is *(surjectively) invariant* if for any (surjective) substitution σ and any $T \in \mathcal{C}$, $\sigma^{-1}T = \{\bar{\alpha} \triangleright \alpha \mid \sigma(\bar{\alpha} \triangleright \alpha) \in T\} \in \mathcal{C}$.

A *closure operator on A* is a mapping $\mathbf{C} : \mathcal{P}(A) \to \mathcal{P}(A)$ such that for any $X, Y \subseteq A$, $X \subseteq \mathbf{C}(X) = \mathbf{C}(\mathbf{C}(X)) \subseteq \mathbf{C}(X \cup Y)$. A set $X \in \mathcal{P}(A)$ such that $\mathbf{C}(X) = X$ is called a *closed set of* $\mathbf{C}$. A closure operator $\mathbf{C}$ is *finitary* if for any $X \subseteq A$, $\mathbf{C}(X) = \bigcup\{\mathbf{C}(Y) \mid Y \subseteq_{\omega} X\}$. The following relations between closure systems and closure operators are well known: 1) if $\mathbf{C}$ is a closure operator on A, then the family of its closed sets is a closure system on A; 2) if $\mathcal{C}$ ia a closure system on A, then the mapping $\mathbf{C}_{\mathcal{C}} : \mathcal{P}(A) \to \mathcal{P}(A)$ defined for each $X \subseteq A$ as $\mathbf{C}_{\mathcal{C}}X := \bigcap\{Y \in \mathcal{C} \mid X \subseteq Y\}$ is a closure operator on A; 3) $\mathcal{C}$ is algebraic iff $\mathbf{C}_{\mathcal{C}}$ is finitary. We use interchangeably the exponential and prefix notations for closure operators, thus $X^{\mathcal{C}} = \mathbf{C}_{\mathcal{C}}X$.

Every closure system $\mathcal{C}$, as a family of subsets ordered under set-inclusion, is a complete lattice. The infimum of a family $\{X_i\}_{i \in I} \subseteq \mathcal{C}$ is its intersection $\bigcap_{i \in I} X_i$, and its supremum is $\bigvee_{i \in I}^{\mathcal{C}} X_i := \mathbf{C}_{\mathcal{C}}(\bigcup_{i \in I} X_i)$; its largest element is A, and its smallest element is $\mathbf{C}_{\mathcal{C}}(\emptyset) = \bigcap \mathcal{C}$.

A *Hilbert system* is a pair $\mathcal{S} = \langle \mathbf{Fm}_{\mathcal{L}}, \mathrm{Th}\,\mathcal{S}\rangle$ such that $\mathrm{Th}\,\mathcal{S} \subseteq \mathcal{P}(\mathrm{Fm}_{\mathcal{L}})$ is an algebraic invariant closure system on $\mathbf{Fm}_{\mathcal{L}}$. A *Gentzen system* is a pair $\mathcal{G} = \langle \mathbf{Fm}_{\mathcal{L}}, \mathrm{Th}\,\mathcal{G}\rangle$ such that $\mathrm{Th}\,\mathcal{G} \subseteq \mathcal{P}(\mathrm{Fm}_{\mathcal{L}}^{+})$ is an algebraic invariant closure system on $\mathbf{Fm}_{\mathcal{L}}^{+}$. For a Hilbert system $\mathcal{S}$ and all $T \in \mathrm{Th}\,\mathcal{S}$, $[T)_{\mathrm{Th}\,\mathcal{S}} := \{U \in \mathrm{Th}\,\mathcal{S} \mid T \subseteq U\}$ denote a principal filter of the lattice $\mathrm{Th}\,\mathcal{S}$ generated by T. If $\mathcal{R}$ is a Hilbert or Gentzen system, we denote $\mathrm{Thm}\,\mathcal{R} := \bigcap \mathrm{Th}\,\mathcal{R}$.

We take a Cantor-style approach towards Gentzen rules: we view a rule not as a "rule"— description of an action, but as a list of all its applications.

A *Gentzen sequent* is a sequence $\bar{s} \triangleright s$ of sequents. A *Gentzen rule* $\bar{s} \vdash s$ is a set of all substitution instances of the Gentzen sequent $\bar{s} \triangleright s$, i.e.,

$$\bar{s} \vdash s := \{\sigma(\bar{s} \triangleright s) \mid \sigma : \mathbf{Fm}_{\mathcal{L}} \to \mathbf{Fm}_{\mathcal{L}}\}.$$

A Gentzen rule $s_1, \ldots, s_n \vdash s$ can also be written as $\dfrac{s_1, \ldots, s_n}{s}$.

Let x, y, z be variables. *Standard* rules (sometimes called *structural*) are rules of the form

$$\begin{array}{lll}
\text{(Ax)} & \vdash \Gamma, x, \Sigma \triangleright x & \text{Axioms} \\
\text{(Ex)} & \Gamma, x, y, \Sigma \triangleright z \vdash \Gamma, y, x, \Sigma \triangleright z & \text{Exchange} \\
\text{(W)} & \Gamma, \Sigma \triangleright y \vdash \Gamma, x, \Sigma \triangleright y & \text{Weakening}
\end{array}$$

$$\text{(Con) } \Gamma, x, x, \Sigma \triangleright y \vdash \Gamma, x, \Sigma \triangleright y \qquad\qquad \text{Contraction}$$
$$\text{(Cut) } \Gamma, x, \Sigma \triangleright y; \ \Theta \triangleright x \vdash \Gamma, \Theta, \Sigma \triangleright y \qquad \text{Cut}$$

where Γ, Σ, Θ range over the set of finite, possibly empty, sequences of variables of $\mathrm{Fm}_{\mathcal{L}}$.

Suppose $\mathcal{G} = \langle \mathbf{Fm}_{\mathcal{L}}, \mathrm{Th}\,\mathcal{G} \rangle$ is a Gentzen system. We say that a Gentzen rule $\bar{s} \vdash s$ *holds in* $\mathcal{G}$ (we write it as $\bar{s} \vdash_{\mathcal{G}} s$) if for every substitution σ and every $\mathcal{G}$-theory T

$$\sigma\{\bar{s}\} \subseteq T \implies \sigma s \in T.$$

3. Closure Relations

Definition 3.1. Let $\mathcal{C}$ be a closure system on $\mathrm{Fm}_{\mathcal{L}}$. Define

$$\mathbf{R}_{\mathcal{L}}\,\mathcal{C} = \{\bar{\alpha} \triangleright \alpha \in \mathrm{Fm}_{\mathcal{L}}^{+} \mid (\forall X \in \mathcal{C})\ \{\bar{\alpha}\} \subseteq X \implies \alpha \in X\}.$$

Definition 3.2. Let $\mathcal{S}$ be a Hilbert system. If $\mathcal{C} \subseteq \mathrm{Th}\,\mathcal{S}$ is an algebraic closure system on $\mathrm{Fm}_{\mathcal{L}}$, then $\mathbf{R}_{\mathcal{L}}\,\mathcal{C}$ is called *a general finite closure relation for* $\mathcal{S}$ or simply *a general closure relation for* $\mathcal{S}$. The set of all general closure relations for $\mathcal{S}$ will be denoted by $\mathbf{Gcr}\,\mathcal{S}$. $\qquad\square$

For every Hilbert system $\mathcal{S}$ of type $\mathcal{L}$ there is a distinguished general closure relation $\mathbf{R}_{\mathcal{L}}\mathrm{Th}\,\mathcal{S}$, which in its turn defines a Gentzen axiomatization for a Gentzen system:

$$\vdash \mathbf{R}_{\mathcal{L}}\mathrm{Th}\,\mathcal{S} := \bigcup\{\vdash \bar{\alpha} \triangleright \alpha \mid \bar{\alpha} \triangleright \alpha \in \mathbf{R}_{\mathcal{L}}\mathrm{Th}\,\mathcal{S}\}.$$

Proposition 3.3. [1, Theorem 2.2.10]
 For any Hilbert system $\mathcal{S}$ of type $\mathcal{L}$

(1) $\mathbf{R}_{\mathcal{L}}\mathrm{Th}\,\mathcal{S} = \{\bar{\alpha} \triangleright \alpha \mid \bar{\alpha} \vdash_{\mathcal{S}} \alpha\}$,
(2) $\mathbf{R}_{\mathcal{L}}\mathrm{Th}\,\mathcal{S}$ *is invariant,*
(3) $\mathbf{Gcr}\,\mathcal{S}$ *can be axiomatized by standard rules and* $\vdash \mathbf{R}_{\mathcal{L}}\mathrm{Th}\,\mathcal{S}$
(4) $\mathbf{Gcr}\,\mathcal{S}$ *forms a Gentzen system on* $\mathbf{Fm}_{\mathcal{L}}$,
(5) $\mathbf{R}_{\mathcal{L}}\mathrm{Th}\,\mathcal{S} = \mathrm{Thm}\,(\mathbf{Gcr}\,\mathcal{S})$.

Closure relations were introduced in [9] as a framework for studying metatheoretical properties of Hilbert systems. The fact that $\mathbf{Gcr}\,S$ form a Gentzen system was first observed also in [9]. The Gentzen system $\mathbf{Gcr}\,S$ formalizes a metalogic over the Hilbert system $\mathcal{S}$. This metalogic is quite weak and equivalent in expressive power to the strict universal Horn logic without equality [4]. Although $\mathbf{Gcr}\,S$ is almost trivial, since can be axiomatized by only taking all proper sequents of $\mathrm{Th}\,\mathcal{S}$ and standard Gentzen rules, by Proposition 3.3(3), it is proved to be useful as a framework for working with other kinds of closure relations like full or axiomatic [1].

We make a distinction between an element a and a vector $\langle a \rangle$ of length one with this element as its only component. This approach requires the following set of technical operators, that would allow us smooth transitions from formulas to sequents and back, and also between the theories of Hilbert and Gentzen systems. Define for every $X \subseteq \mathrm{Fm}_{\mathcal{L}}$ and every $\mathcal{A} \subseteq \mathrm{Fm}_{\mathcal{L}}^+$

$$\triangleright X := \{ \triangleright \alpha \mid \alpha \in X \},$$

$$\mathrm{Thm}\, \mathcal{A} := \{ \alpha \in \mathrm{Fm}_{\mathcal{L}} \mid \triangleright \alpha \in \mathcal{A} \},$$

$$\Theta\, \mathcal{A} := \{ \triangleright \alpha \in \mathrm{Fm}_{\mathcal{L}} \mid \triangleright \alpha \in \mathcal{A} \}.$$

Thus we obtain operators

$$(\triangleright) : \mathcal{P}(\mathrm{Fm}_{\mathcal{L}}) \to \mathcal{P}(\mathrm{Fm}_{\mathcal{L}}^1),$$

$$\mathrm{Thm} : \mathcal{P}(\mathrm{Fm}_{\mathcal{L}}^+) \to \mathcal{P}(\mathrm{Fm}_{\mathcal{L}}),$$

$$\Theta : \mathcal{P}(\mathrm{Fm}_{\mathcal{L}}^+) \to \mathcal{P}(\mathrm{Fm}_{\mathcal{L}}^1).$$

Mnemonically, the Greek letter Θ above stands for "Theorems".

Reminder. In the following proofs we rely heavily on, so called, "exponential" notation for closures of sets. Namely, if $\mathcal{C}$ is a closure system on some set X, then for all $Y \subseteq X$:

$$Y^{\mathcal{C}} = (Y)^{\mathcal{C}} := \bigcap\nolimits_{Y \subseteq F \in \mathcal{C}} F.$$

Definition 3.4. For a Hilbert system $\mathcal{S}$, define the set of *axiomatic closure relations of* $\mathcal{S}$ as follows:

$$\mathbf{Acr}\, \mathcal{S} := \{ (\triangleright T)^{\mathbf{Gcr}\, \mathcal{S}} \mid T \in \mathrm{Th}\, \mathcal{S} \}.$$

An element of $\mathbf{Acr}\, \mathcal{S}$ is called *an axiomatic closure relation for* $\mathcal{S}$. $\qquad\square$

Note, that, in the definition above, the set $\triangleright T$ contains sequents of the form $\triangleright \alpha$, $\alpha \in T$, where $T \subseteq \mathrm{Fm}_{\mathcal{L}}$, and we take the closure of $\triangleright T$ in the family of Gentzen theories, each of them is a set of sequents itself.

Proposition 3.5. *For every Hilbert system $\mathcal{S}$ of type $\mathcal{L}$*

(1) $\mathbf{Acr}\, \mathcal{S} \subseteq \mathbf{Gcr}\, \mathcal{S}$,
(2) $\mathcal{A} \in \mathbf{Acr}\, \mathcal{S} \implies \mathcal{A} = (\Theta\, \mathcal{A})^{\mathbf{Gcr}\, \mathcal{S}}$,
(3) $\mathbf{Acr}\, \mathcal{S} = \{ (\triangleright X)^{\mathbf{Gcr}\, \mathcal{S}} \mid X \subseteq \mathrm{Fm}_{\mathcal{L}} \}$,
(4) $\mathbf{Acr}\, \mathcal{S} = \{ \mathbf{R}_{\mathcal{L}}[T]_{\mathrm{Th}\, \mathcal{S}} \mid T \in \mathrm{Th}\, \mathcal{S} \}$.

8

(5) *For every $X \subseteq \mathrm{Fm}_{\mathcal{L}}$,*

$$\bar{\alpha} \triangleright \alpha \in (\triangleright X)^{\mathbf{Gcr}\,\mathcal{S}} \iff \alpha \in \{\bar{\alpha}\}^{\mathcal{S}} \vee X^{\mathcal{S}} \iff X, \bar{\alpha} \vdash_{\mathcal{S}} \alpha.$$

Proof. (1) By definition.

(2) If $\mathcal{A} \in \mathbf{Acr}\,\mathcal{S}$, then $\mathcal{A} = (\triangleright T)^{\mathbf{Gcr}\,\mathcal{S}}$ for some $T \in \mathrm{Th}\,\mathcal{S}$. Then

$$\triangleright T \subseteq \Theta\,\mathcal{A} \subseteq \mathcal{A} \implies \mathcal{A} = (\triangleright T)^{\mathbf{Gcr}\,\mathcal{S}} \subseteq (\Theta\,\mathcal{A})^{\mathbf{Gcr}\,\mathcal{S}} \subseteq \mathcal{A}^{\mathbf{Gcr}\,\mathcal{S}} = \mathcal{A}$$

$$\implies \mathcal{A} = (\Theta\,\mathcal{A})^{\mathbf{Gcr}\,\mathcal{S}}.$$

(3) If $\mathcal{A} \in \mathbf{Acr}\,\mathcal{S}$, then $\mathcal{A} = (\Theta\,\mathcal{A})^{\mathbf{Gcr}\,\mathcal{S}} = (\triangleright \mathrm{Thm}\,\mathcal{A})^{\mathbf{Gcr}\,\mathcal{S}}$. For the other direction, suppose $\mathcal{A} = (\triangleright X)^{\mathbf{Gcr}\,\mathcal{S}}$, for some $X \subseteq \mathrm{Fm}_{\mathcal{L}}$. Then $\mathcal{A} = (\Theta\,\mathcal{A})^{\mathbf{Gcr}\,\mathcal{S}} = (\triangleright \mathrm{Thm}\,\mathcal{A})^{\mathbf{Gcr}\,\mathcal{S}}$, because

$$(\supseteq)\ \Theta\,\mathcal{A} \subseteq \mathcal{A} \implies (\Theta\,\mathcal{A})^{\mathbf{Gcr}\,\mathcal{S}} \subseteq \mathcal{A}^{\mathbf{Gcr}\,\mathcal{S}} = \mathcal{A},$$

$$(\subseteq)\ \mathcal{A} = (\triangleright X)^{\mathbf{Gcr}\,\mathcal{S}} \implies \triangleright X \subseteq \mathcal{A} \implies \triangleright X \subseteq \Theta\,\mathcal{A}$$

$$\implies \mathcal{A} = (\triangleright X)^{\mathbf{Gcr}\,\mathcal{S}} \subseteq (\Theta\,\mathcal{A})^{\mathbf{Gcr}\,\mathcal{S}}.$$

(4) Suppose $\mathcal{A} \in \mathbf{Acr}\,\mathcal{S}$. Then, by (3), $\mathcal{A} = (\triangleright T)^{\mathbf{Gcr}\,\mathcal{S}}$, where $T = \mathrm{Thm}\,\mathcal{A} \in \mathrm{Th}\,\mathcal{S}$. Let $\mathcal{C} = [T)_{\mathrm{Th}\,\mathcal{S}}$. Being a general closure relation for $\mathcal{S}$, $\mathcal{A} = \mathbf{R}_{\mathcal{L}}\,\mathcal{D}$, for some algebraic closure system $\mathcal{D} \subseteq \mathrm{Th}\,\mathcal{S}$. Then $\mathcal{A} = \mathbf{R}_{\mathcal{L}}\,\mathcal{C}$, because

$$(\supseteq)\ T \overset{(3)}{=} \bigcap \mathcal{D} \implies \mathcal{D} \subseteq [T)_{\mathrm{Th}\,\mathcal{S}} = \mathcal{C} \implies \mathbf{R}_{\mathcal{L}}\,\mathcal{C} \subseteq \mathbf{R}_{\mathcal{L}}\,\mathcal{D} = \mathcal{A},$$

$$(\subseteq)\ \Theta\,\mathbf{R}_{\mathcal{L}}\,\mathcal{D} = \triangleright(\bigcap \mathcal{D}) = \triangleright T = \triangleright(\bigcap \mathcal{C}) = \Theta\,\mathbf{R}_{\mathcal{L}}\,\mathcal{C} \implies \triangleright T \subseteq \mathbf{R}_{\mathcal{L}}\,\mathcal{C}$$

$$\implies \mathcal{A} = (\triangleright T)^{\mathbf{Gcr}\,\mathcal{S}} \subseteq \mathbf{R}_{\mathcal{L}}\,\mathcal{C}.$$

(5) $\bar{\alpha} \triangleright \alpha \in (\triangleright T)^{\mathbf{Gcr}\,\mathcal{S}} \overset{(3)}{=} \mathbf{R}\,[T)_{\mathrm{Th}\,\mathcal{S}}$

$$\iff \alpha \in \{\bar{\alpha}\}^{[T)_{\mathrm{Th}\,\mathcal{S}}} = (T \cup \{\bar{\alpha}\})^{\mathcal{S}} = T \vee \{\bar{\alpha}\}^{\mathcal{S}} \iff T, \bar{\alpha} \vdash_{\mathcal{S}} \alpha. \qquad \square$$

Lemma 3.6. $\mathbf{Acr}\,\mathcal{S}$ *is a closure system iff for all families* $\{\mathcal{A}_i\}_{i \in I} \subseteq$ $\mathbf{Acr}\,\mathcal{S}$

$$\bigcap_{i \in I} \mathcal{A}_i = (\bigcap_{i \in I} \Theta\,\mathcal{A}_i)^{\mathbf{Gcr}\,\mathcal{S}}.$$

Proof. It follows directly from the implications

$$(\Rightarrow)\ \Theta(\bigcap_{i \in I} \mathcal{A}_i) = \bigcap_{i \in I} \Theta\,\mathcal{A}_i$$

$$\implies \bigcap_{i \in I} \mathcal{A}_i \overset{3.5(2)}{=} (\Theta(\bigcap_{i \in I} \mathcal{A}_i))^{\mathbf{Gcr}\,\mathcal{S}} = (\bigcap_{i \in I} \Theta\,\mathcal{A}_i)^{\mathbf{Gcr}\,\mathcal{S}}.$$

$$(\Leftarrow)\ \bigcap_{i \in I} \mathcal{A}_i = (\bigcap_{i \in I} \Theta\,\mathcal{A}_i)^{\mathbf{Gcr}\,\mathcal{S}} \overset{3.5(3)}{\in} \mathbf{Acr}\,\mathcal{S}. \qquad \square$$

4. The Deduction-Detachment Theorem

The following is a standard definition:

Definition 4.1. A Hilbert system $\mathcal{S}$ admits a *multiterm deduction-detachment theorem* (DDT_Δ) with respect to a finite (may be empty) set $\Delta(x, y)$ of formulas of two variables if the following holds

(1) $x, \Delta(x, y) \vdash_{\mathcal{S}} y$ $/ \Delta$-detachment,

(2) $\dfrac{\Gamma, \alpha \vdash_{\mathcal{S}} \beta}{\Gamma \vdash_{\mathcal{S}} \Delta(\alpha, \beta)}$,   for all $\alpha, \beta \in \mathrm{Fm}_{\mathcal{L}}$, $/ \Delta$-deduction. $\square$

Lemma 4.2. *Suppose* $\mathbf{Acr}\,\mathcal{S}$ *for some Hilbert system* $\mathcal{S}$ *is an invariant closure system, and let* $\Delta(x, y)$ *be a nonempty set of formulas of two variables. Then* $\mathcal{S}$ *admits* DDT_Δ *iff*

$$\{x \triangleright y\}^{\mathbf{Acr}\,\mathcal{S}} = (\triangleright \Delta(x, y))^{\mathbf{Acr}\,\mathcal{S}}. \qquad (*)$$

Proof. $(\Rightarrow)$ The proof is straightforward.
$(\Leftarrow)$ The statement follows from the implications:

$$x \triangleright y \in (\triangleright \Delta(x, y))^{\mathbf{Acr}\,\mathcal{S}} \overset{3.5}{=} (\triangleright \Delta(x, y))^{\mathbf{Gcr}\,\mathcal{S}}$$

$$\overset{3.5(5)}{\Longrightarrow} y \in \{x\}^{\mathcal{S}} \vee (\Delta(x, y))^{\mathcal{S}} \implies x, \Delta(x, y) \vdash_{\mathcal{S}} y. \quad / \Delta\text{-detachment}$$

$$\Gamma, \alpha \vdash_{\mathcal{S}} \beta \overset{3.5(5)}{\Longrightarrow} \alpha \triangleright \beta \in (\triangleright \Gamma)^{\mathbf{Gcr}\,\mathcal{S}} = (\triangleright \Gamma)^{\mathbf{Acr}\,\mathcal{S}}$$

$$\implies (\triangleright \Delta(\alpha, \beta))^{\mathbf{Gcr}\,\mathcal{S}} = (\triangleright \Delta(\alpha, \beta))^{\mathbf{Acr}\,\mathcal{S}} \overset{(*)}{=} \{\alpha \triangleright \beta\}^{\mathbf{Acr}\,\mathcal{S}} \subseteq (\triangleright \Gamma)^{\mathbf{Gcr}\,\mathcal{S}}$$

$$\implies \Delta(\alpha, \beta) \subseteq \Gamma^{\mathcal{S}} \overset{3.5(5)}{\Longrightarrow} \Gamma \vdash_{\mathcal{S}} \Delta(\alpha, \beta). \qquad / \Delta\text{-deduction} \qquad \square$$

Examples. 1) Consider the normal modal logic $S4$. Let $\mathrm{Th}\,S4_{\vdash}$ be the family of sets of modal formulas that are closed under modus ponens and each contains all theorems of $S4$. Then, by deduction theorem for $S4$,

$$\{x \triangleright y\}^{\mathbf{Acr}\,S4_{\vdash}} = \{\Box x \to y\}^{\mathbf{Gcr}\,S4_{\vdash}} = \{\Box x \to y\}^{\mathbf{Acr}\,S4_{\vdash}}.$$

2) The *inconsistent* Hilbert system $\mathcal{S} = \langle \mathbf{Fm}_{\mathcal{L}}, \{\mathrm{Fm}_{\mathcal{L}}\} \rangle$ over the language $\mathcal{L}$ admits DDT_Δ with respect to any finite set $\Delta \subseteq \mathrm{Fm}_{\mathcal{L}}$ of formulas, because

$$(\triangleright \Delta)^{\mathbf{Acr}\,\mathcal{S}} = (\triangleright \Delta)^{\mathbf{Gcr}\,\mathcal{S}} = \mathrm{Fm}_{\mathcal{L}}^{+} = \{x \triangleright y\}^{\mathbf{Acr}\,\mathcal{S}}.$$

3) We also define that the *almost inconsistent* Hilbert system $\mathcal{S} = \langle \mathbf{Fm}_{\mathcal{L}}, \{\emptyset, \mathrm{Fm}_{\mathcal{L}}\} \rangle$ over $\mathcal{L}$ admits $\mathrm{DDT}_\emptyset$, because

$$\{x \triangleright y\}^{\mathbf{Acr}\,\mathcal{S}} = \mathbf{R}_{\mathcal{L}}\{\emptyset, \mathrm{Fm}_{\mathcal{L}}\} = \mathrm{Fm}_{\mathcal{L}}^{+} \setminus \mathrm{Fm}_{\mathcal{L}}^{1} = (\emptyset)^{\mathbf{Gcr}\,\mathcal{S}} = (\emptyset)^{\mathbf{Acr}\,\mathcal{S}}.$$

Theorem 4.3. *Let $\mathcal{S}$ be a Hilbert system with theorems. Then $\mathbf{Acr}\,\mathcal{S}$ forms a Gentzen system iff $\mathcal{S}$ admits a multiterm deduction-detachment theorem.*

Proof. In view of the remarks above, it suffices to prove the theorem for $\mathcal{S}$ that is not inconsistent.

($\Rightarrow$) Suppose $\mathbf{Acr}\,\mathcal{S}$ is a closure system, then there is a closure of the set $\{x \triangleright y\}$ in $\mathbf{Acr}\,\mathcal{S}$. If $\{x \triangleright y\}^{\mathbf{Acr}\,\mathcal{S}} = (\emptyset)^{\mathbf{Gcr}\,\mathcal{S}}$, then

$$\{x \triangleright y\}^{\mathbf{Acr}\,\mathcal{S}} = (\emptyset)^{\mathbf{Gcr}\,\mathcal{S}} \implies x \triangleright y \in (\emptyset)^{\mathbf{Gcr}\,\mathcal{S}} \overset{3.5(5)}{\implies} x \vdash_{\mathcal{S}} y,$$

so $\mathcal{S}$ is either inconsistent or almost inconsistent, a contradiction with the assumption. Thus $\{x \triangleright y\}^{\mathbf{Acr}\,\mathcal{S}} = (\triangleright T)^{\mathbf{Gcr}\,\mathcal{S}}$, for some $T \in \mathrm{Th}\,\mathcal{S}$, such that $T \neq \mathrm{Thm}\,\mathcal{S}$, because $(\emptyset)^{\mathbf{Gcr}\,\mathcal{S}} = (\triangleright \mathrm{Thm}\,\mathcal{S})^{\mathbf{Gcr}\,\mathcal{S}}$. Since $\{x \triangleright y\}^{\mathbf{Acr}\,\mathcal{S}}$ is compact in $\mathbf{Acr}\,\mathcal{S}$, there is a finite subset $\mathcal{O} \subseteq T$, such that $\{x \triangleright y\}^{\mathbf{Acr}\,\mathcal{S}} = \mathcal{O}^{\mathbf{Gcr}\,\mathcal{S}}$. Suppose σ is any substitution such that $\sigma\{x,y\} = \{x,y\}$ and $\sigma(\mathrm{Var}\setminus\{x,y\}) \subseteq \{x,y\}$ and let $\Delta(x,y) = \sigma\mathcal{O}$. Since $\mathbf{Acr}\,\mathcal{S}$ forms a Gentzen system, it is invariant under inverse substitutions, therefore

$$\{x \triangleright y\}^{\mathbf{Acr}\,\mathcal{S}} = \{\sigma x \triangleright \sigma y\}^{\mathbf{Acr}\,\mathcal{S}} = (\sigma\mathcal{O})^{\mathbf{Acr}\,\mathcal{S}} = (\triangleright\Delta(x,y))^{\mathbf{Acr}\,\mathcal{S}}.$$

So, by Lemma 4.2, $\mathcal{S}$ admits DDT_Δ.

($\Leftarrow$) Suppose $\mathcal{S}$ has DDT_Δ, where $\Delta \neq \emptyset$. Δ can be viewed as a function $\Delta : \mathrm{Fm}_{\mathcal{L}}^2 \to \mathcal{P}(\mathrm{Fm}_{\mathcal{L}})$. Furthermore it can be extended to a function from $\mathrm{Fm}_{\mathcal{L}}^+$ to $\mathcal{P}(\mathrm{Fm}_{\mathcal{L}})$ inductively as follows

$$\Delta(\triangleright\alpha) := \alpha,$$
$$\Delta(\alpha_0,\ldots,\alpha_n \triangleright \alpha) := \Delta(\alpha_0,\ldots,\alpha_{n-1} \triangleright \Delta(\alpha_n,\alpha))$$
$$:= \bigcup_{\delta\in\Delta}\{\Delta(\alpha_0,\ldots,\alpha_{n-1} \triangleright \delta(\alpha_n,\alpha))\},$$

and further, in the usual way, to a complex function $\Delta : \mathcal{P}(\mathrm{Fm}_{\mathcal{L}}^+) \to \mathcal{P}(\mathrm{Fm}_{\mathcal{L}})$. Thus, for every $\mathcal{A} \in \mathbf{Acr}\,\mathcal{S}$, the following holds

(1) $\triangleright\alpha \in \mathcal{A} \iff \Delta(\triangleright\alpha) \overset{def}{=} \alpha \in \mathrm{Thm}\,\mathcal{A}$

(2) $\bar{\alpha}, \alpha_{|\bar{\alpha}|} \triangleright \alpha \in \mathcal{A} \overset{3.5(5)}{\iff} \mathrm{Thm}\,\mathcal{A}, \bar{\alpha}, \alpha_{|\bar{\alpha}|} \vdash_{\mathcal{S}} \alpha \overset{4.1}{\iff} \mathrm{Thm}\,\mathcal{A}, \bar{\alpha} \vdash_{\mathcal{S}} \Delta(\alpha_{|\bar{\alpha}|},\alpha)$

$\overset{3.5(5)}{\iff} \bar{\alpha} \triangleright \Delta(\alpha_{|\bar{\alpha}|},\alpha) \subseteq \mathcal{A} \iff \ldots \iff \triangleright\Delta(\bar{\alpha}, \alpha_{|\bar{\alpha}|} \triangleright \alpha) \subseteq \mathcal{A}$

$\iff \Delta(\bar{\alpha}, \alpha_{|\bar{\alpha}|} \triangleright \alpha) \subseteq \mathrm{Thm}\,\mathcal{A}.$

In other words: $\bar{\alpha} \triangleright \alpha \in \mathcal{A} \iff \Delta(\bar{\alpha} \triangleright \alpha) \in \mathrm{Thm}\,\mathcal{A}.$ $\qquad (*)$

Therefore, for every family $\{\mathcal{A}_i\}_{i\in I} \subseteq \mathbf{Acr}\,\mathcal{S}$,

$$\bar{\alpha}\triangleright\alpha \in \bigcap_{i\in I}\mathcal{A}_i \iff (\forall i \in I)\ \bar{\alpha}\triangleright\alpha \in \mathcal{A}_i$$

$$\overset{(*)}{\iff} (\forall i \in I)\ \Delta(\bar{\alpha}\triangleright\alpha) \subseteq \mathrm{Thm}\,\mathcal{A}_i \iff \Delta(\bar{\alpha}\triangleright\alpha) \subseteq \bigcap_{i\in I}\mathrm{Thm}\,\mathcal{A}_i$$

$$\overset{(*)}{\iff} \bar{\alpha}\triangleright\alpha \in (\triangleright(\bigcap_{i\in I}\mathrm{Thm}\,\mathcal{A}_i))^{\mathbf{Gcr}\,\mathcal{S}} = (\bigcap_{i\in I}\Theta\mathcal{A}_i)^{\mathbf{Gcr}\,\mathcal{S}} \in \mathbf{Acr}\,\mathcal{S}.$$

Thus, by Lemma 3.6, $\mathbf{Acr}\,\mathcal{S}$ is closed under arbitrary intersections, hence a closure system.

Now suppose $\mathcal{A} \in \mathbf{Acr}\,\mathcal{S}$ and σ is any substitution. Then

$$\bar{\alpha}\triangleright\alpha \in \sigma^{-1}\mathcal{A} \iff \sigma(\bar{\alpha}\triangleright\alpha) \in \mathcal{A}$$

$$\iff \Delta(\sigma\bar{\alpha}\triangleright\sigma\alpha) = \sigma\Delta(\bar{\alpha}\triangleright\alpha) \subseteq \mathrm{Thm}\,\mathcal{A}$$

$$\iff \Delta(\bar{\alpha}\triangleright\alpha) \subseteq \sigma^{-1}(\mathrm{Thm}\,\mathcal{A}) \iff \bar{\alpha}\triangleright\alpha \in (\sigma^{-1}\Theta\mathcal{A})^{\mathbf{Gcr}\,\mathcal{S}} \in \mathbf{Acr}\,\mathcal{S}.$$

Thus, in addition to being a closure system, $\mathbf{Acr}\,\mathcal{S}$ is invariant. Finally, $\mathbf{Acr}\,\mathcal{S}$ is always algebraic — it follows from [1, Lemma 4.1.2]. $\square$

The characterization provided by Theorem 4.3 is closely related to some known partial characterizations of the deduction-detachment theorem. For instance, J. Czelakowski [5, Theorem 2.6.8] have shown that a protoalgebraic deductive system $\mathcal{S}$ admits a multiterm deduction-detachment theorem iff the lattice $\mathrm{Th}\,\mathcal{S}$ is infinitely meet-distributive over its compact elements, where

Definition 4.4. A complete lattice L is *infinitely meet-distributive over its compact elements* if for every compact element $c \in L$ and every family $\{a_i\}_{i\in I} \subseteq L$

$$c \vee \left(\bigwedge_{i\in I}a_i\right) = \bigwedge_{i\in I}(c \vee a_i). \qquad \square$$

The relation that this condition has with Theorem 4.3 is established in the following theorem.

Theorem 4.5. $\mathbf{Acr}\,\mathcal{S}$ *is a closure system iff the lattice* $\mathrm{Th}\,\mathcal{S}$ *is infinitely meet-distributive over its compact elements.*

Proof. In this theorem $\vee$ denotes the join in the complete lattice $\mathrm{Th}\,\mathcal{S}$. ($\Rightarrow$) Suppose $\mathbf{Acr}\,\mathcal{S}$ is closed under arbitrary intersections. Let $\{T_i\}_{i\in I} \subseteq \mathrm{Th}\,\mathcal{S}$

12

and $\{\bar{\alpha}\} \subseteq_\omega \mathrm{Fm}_\mathcal{L}$. Then for any $\alpha \in \mathrm{Fm}_\mathcal{L}$

$$\alpha \in \bigcap_{i \in I}(\{\bar{\alpha}\}^\mathcal{S} \vee T_i) \iff (\forall i \in I)\ \alpha \in \{\bar{\alpha}\}^\mathcal{S} \vee T_i$$

$$\overset{3.5(5)}{\iff} (\forall i \in I)\ \bar{\alpha} \triangleright \alpha \in (\triangleright T_i)^{\mathbf{Gcr}\,\mathcal{S}} \iff \bar{\alpha} \triangleright \alpha \in \bigcap_{i \in I}(\triangleright T_i)^{\mathbf{Gcr}\,\mathcal{S}}$$

$$\overset{3.6}{\iff} \bar{\alpha} \triangleright \alpha \in \left(\bigcap_{i \in I}(\triangleright T_i)\right)^{\mathbf{Gcr}\,\mathcal{S}} = \left(\triangleright\left(\bigcap_{i \in I} T_i\right)\right)^{\mathbf{Gcr}\,\mathcal{S}}$$

$$\overset{3.5(5)}{\iff} \alpha \in \{\bar{\alpha}\}^\mathcal{S} \vee \left(\bigcap_{i \in I} T_i\right).$$

Thus $\bigcap_{i \in I}(\{\bar{\alpha}\}^\mathcal{S} \vee T_i) = \{\bar{\alpha}\}^\mathcal{S} \vee (\bigcap_{i \in I} T_i)$.

($\Leftarrow$) By contradiction. Suppose $\mathrm{Th}\,\mathcal{S}$ is infinitely meet-distributive over compact elements, but there is a family $\{\mathcal{A}_i\}_{i \in I} \subseteq \mathbf{Acr}\,\mathcal{S}$ and a sequent $\bar{\alpha} \triangleright \alpha$ such that $\bar{\alpha} \triangleright \alpha \in \bigcap_{i \in I} \mathcal{A}_i$ and $\bar{\alpha} \triangleright \alpha \notin (\bigcap_{i \in I} \Theta\, \mathcal{A}_i)^{\mathbf{Gcr}\,\mathcal{S}}$. Then

1) $\bar{\alpha} \triangleright \alpha \notin (\bigcap_{i \in I} \Theta\, \mathcal{A}_i)^{\mathbf{Gcr}\,\mathcal{S}} \overset{3.5(5)}{\iff} \alpha \notin \{\bar{\alpha}\}^\mathcal{S} \vee (\bigcap_{i \in I} \mathrm{Thm}\,\mathcal{A}_i)$,

2) $\bar{\alpha} \triangleright \alpha \in \bigcap_{i \in I} \mathcal{A}_i \iff (\forall i \in I)\ \bar{\alpha} \triangleright \alpha \in \mathcal{A}_i$

$\iff (\forall i \in I)\ \alpha \in \{\bar{\alpha}\}^\mathcal{S} \vee \mathrm{Thm}\,\mathcal{A}_i \iff \alpha \in \bigcap_{i \in I}(\{\bar{\alpha}\}^\mathcal{S} \vee \mathrm{Thm}\,\mathcal{A}_i).$

But, by assumption, $\{\bar{\alpha}\}^\mathcal{S} \vee (\bigcap_{i \in I} \mathrm{Thm}\,\mathcal{A}_i) = \bigcap_{i \in I}(\{\bar{\alpha}\}^\mathcal{S} \vee \mathrm{Thm}\,\mathcal{A}_i)$, a contradiction. $\qquad\square$

Remarks.

1. The fact, that the set of axiomatic closure relations is closed under finite intersections if and only if the lattice of all $\mathcal{S}$-theories is distributive, was first observed in [8].

2. Note that $\mathrm{Th}\,\mathcal{S}$ is always infinitely join-distributive over its compact elements, i.e.,

$$\{\bar{\alpha}\}^\mathcal{S} \cap \left(\bigvee_{i \in I} \mathrm{Thm}\,\mathcal{A}_i\right) = \bigvee_{i \in I}(\{\bar{\alpha}\}^\mathcal{S} \cap \mathrm{Thm}\,\mathcal{A}_i),$$

by Proposition 2.5.1[5], since $\mathrm{Th}\,\mathcal{S}$ is algebraic.

3. A number of prominent Hilbert systems fail to admit DDT. For instance, the deductive system of the smallest modal propositional logic K does not admit DDT, because the lattice $\mathrm{Th}\,K_\vdash$ is not infinite meet-distributive over its compact elements, even though it is distributive (see [5] for some details). Thus $\mathbf{Acr}\,K_\vdash$ is closed under finite intersections, but not under arbitrary intersections. On the other hand, the deductive system for Anderson and Belnap's logic of relevant implication R does not admit DDT for a different reason — $\mathbf{Acr}\,R_\vdash$ is not invariant ([10,5] and Dziobiak, unpublished).

5. DDT and Protoalgebraic Hilbert Systems

The deduction-detachment theorem was originally introduced by Tarski and Herbrand (independently) to investigate algebraizability phenomenon of the classical logic. Therefore it is particularly interesting to investigate DDT in the context of protoalgebraic Hilbert systems, which were specifically introduced to capture weak forms of algebraizability phenomenon [2].

Definition 5.1. A Hilbert system $\mathcal{S}$ is protoalgebraic if there exists a non empty finite set $\Delta(x,y)$ of formulas of two variables, such that

$\quad$ (1) $\quad \vdash_{\mathcal{S}} \Delta(x,x)$

$\quad$ (2) $\quad x, \Delta(x,y) \vdash_{\mathcal{S}} y \qquad\qquad$ / Δ-detachment $\qquad\qquad\square$

The class of protoalgebraic Hilbert systems contains many well-known propositional deductive systems, such as deductive systems of intuitionistic, modal and relevant logics and, of course, that of the classical logic. It is easy to see that DDT_Δ implies protoalgebraicity with the same set of formulas Δ. But protoalgebraicity is strictly weaker than DDT, as the above-mentioned examples of deductive systems for K and R show.

$\quad$ For protoalgebraic systems, the conditions for DDT can be weakened, as it was shown by J. Czelakowski [5]. Further we will show how some of J. Czelakowski's results can be derived using our approach.

Lemma 5.2. *If $\mathcal{S}$ is protoalgebraic, then $\mathbf{Acr}\,\mathcal{S}$ is surjectively invariant.*

Proof. Suppose $\mathcal{A} \in \mathbf{Acr}\,\mathcal{S}$ and σ is a surjective substitution. By Lemma 3.5(4), $\mathcal{A} = \mathbf{R}_{\mathcal{L}}[T)_{\mathrm{Th}\,\mathcal{S}}$ for some $T \in \mathrm{Th}\,\mathcal{S}$. Then, by characteristic property of protoalgebraic deductive systems [9, Theorem 2.6]:

$$\sigma^{-1}[T)_{\mathrm{Th}\,\mathcal{S}} = [\sigma^{-1}T)_{\mathrm{Th}\,\mathcal{S}}, \qquad\qquad (*)$$

therefore

$$\bar{\alpha} \rhd \alpha \in \sigma^{-1}\mathcal{A} \iff \sigma\bar{\alpha} \rhd \sigma\alpha \in \mathcal{A}$$

$$\overset{3.5(4)}{\iff} (\forall F \in [T)_{\mathrm{Th}\,\mathcal{S}})\,(\{\sigma\bar{\alpha}\} \subseteq F \implies \sigma\alpha \in F)$$

$$\iff (\forall F \in [T)_{\mathrm{Th}\,\mathcal{S}})\,(\{\bar{\alpha}\} \subseteq \sigma^{-1}F \implies \alpha \in \sigma^{-1}F)$$

$$\iff (\forall G \in \sigma^{-1}[T)_{\mathrm{Th}\,\mathcal{S}})\,(\{\bar{\alpha}\} \subseteq G \implies \alpha \in G)$$

$$\overset{(*)}{\iff} (\forall G \in [\sigma^{-1}T)_{\mathrm{Th}\,\mathcal{S}})\,(\{\bar{\alpha}\} \subseteq G \implies \alpha \in G)$$

i.e., $\sigma^{-1}\mathcal{A} = \mathbf{R}_{\mathcal{L}}[\sigma^{-1}T)_{\mathrm{Th}\,\mathcal{S}} \in \mathbf{Acr}\,\mathcal{S}$. $\qquad\qquad\square$

$\quad$ The following lemma is a close analogue of Lemma 2.3 [6].

14

Lemma 5.3. *Suppose $\mathcal{R}$ is a finitary closure system on $\mathrm{Fm}_{\mathcal{L}}^{+}$. Then $\mathcal{R}$ is invariant, if it is surjectively invariant.*

Proof. Based on correspondence between closure systems and consequence relations, it suffices to show that for all $X, Y \subseteq \mathrm{Fm}_{\mathcal{L}}^{+}$ and all substitution σ

$$X^{\mathcal{R}} \subseteq Y^{\mathcal{R}} \implies (\sigma X)^{\mathcal{R}} \subseteq (\sigma Y)^{\mathcal{R}}.$$

Since $\mathcal{R}$ is an algebraic closure system, for every sequent $s \in X$, there is a finite subset $Y_s \subseteq_\omega Y$, such that $\{s\}^{\mathcal{R}} \subseteq Y_s^{\mathcal{R}}$. Since $Y_s \cup \{s\}$ is finite, then $\mathrm{Var}(Y_s \cup \{s\})$ is finite, therefore there is a surjective substitution σ_s, such that $\sigma_s s = \sigma s$, $\sigma_s Y_s = \sigma Y_s$. Thus

$$(\sigma X)^{\mathcal{R}} = \left(\textstyle\bigcup_{s \in X} \sigma_s s\right)^{\mathcal{R}} \subseteq \left(\textstyle\bigcup_{s \in X} \sigma_s Y_s\right)^{\mathcal{R}} = \left(\textstyle\bigcup_{s \in X} \sigma Y_s\right)^{\mathcal{R}} \subseteq (\sigma Y)^{\mathcal{R}}. \quad \square$$

Corollary 5.4. [5, Theorem 2.6.8(i),(iii)] *Let $\mathcal{S}$ be a protoalgebraic Hilbert system. If the lattice $\mathrm{Th}\,\mathcal{S}$ is infinitely meet-distributive over its compact elements, then $\mathcal{S}$ admits a multiterm deduction-detachment theorem.*

Proof. By Lemma 5.2, since $\mathcal{S}$ is protoalgebraic, $\mathbf{Acr}\,\mathcal{S}$ is surjectively invariant. On the other hand, since $\mathrm{Th}\,\mathcal{S}$ is infinitely meet-distributive over its compact elements, then $\mathbf{Acr}\,\mathcal{S}$ is a closure system on $\mathrm{Fm}_{\mathcal{L}}^{+}$, by Theorem 4.5. Thus, by Lemma 5.3, $\mathbf{Acr}\,\mathcal{S}$ is invariant. Altogether, we have that $\mathbf{Acr}\,\mathcal{S}$ is an algebraic, invariant, closure system on $\mathrm{Fm}_{\mathcal{L}}^{+}$, hence it is a Gentzen system, therefore, by Theorem 4.3, $\mathcal{S}$ admits a multiterm deduction-detachment theorem. $\quad \square$

Another case, where the characterization of DDT given by Theorem 4.3, can be enlightening, is related to the question of existence of fully adequate Gentzen systems for weakly algebraizable Hilbert systems. Some definitions are in place.

Definition 5.5. For a Hilbert system $\mathcal{S}$, define the set of *full closure relations of $\mathcal{S}$* as follows:

$$\mathbf{Fcr}\,\mathcal{S} := \{(\theta)^{\mathbf{Gcr}\,\mathcal{S}} \mid \theta \in \mathrm{Con}\,\mathbf{Fm}_{\mathcal{L}}\}.$$

An element of $\mathbf{Fcr}\,\mathcal{S}$ is called *a full closure relation for $\mathcal{S}$*. $\quad \square$

Note, that the above definition is correct: $\theta \in \mathrm{Con}\,\mathbf{Fm}_{\mathcal{L}}$ contains sequents of the form $\alpha \triangleright \beta \overset{def}{=} \langle \alpha, \beta \rangle$, for every $\langle \alpha, \beta \rangle \in \theta$, and we take the closure of θ in the closure system of Gentzen theories.

More traditionally is to define full closure relations through closure systems of compatible $\mathcal{S}$-theories.

An $\mathcal{S}$-theory T is *compatible* with a congruence $\theta \in \mathrm{Con}\,\mathbf{Fm}_{\mathcal{L}}$, if $\langle \alpha, \beta \rangle \in \theta$ and $\alpha \in T$ imply $\beta \in T$. Abusing our notation, we denote the set of all compatible with θ theories by $[\theta)_{\mathrm{Th}\,\mathcal{S}}$. It can be shown [1, Proposition 4.2.2] that

$$\mathbf{Fcr}\,\mathcal{S} := \{\mathbf{R}_{\mathcal{L}}[\theta)_{\mathrm{Th}\,\mathcal{S}} \mid \theta \in \mathrm{Con}\,\mathbf{Fm}_{\mathcal{L}}\}.$$

Full closure relations had previously got more attention than axiomatic closure relations, mainly in connection with algebraizability phenomenon beyond the class of protoalgebraic systems (see [7] for details). If, for a Hilbert system $\mathcal{S}$ with theorems, $\mathbf{Fcr}\,\mathcal{S}$ forms a Gentzen system, it is called the *fully adequate Gentzen system for $\mathcal{S}$* [7,9]. For weakly algebraizable Hilbert systems there is an especially elegant criterion for the existence of the fully adequate Gentzen system.

Corollary 5.6. [9, Corollary 5.7] *Let $\mathcal{S}$ be a weakly algebraizable Hilbert system. Then $\mathcal{S}$ has a fully adequate Gentzen system iff it admits a multi-term deduction-detachment theorem.*

Proof. By Theorem 3.6(i),(iv)[6], for a weakly algebraizable Hilbert system $\mathcal{S}$, there is one-to-one correspondence between families of closure systems $\{[\theta)_{\mathrm{Th}\,\mathcal{S}} \mid \theta \in \mathrm{Con}\,\mathbf{Fm}_{\mathcal{L}}\}$ and $\{[T)_{\mathrm{Th}\,\mathcal{S}} \mid T \in \mathrm{Th}\,\mathcal{S}\}$, therefore

$$\mathbf{Fcr}\,\mathcal{S} = \{\mathbf{R}_{\mathcal{L}}[\theta)_{\mathrm{Th}\,\mathcal{S}} \mid \theta \in \mathrm{Con}\,\mathbf{Fm}_{\mathcal{L}}\} = \{\mathbf{R}_{\mathcal{L}}[T)_{\mathrm{Th}\,\mathcal{S}} \mid T \in \mathrm{Th}\,\mathcal{S}\} = \mathbf{Acr}\,\mathcal{S}.$$

The rest follows directly from Theorem 4.3. $\qquad\qquad\square$

References

1. S. Babyonyshev, Metatheories of deductive systems, *Ph.D. Thesis*, Iowa State University, 2004. (available at http://www.geocities.com/bsv70/thesis.pdf)
2. W. Block and D. Pigozzi, Protoalgebraic logics, *Studia Logika*, 45:337–369 (1986).
3. W. Block and D. Pigozzi, Abstract algebraic logic and the deduction theorem, *The Bulletin of Symbolic Logic*, to appear.
4. S.L. Bloom, Some theorems on structural consequence operations, *Studia Logica* 34:1–9 (1975).
5. J. Czelakowski, Protoalgebraic logics, *Kluwer Academic Publishers*, Dordrecht, 2001.
6. J. Czelakowski and R. Jansana, Weakly algebraizable logics, *The Journal of Symbolic Logic*, v. 65, 2:641–668 (2000).
7. J.M. Font and R. Jansana, A general algebraic semantics for sentential logics, *Springer-Verlag, Number 7 in Lecture Notes in Logic*, 1996.
8. J.M. Font, R. Jansana and D. Pigozzi, Fully adequate Gentzen systems and closure properties of the class of full models, *Manuscript*, 1999.

9. J.M. Font, R. Jansana and D. Pigozzi, Fully adequate Gentzen systems and the deduction theorem, *Reports on mathematical logic*, 35:115–165 (2001).

10. L.L. Maximova, Formal proofs in the calculus of strong implication (in Russian), *Algebra i Logika*, v. 5, 6:33–38 (1966).

COMPUTABLE NUMBERINGS IN THE HIERARCHY OF ERSHOV*

S. A. BADAEV and Zh. T. TALASBAEVA

Kazakh National University,
Al-Farabi ave., 71,
Almaty, 050038, Kazakhstan
E-mail: badaev@kazsu.kz, star@kazsu.kz

The paper is aimed to show some new results of the authors on Rogers semilattices for the families of sets in the hierarchy of Ershov and to discuss similarities and differences of the algebraic properties of these semilattices in contrary with the semilattices of computable numberings for the families of the arithmetical sets.

1. Background and preliminaries

A lot of known notions of computability became a special case of the general approach suggested by S. Goncharov and A. Sorbi [1]. This approach activated the study of computability at several domains in logic and computer science. Among these areas of research one can find investigations on computable numberings in arithmetical hierarchy [2–4] and hierarchy of Ershov [5–7] century. These investigations are usually made along the results known in the classical case of computable numbering of the families of c.e. sets. Class of c.e. sets forms level 1 in both arithmetical and Ershov's hierarchies. We will separate this classical case from the other cases in these hierarchies assuming that only levels over level 1 are considered if we spoke on the families in the arithmetical or Ershov's hierarchies.

Computable numberings of a family of c.e. sets can be considered as uniform enumeration procedures for the sets of the family. Monotonicity is the most significant feature of such computations. This means that every number (treated as a piece of information) enumerated in any set via uniform enumeration procedure never leaves this set in future. Thus, information accumulated in every set of the family is growing more and more, i.e. monotonously with respect to growth of time. In contrary with the classical case, during computations via computable numbering of a family in

---

*This work was partially supported by grant INTAS–00–499.

the arithmetical or Ershov's hierarchies, any number could enter a set and, later, it could leave the set, and after that again enter it and *etc.* In the case of the hierarchy of Ershov, number of these 'enter-leaves' is bounded by the level of hierarchy in which considered family is involved while, in the case of the arithmetical hierarchy, it is usually required for a number of 'enter-leaves' to be finite only. Besides, in the latter case one can use jumps of empty sets as oracles in the computations.

We refer to handbooks [8, 9] for the notions and standard notations on computability theory. Undefined notions of the theory of numberings can be found in [10–12]. Basic notions and facts on the hierarchy of Ershov one can find in [13]. We use notation $c(x, y)$ for pairing function, $l(n)$ and $r(n)$ denote correspondingly left and right components of the n-th pair.

The notion of computable numbering for a family $\mathcal{A}$ from class Σ_n^i of level n in Ershov's ($i = -1$) or arithmetical ($i = 0$) hierarchy may be deduced [1, 12] from the approach of Goncharov–Sorbi as follows. Numbering $\alpha : \omega \mapsto \mathcal{A}$ is $\Sigma_n^i - computable$ if $\{\langle x, m \rangle \ : \ x \in \alpha(m)\} \in \Sigma_n^i$. Note that relation $\{\langle x, m \rangle \ : \ x \in \alpha(m)\} \in \Sigma_n^{-1}$ is equivalent to relation "$\{\langle x, m \rangle \ : \ x \in \alpha(m)\}$ is $n-$ computably enumerable" in a sense of Putnam [14].

If there exists a computable function f which translates indices of the sets in numbering $\alpha : \omega \mapsto \mathcal{A}$ into the indices of the same sets in numbering $\beta : \omega \mapsto \mathcal{A}$, i.e. $\alpha(x) = \beta(f(x))$ for all x, then α is called *reducible* to β (symbolically, $\alpha \leqslant \beta$). Numberings which are reducible to each other are called *equivalent. Rogers semilattice* $\mathcal{R}_n^i(\mathcal{A})$ of a family $\mathcal{A} \subseteq \Sigma_n^i$ is a quotient structure of all Σ_n^i- computable numberings of the family $\mathcal{A}$ modulo equivalence of the numberings ordered by the relation induced by reducibility of numberings. $\mathcal{R}_n^i(\mathcal{A})$ allows one to measure different computations of a given family $\mathcal{A}$. Rogers semilattices $\mathcal{R}_n^i(\mathcal{A})$ are used also as a tool to classify properties of Σ_n^i- computable numberings for different families $\mathcal{A}$.

Historically, the first two problems on Rogers semilattices of the families of c.e. sets were raised by Ershov: What is a possible cardinality of a Rogers semilattice? Can a Rogers semilattice be a lattice?

There is a sharp distinction in Rogers semilattices of the families of $\Sigma_n^{-1}-$ sets for $n > 1$ and Rogers semilattices of Σ_n^0- sets for $n > 0$. In 1971, Khutoretsky [15] proved that no non-trivial Rogers semilattice $\mathcal{R}_1^0(\mathcal{A})$ of a family $\mathcal{A}$ of c.e.e. sets could be decomposed into union of disjoint principal ideal and principal filter. As a straightforward consequence he concluded that if $\mathcal{R}_1^0(\mathcal{A})$ contains two elements then it is infinite. The Rogers semilattices of Σ_n^0- computable numberings for $n > 1$ also consists either of a single or infinitely many elements too [1]. Unexpectedly, the statement of decomposition of Rogers semilattices holds for some families in Ershov's hierarchy [7]. It should be noted that the latter result says nothing on a possible cardinality of Rogers semilattices in the case of Ershov's hierarchy. Unfortunately, the methods employed in the theorems of Khutoretsky, and Goncharov–Sorbi are of no use in the case of computability in the hierarchy of Ershov. Non-monotonicity of computations in this hierarchy prevents anybody to use these methods for resolution the problem of cardinality as well as many other problems. The question on a cardinality of Rogers semilattices in the case of Ershov's hierarchy becomes actual and seems to be non-trivial. Our paper can be thought as the first step towards a resolution for this natural question.

2. One-element Rogers semilattices

It is well-known [5] that there exists an infinite family $\mathcal{A}$ of c.e. sets with one-element Rogers semilattice $\mathcal{R}_2^{-1}(\mathcal{A})$. We suggest a bit another construction which gives examples of the families of an arbitrary countable cardinality with one-element Rogers semilattices.

Theorem 2.1. *For every $n \in \omega \cup \{\omega\}$, $n \neq 0$, there exists a family $\mathcal{A}$ consisting of n c.e. sets whose Rogers semilattice $\mathcal{R}_2^{-1}(\mathcal{A})$ is one-element.*

Proof. At first, we will build computable Friedberg numbering α of an infinite family $\mathcal{A} \leftrightharpoons \alpha(\omega)$ of the c.e. sets such that, for every k,

$$\text{if } \pi_k \text{ is a numbering of } \mathcal{A} \text{ then } \pi_k \leqslant \alpha.$$

Here and furthermore, $\{\pi_k\}_{k \in \omega}$ stands for any computable numbering of the class of all Σ_2^{-1}– computable numberings of the families of Σ_2^{-1}– sets. Without lost of generality, we presuppose that $\pi_0(0) = \emptyset$.

We will also build a supplementary sequence $\{g_k\}_{k \in \omega}$ of partial computable functions. Function g_k is aimed to make reduction of numbering π_k to numbering α if π_k is a numbering of the family $\mathcal{A}$.

Let $a(k, x, i)$ be any computable one-to-one function and let $\{\pi_k^t(x)\}$ be a strong triple array of finite sets which uniformly approximates computations in numberings $\pi_k, k \in \omega$. We presuppose that $\pi_0^t(0) = \emptyset$ for all t.

The construction

Stage 0. Let $\alpha^0(e) = \emptyset$ and let $g_k^0(e)$ be undefined for all k and e. Go to the next stage.

20

Stage $t+1$. Let $m = l(t)$ and consider numbers k, x, i such that $c(c(k, x), i) = m$. Carry out instructions of cases 1,2.

Case 1. If t is the least number such that $l(t) = m$ then enumerate number $a(k, x, i)$ into the set $\alpha(i)$. Go to *End of stage.*

Case 2. If there exists $t' < t$ such that $l(t') = m$ then carry out one the following mutually exclusive subcases 2.1–2.3.

Subcase 2.1. $a(k, x, i) \in \pi_k^{t+1}(x)$ and $g_k^t(x)$ is undefined. Let $g_k^{t+1}(x) = i$ and go to *End of stage.*

Subcase 2.2. $g_k^t(x) = i$ and $a(k, x, i) \notin \pi_k^{t+1}(x)$. Enumerate $a(k, x, i)$ into $\alpha(j)$ for every $j \neq i$. Go to *End of stage.*

Subcase 2.3. Subcases 2.1, 2.2 do not hold. Go to *End of stage.*

End of stage. For every m, y, z, if $g_m^{t+1}(y)$ and $\alpha^{t+1}(z)$ have not been explicitly defined above then let $g_m^{t+1}(y) = g_m^t(y)$ and let $\alpha^{t+1}(z) = \alpha^t(z)$.

Properties of the construction

1. For every k, x, i, number $a(k, x, i)$ is contained in the set $\alpha(i)$.

2. For every k, x, i, number $a(k, x, i)$ belongs to either one or all the sets of the family $\mathcal{A}$.

3. For every i, number $a(0, 0, i)$ is contained in the set $\alpha(i)$ only. In particular, $\mathcal{A}$ is infinite family and α is Friedberg numbering.

4. For every k, x, t, if $g_k^t(x)$ is defined then $g_k^s(x)$ is also defined and $g_k^s(x) = g_k^t(x)$ for all $s \geq t$. In particular, every g_k is partially computable function.

Properties 1–4 are evident.

5. For every k, x, i, number $a(k, x, i)$ belongs to all sets of $\mathcal{A}$ if and only if $g_k(x)$ is defined, $g_k(x) = i$, and $a(k, x, i) \notin \pi_k(x)$.

Proof. First of all, note that number $a(k, x, i)$ can be enumerated into all sets of the family $\mathcal{A}$ if and only if *Subcase* 2.2 holds at some stage $t_0 + 1$ such that $l(t_0) = c(c(k, x), i)$.

$\Rightarrow$ Let $l(t_0) = c(c(k, x), i)$ and let $t_0 + 1$ be the stage at which *Subcase* 2.2 holds. Then $g_k^{t_0}(x) = i$ and $a(k, x, i) \notin \pi_k^{t_0+1}(x)$. Let $t_1 + 1 < t_0 + 1$ be the stage at which $g_k(x)$ had been defined by *Subcase* 2.1. Then $a(k, x, i) \in \pi_k^{t_1+1}(x)$. Since $\pi_k(x) \in \Sigma_2^{-1}$ it follows that $a(k, x, i) \notin \pi_k(x)$.

$\Leftarrow$ Let $t_1 + 1$ be the stage at which $g_k(x)$ had been let equal to i. By construction, $l(t_1) = c(c(k, x), i)$ and *Subcase* 2.1 holds at stage $t_1 + 1$. Then $a(k, x, i) \in \pi_k^{t_1+1}(x)$. Let s be the stage at which $a(k, x, i)$ leaves $\pi_k(x)$. Let t_2 be the least number such that $t_2 \geq s$ and $l(t_2) = c(c(k, x), i)$. Then *Subcase* 2.2 holds at stage $t_2 + 1$.

6. For every k, if π_k is a numbering of $\mathcal{A}$ then g_k is total function.

Proof. If π_k is a numbering of $\mathcal{A}$ then for every x there exists i such that $\alpha(i) = \pi_k(x)$. By property 1, $a(k,x,i) \in \alpha(i)$ and, hence, $a(k,x,i) \in \pi_k(x)$. Let s be the least number such that $a(k,x,i) \in \pi_k^s(x)$. Clearly, $a(k,x,i) \in \pi_k^t(x)$ for all $t \geq s$ since $\pi_k(x) \in \Sigma_2^{-1}$ and $a(k,x,i) \in \pi_k(x)$. Then $g_k(x)$ will be defined by *Subcase* 2.1 at the least stage $t_0 + 1 \geq s$ such that $l(t_0) = c(c(k,x),i)$.

7. For every k, if π_k is a numbering of the family $\mathcal{A}$ then $\pi_k(x) = \alpha(g_k(x))$ for all x.

Proof. Let π_k be any numbering of $\mathcal{A}$ and let x be an arbitrary number. By property 6, $g_k(x) = i$ for some i. By property 1, $a(k,x,i) \in \alpha(i)$. By properties 2,3, we have two alternatives: either $a(k,x,i) \notin \alpha(j)$ for all $j \neq i$ or $a(k,x,i) \in \alpha(s)$ for all s. For the first alternative, equality $\pi_k(x) = \alpha(g_k(x))$ is obvious since $a(k,x,i) \in \pi_k(x)$ too. If the second alternative takes place then $a(k,x,i) \notin \pi_k(x)$. Hence, $\pi_k(x) \notin \mathcal{A}$. A contradiction. Thus, only first alternative is valid.

The statement of theorem for an existence of infinite family is a straightforward consequence of properties 3, 7.

Let now n be any positive number. To build a family $\mathcal{A}$ consisting of n elements we adopt the construction above by changing only instructions of *Case* 1 as follows.

Case 1. If t is the least number such that $l(t) = m$ then enumerate number $a(k,x,i)$ into set $\alpha(i)$. Besides, if $i \geq n-1$ then enumerate number $a(k,x,i)$ into $\alpha(j)$ for all $j \geq n-1, j \neq i$. Go to *End of stage*.

New version of *Case* 1 implies that property 3 of the modified construction should be changed in a following way.

3. For every $q > n-1$, $\alpha(q) = \alpha(n-1)$. For every $i \leq n-1$, number $a(0,0,i)$ is contained in the set $\alpha(i)$ only. In particular, $\mathcal{A}$ consists of n elements and α is decidable numbering.

It is easy to check that all other properties of the modified construction as well as their proofs are the same as the old ones. And again the statement of theorem is a straightforward consequence of properties $3, 7$. $\square$

The families constructed in theorem 2.1 are effectively discrete. According to well-known fact of the theory of numberings [10, 11] (see also [12] for details and background), if a family of c.e. sets is effectively discrete then its Rogers semilattice consists of a single element. It is obvious that the family $\mathcal{B} \leftrightharpoons \{\{x\} : x \in \omega\}$ is effectively discrete. Nevertheless, Rogers semilattice $\mathcal{R}_2^{-1}(\mathcal{B})$ is infinite. It is not hard to prove this directly or show it with Corollary 3.3(6) below.

Let a family $\mathcal{C} \subseteq \Sigma_n^0$ be finite and non-discrete. Then $\mathcal{C}$ contains a pair of embedded sets. This implies that $\mathcal{R}_n^0(\mathcal{C})$ is infinite (see [10–12] for $n = 1$ and see [1] for $n > 1$). And, again, we have rather different result in the case of Ershov's hierarchy.

Theorem 2.2. *There exists a family consisting of two embedded Σ_2^{-1} – sets whose Rogers semilattice is one-element.*

Proof. Like the proof of Theorem 2.1, we will build by stages decidable numbering α of the family $\mathcal{A} = \alpha(\omega)$ such that for every k

$$\text{if } \pi_k \text{ is a numbering of } \mathcal{A} \text{ then } \pi_k \leqslant \alpha.$$

We will also build a supplementary sequence $\{g_k\}_{k \in \omega}$ of partially computable functions. Let $a(k, x, i)$ be any computable injective function.

The construction

Stage 0. Let $\alpha^0(e) = \emptyset$ and let $g_k^0(e)$ be undefined for all k and e. Go to the next stage.

Stage $t+1$. Let $m = l(t)$, $k = l(m)$, $x = r(m)$ and carry out the instructions of the following 3 cases.

Case 1. If t is the least number such that $l(t) = m$ then enumerate $a(k, x, 0)$ into $\alpha(0)$ and enumerate $a(k, x, 0)$ and $a(k, x, 1)$ into $\alpha(1)$. Go to *End of stage*.

Case 2. If there exists $t' < t$ such that $l(t') = m$ and $g_k^t(x)$ is undefined then carry out one of the following 4 mutually exclusive subcases.

Subcase 2.1. If $\{a(k, x, 0), a(k, x, 1)\} \subseteq \pi_k^{t+1}(x)$ then let $g_k^{t+1}(x) = 1$. Go to *End of stage*.

Subcase 2.2. If $a(k, x, 0) \in \alpha^t(0) \cap \pi_k^{t+1}(x)$ and $a(k, x, 1) \notin \pi_k^{t+1}(x)$ then extract $a(k, x, 0)$ from $\alpha(0)$. Go to *End of stage*.

Subcase 2.3. If $a(k, x, 0) \notin \alpha^t(0) \cup \pi_k^{t+1}(x)$ and $a(k, x, 0) \in \pi_k^s(x)$ for some $s \leq t$ then let $g_k^{t+1}(x) = 0$. Go to *End of stage*.

Subcase 2.4. Subcases 2.1–2.3 do not hold. Go to *End of stage*.

Case 3. If $g_k(x)$ is defined then carry out the instructions of the following two subcases.

Subcase 3.1. If $g_k(x) = 1$ and $a(k, x, 1) \notin \pi_k^{t+1}(x)$ then enumerate number $a(k, x, 1)$ into set $\alpha(0)$. Go to *End of stage*.

Subcase 3.2. Subcase 3.1 fails. Go to *End of stage*.

End of stage. For all m, y, z, if $g_m^{t+1}(y)$ or $\alpha^{t+1}(z)$ have not been explicitly defined above then let $g_m^{t+1}(y) = g_m^t(y)$, let $\alpha^{t+1}(z) = \alpha^t(z)$ for $z \leq 1$, and let $\alpha^{t+1}(z) = \alpha^{t+1}(1)$ for $z > 1$. Go to the next stage.

Properties of the construction

1. For all $i \geq 1$, $\alpha(i) = \alpha(1)$. In particular, $\mathcal{A} = \{\alpha(0), \alpha(1)\}$.

2. $\mathcal{A} \subseteq \Sigma_2^{-1}$ and α is Σ_2^{-1}–computable numbering of $\mathcal{A}$.

3. For every $y \in \alpha(0) \cup \alpha(1)$, there exist $i \leq 1$ and $k, x \in \omega$ such that $y = a(k, x, i)$.

4. For all k, x, the set $\alpha(0) \cap \{a(k, x, 0), a(k, x, 1)\}$ may be equal to one of the sets $\emptyset$, $\{a(k, x, 0)\}$, or $\{a(k, x, 0), a(k, x, 1)\}$, while the intersection $\alpha(1) \cap \{a(k, x, 0), a(k, x, 1)\}$ is always equal to $\{a(k, x, 0), a(k, x, 1)\}$.

5. For every k, x, t, if $g_k^t(x)$ is defined then $g_k^s(x)$ is also defined and $g_k^s(x) = g_k^t(x)$ for all $s \geq t$. In particular, every g_k is partially computable function.

Properties 1–5 are evident.

6. For every k, if π_k is a numbering of $\mathcal{A}$ then the function g_k is total.

Proof. Let π_k be any numbering of $\mathcal{A}$ and let x be any number. Denote $c(k, x)$ by m and let $T_m = \{t + 1 \mid l(t) = m\}$. By construction, $g_k(x)$ can be defined only at stages from T_m in a result of carrying *Subcase* 2.1 or *Subcase* 2.3. Let $t_0 + 1$ be the least number of T_m. *Case* 1 takes place only once at stages from T_m, namely at stage $t_0 + 1$. At this stage, $a(k, x, 0)$ is enumerated into $\alpha(0)$, and numbers $a(k, x, 0), a(k, x, 1)$ are enumerated into $\alpha(1)$. Let $t_1 + 1 \in T_m$ be the least stage such that $t_1 > t_0$ and

$$\alpha^s(0) \cap \{a(k, x, 0), a(k, x, 1)\} = \alpha^{t_1}(0) \cap \{a(k, x, 0), a(k, x, 1)\}, \quad (1)$$
$$\alpha^s(1) \cap \{a(k, x, 0), a(k, x, 1)\} = \alpha^{t_1}(1) \cap \{a(k, x, 0), a(k, x, 1)\}, \quad (2)$$
$$\pi_k^s(x) \cap \{a(k, x, 0), a(k, x, 1)\} = \pi_k^{t_1}(x) \cap \{a(k, x, 0), a(k, x, 1)\} \quad (3)$$

for all $s \geq t_1$. An existence of such $t_1 + 1$ follows from property 2.

If we have to prove nothing. So, we suppose that $g_k^{t_1}(x)$ is undefined.

By property 1, either $\pi_k(x) = \alpha(0)$ or $\pi_k(x) = \alpha(1)$. At first, suppose that $\pi_k(x) = \alpha(0)$. If $a(k, x, 1) \in \alpha^{t_1+1}(0)$ then at some stage $t_2 + 1 \in T_m$, where $t_0 < t_2 \leq t_1$, number $a(k, x, 1)$ is enumerated into $\alpha(0)$ by the instructions of *Subcase* 3.1. Then $g_k^{t_2}(x)$ is defined and we have to prove nothing. Thus, we can suppose that $a(k, x, 1) \notin \alpha^{t_1+1}(0)$ and that $g_k^{t_1}(x)$ is undefined.

Obviously, *Case* 1 and *Subcase* 2.1 do not hold at stage t_1+1. *Subcase* 2.2 is also impossible at stage $t_1 + 1$. Otherwise, $a(k, x, 0)$ have to be extracted from $\alpha(0)$ at stage $t_1 + 1$ in contradiction with (1). It is sufficient to show that the conditions of *Subcase* 2.3 are valid at stage $t_1 + 1$. We will do this by contradiction. Assume that $a(k, x, 0) \notin \pi_k^s(x)$ for all $s \leq t_1$. Then $a(k, x, 0) \notin \pi_k^t(x)$ for all t because of (3). Therefore, *Subcase* 2.2 is not impossible before stage $t_1 + 1$. This implies $a(k, x, 0) \in \alpha(0) \setminus \pi_k(x)$, a contradiction. So, we can assume that $a(k, x, 0)$ was enumerated into $\pi_k(x)$ at some stage $s \leq t_1$ and $a(k, x, 0) \in \alpha^t(0) \cup \pi_k^{t+1}(x)$. Since *Subcase* 2.2 is impossible at stage $t_1 + 1$ it follows that either $a(k, x, 0) \in \alpha^{t_1}(0) \setminus \pi_k^{t_1+1}(x)$ or $a(k, x, 0) \in \pi_k^{t_1+1}(x) \setminus \alpha^{t_1}(0)$. Hence, $a(k, x, 0) \in (\alpha(0) \setminus \pi_k(x)) \cup (\pi_k(x) \setminus \alpha(0))$ because of (1) and (3). A contradiction. Thus, if $g_k^{t_1}(x)$ is undefined then $g_k^{t_1+1}(x) = 0$ according with the instructions of *Subcase* 2.3.

Let now $\pi_k(x)$ be equal to $\alpha(1)$. By property 4 and equality (2), $\{a(k, x, 0), a(k, x, 1)\} \subseteq \pi_k^{t_1+1}(x)$. If $g_k^{t_1}(x)$ is still undefined then *Subcase* 2.1 holds at stage $t_1 + 1$ and we define $g_k^{t_1+1}(x) = 1$.

7. For any k, if π_k is a numbering of $\mathcal{A}$ then $\pi_k(x) = \alpha(g_k(x))$ for all x.

Proof. Let π_k be any numbering of $\mathcal{A}$ and let x be an arbitrary number. Let m, T_m, and $t_0 + 1$ be chosen as in the proof of property 6. By property 6, $g_k(x)$ is defined at some stage $t_1 + 1 \in T_m$.

By construction, either $g_k(x) = 0$ or $g_k(x) = 1$. At first, assume that $g_k(x) = 0$ and prove equality $\pi_k(x) = \alpha(0)$. Since $g_k^{t_1+1}(x) = 0$ it follows that only *Subcase* 2.3 could take place at stage $t_1 + 1$. Hence, $a(k, x, 0) \notin \pi_k^{t_1+1}(x)$ and $a(k, x, 0) \in \pi_k^s(x)$ for some $s \leq t_1$. It follows that $a(k, x, 0) \notin \pi_k(x)$ since $\pi_k(x) \in \Sigma_2^{-1}$. Property 4 implies inequality $\pi_k(x) \neq \alpha(1)$. Therefore, $\pi_k(x) = \alpha(0)$.

Assume now $g_k(x) = 1$ and prove that $\pi_k(x) = \alpha(1)$. Since $g_k^{t_1+1}(x) = 1$ it follows that only *Subcase* 2.1 could take place at stage $t_1 + 1$. Hence, $a(k, x, 1) \in \pi_k^{t_1+1}(x)$. Note that, by construction, number $a(k, x, 1)$ can be enumerated into $\alpha(0)$ if and only if *Subcase* 3.1 holds at some stage from T_m followed stage $t_1 + 1$.

We have only two alternatives for *Subcase* 3.1:

(i) *Subcase* 3.1 never holds at stages $t + 1 > t_1 + 1$ from T_m. Then, for all these stages, $a(k, x, 1) \in \pi_k^{t+1}(x)$ and $a(k, x, 1)$ is not enumerated into $\alpha^{t+1}(0)$. Thus, $a(k, x, 1) \in \pi_k(x) \setminus \alpha(0)$. Therefore, $\pi_k(x) = \alpha(1)$.

(ii) There exists stage $t_2 + 1 \in T_m$ such that $t_2 + 1 > t_1 + 1$ and *Subcase* 3.1 holds at stage $t_2 + 1$. Then $a(k, x, 1) \notin \pi_k(x)$ but $a(k, x, 1) \in \alpha(0) \cap \alpha(1)$. Therefore, $\pi_k(x) \notin \mathcal{A}$. A contradiction. Hence, alternative (ii) is impossible.

8. $\alpha(0) \subset \alpha(1)$.

Proof. By properties 3,4, $\alpha(0) \subseteq \alpha(1)$ since the function $a(k, x, i)$ is injective. Remind that $\pi_0^t(0) = \emptyset$ for all t. It implies immediately that *Subcases* 2.1, 3.1 never hold at stages $t + 1$ such that $l(t) = c(0, 0)$. Therefore, $a(0, 0, 1) \in \alpha(1) \setminus \alpha(0)$. $\qquad\square$

3. Infinite Rogers semilattices

Let A and B be any sets. For an arbitrary non-trivial subset R of ω, we will denote by ν_R a numbering of the family $\mathcal{A} = \{A, B\}$ defined as

$$\nu_R(x) = \begin{cases} A, & \text{if } x \in R \\ B, & \text{if } x \notin R \end{cases}$$

for each $x \in \omega$. For every sets R, Q, we have obviously $\nu_R \leqslant \nu_Q$ if and only if $R \leqslant_m Q$. Besides, $\nu_R \oplus \nu_Q \equiv \nu_{R \oplus Q}$. Thus, mapping ν induces isomorphism of upper semilattice of $\mathbf{m}-$ degrees onto upper semilattice of all numberings of $\mathcal{A}$ modulo equivalence of numberings. This is well-known fact of general theory of numberings [10], in which a nature of the elements A, B of the family $\mathcal{A}$ does not matter at all.

But if the family $\mathcal{A}$ consists of embedded c.e. sets $B \subset A$ and R is c.e. too then ν_R is classically computable numbering and mapping ν induces isomorphism of upper semilattice $\mathbf{L}_{\mathbf{m}}^{\mathbf{0}}$ of $\mathbf{m}-$ degrees of non-trivial c.e. sets onto Rogers semilattice $\mathcal{R}_1^0(\mathcal{A})$ [10]. Computability of numbering ν_R is a straightforward with monotonicity of the following uniform procedure: for each x enumerate numbers of B into $\nu_R(x)$ until x have been appeared in some fixed enumeration of R, and if this occurs then enumerate numbers of A into $\nu_R(x)$.

If we consider $\Sigma_2^{-1}-$ sets A and B and try to switch from enumeration of B to enumeration of A during building $\nu_R(x)$ we could meet incorrect result. For instance, let number $a \in A \setminus B$ have been appeared in enumeration of B. Then it have to be extracted from B later, say, at stage s. For all $x \in R$ such that x is enumerated in R after stage s, switching will result either $a \notin \nu_R(x)$ (and, hence, $\nu_R(x) \neq A$) or $\nu_R(x) \notin \Sigma_2^{-1}$. Besides, if $A \setminus B$ contains two or more elements then $\nu_R(x) \neq B$ too. Searching for the conditions to avoid incorrect switching lead us to the following statement.

Theorem 3.1. *Let n be non-zero even number, $A, B \in \Sigma_n^{-1}$, and $\mathcal{A} = \{A, B\}$. Besides, let $B \in \Sigma_{n-1}^{-1}$ if n is odd. Let R be an arbitrary non-trivial c.e. set. If there exist c.e. sets B_0, C and there exists $\Sigma_{n-1}^{-1}-$ set B_1 such that*

(i) $B = B_0 \setminus B_1$, $B_1 \cap A = \emptyset$,
(ii) $C \cap A = \emptyset$,
(iii) $C \supseteq B \setminus A$

then ν_R is $\Sigma_n^{-1}-$ computable numbering of the family $\mathcal{A}$.

Proof. We will build some numbering α by suggesting a uniformly computable enumeration procedure for the sets $\alpha(x), x \in \omega$. We fix arbitrary number x and describe the set $\alpha^t(x)$ for any stage t.

Let A_0 be any c.e. set and let A_1 be $\Sigma_{n-1}^{-1}-$ set such that $A_0 \setminus A_1 = A$. Fix any computable approximations $\{A_0^t\}_{t\in\omega}$, $\{A_1^t\}_{t\in\omega}$, $\{B_0^t\}_{t\in\omega}$, $\{B_1^t\}_{t\in\omega}$, $\{C^t\}_{t\in\omega}$, $\{R^t\}_{t\in\omega}$ of the sets A_0, A_1, B_0, B_1, C, R correspondingly.

The construction

Stage 0. Let $\alpha^0(x) = \emptyset$. Go to the next stage.

Stage $t + 1$. If $x \notin R^t$ then let $\alpha^{t+1}(x) = B_0^t \setminus B_1^t$. If $x \in R^t$ then let

$$\alpha^{t+1}(x) = [\bigcup_{i\leq s}\alpha^i(x) \setminus (B_1^t \cup C^t)] \cup [(A_0^t \setminus \bigcup_{i\leq s}\alpha^i(x)) \setminus (A_1^t \setminus \bigcup_{i\leq s}\alpha^i(x))].$$

Here s is the least stage such that $x \in R^s$. Go to the next stage.

Note that, for each sequence $\{X^t\}_{t\in\omega}$ of sets considered in the construction, $\underline{\lim}\, X^t = \overline{\lim}\, X^t$ since, for every number x, there exists stage t_x such that either $x \in X^t$ for all $t > t_x$ or $x \notin X^t$ for all $t > t_x$. Thus, we can use operator lim instead of operators $\underline{\lim}$ and $\overline{\lim}$. Remind [16], that

$$\underline{\lim}\, X^t = \bigcup_{i\geq 0}\bigcap_{j\geq 0} X^{i+j} \quad \text{and} \quad \overline{\lim}\, X^t = \bigcap_{i\geq 0}\bigcup_{j\geq 0} X^{i+j}.$$

We will show that $\alpha(x) = \nu_R(x)$ for all x and that α is $\Sigma_n^{-1}-$ computable numbering. If $x \notin R$ then

$$\alpha(x) = \lim_{t\to\infty}(B_0^t \setminus B_1^t) = \lim_{t\to\infty} B_0^t \setminus \lim_{t\to\infty} B_1^t = B_0 \setminus B_1 = B,$$

and therefore, $\alpha(x) = \nu_R(x)$.

Let now $x \in R$ and show that $\alpha(x) = A$. Let s be the least stage such that $x \in R^s$. By construction,

$$\alpha(x) = \lim_{t\to\infty}\{[\bigcup_{i\leq s}\alpha^i(x) \setminus (B_1^t \cup C^t)] \cup [(A_0^t \setminus \bigcup_{i\leq s}\alpha^i(x)) \setminus (A_1^t \setminus \bigcup_{i\leq s}\alpha^i(x))]\}.$$

We apply elementary properties of lim operator and set-theoretic operations to get the following equalities:

$$\alpha(x) = [\bigcup_{i\leq s}\alpha^i(x) \setminus (B_1 \cup C)] \cup [(A_0 \setminus \bigcup_{i\leq s}\alpha^i(x)) \setminus (A_1 \setminus \bigcup_{i\leq s}\alpha^i(x))]$$

$$= [\bigcup_{i\leq s}\alpha^i(x) \setminus (B_1 \cup C)] \cup [A \setminus \bigcup_{i\leq s}\alpha^i(x)].$$

To show that $\alpha(x) = A$ it is sufficient to prove now that

$$\bigcup_{i \leq s} \alpha^i(x) \setminus (B_1 \cup C) = \bigcup_{i \leq s} \alpha^i(x) \cap A.$$

Obviously,

$$\bigcup_{i \leq s} \alpha^i(x) \setminus (B_1 \cup C) = [(\bigcup_{i \leq s} \alpha^i(x) \cap A) \setminus (B_1 \cup C)]$$

$$\cup [(\bigcup_{i \leq s} \alpha^i(x) \setminus A) \setminus (B_1 \cup C)]$$

$$= [\bigcup_{i \leq s} \alpha^i(x) \cap A \cap \overline{B_1} \cap \overline{C}] \cup [\bigcup_{i \leq s} \alpha^i(x) \cap \overline{A} \cap \overline{B_1} \cap \overline{C}].$$

Hypothesizes (i), (ii) of theorem imply that $A \cap \overline{B_1} \cap \overline{C} = A$ and, hence, we need only to show that

$$\bigcup_{i \leq s} \alpha^i(x) \cap \overline{A} \cap \overline{B_1} \cap \overline{C} = \emptyset.$$

It is easy to see the latter with evident inclusions $\alpha^i(x) \subseteq B_0$ for all $i \leq s$ and hypothesis (iii) of theorem:

$$\bigcup_{i \leq s} \alpha^i(x) \cap \overline{A} \cap \overline{B_1} \cap \overline{C} \subseteq B_0 \cap \overline{A} \cap \overline{B_1} \cap \overline{C} = B \cap \overline{A} \cap \overline{C} = (B \setminus A) \cap \overline{C} = \emptyset.$$

Thus, $\nu_R(x) = \alpha(x)$ for every x. Procedure of building the sets $\alpha^t(x)$ is uniformly computable on x, t. Therefore, we have to check thoroughly that, for every y, number of 'enter–leaves' y into–from set $\alpha(x)$ does not exceed n by the end of the construction.

To be precise in the evaluating a number of the 'enter-leaves' we should note the following two remarks concerning the computation procedures in our construction.

(i) How should we execute the procedures defined by expressions of the type $X^{t+1} = Y^{t+1} \setminus Z^{t+1}$? Here $\{Y^t\}$ and $\{Z^t\}$ are given computable approximations of some sets while an approximation $\{X^t\}$ is creating in our construction. Trivially, if some number x appears Y^{t+1}, i.e. $x \in Y^{t+1}$, and $x \notin Z^{t+1}$ then we enumerate x into X^{t+1} even if $x \in X^t$. Evidently, if $x \notin Y^{t+1}$ we do not enumerate x into X^{t+1}. And what about the case if $x \in Y^{t+1}$, $x \notin Y^{t+1}$, and $x \in Z^{t+1}$? In this case we do not touch x at stage $t + 1$ at all. By the other words, number of the 'enter-leaves' done for x at stage $t + 1$ is equal to 0.

(ii) For every k, x, t and any $X \in \Sigma_k^{-1}$, if $x \in X^t$ then number of the 'enter-leaves' for x during computation of X by the end of stage t does not exceed

k in the case of odd k and it is less than or equal to $k-1$ in the case of even k.

Let x, y be the arbitrary numbers. If $x \notin R$ then behavior of y with respect to $\alpha(x)$ is completely controlled by the enumeration procedure for B. Therefore, total number of the 'enter-leaves' for y by the end of the construction does not exceed n.

If $x \in R^s$, where s is the least stage when x has been enumerated in R, and $y \in \bigcup_{i \leq s} \alpha^i(x)$ then number of the 'enter-leaves' for y by the end of stage s does not exceed $n-1$ because of (ii). By construction, after stage s number y may be extracted from $\alpha(x)$ but never enumerated into $\alpha(x)$ again. Therefore, total number of the 'enter-leaves' for y is less or equal to n. In opposite, if $y \notin \bigcup_{i \leq s} \alpha^i(x)$ then number of the 'enter-leaves' for y by the end of stage s is equal to 0 because of (i). By construction, after stage s, y can enter or leave $\alpha(x)$ according to enumeration procedure for A only. Therefore, in all cases, total number of the 'enter–leaves' for y into–from the set $\alpha(x)$ does not exceed n. $\qquad\square$

We can construct infinitely many pairwise incomparable numberings ν_R of the family A by means of varying R over an infinite family of pairwise $\mathbf{m}$–incomparable c.e. sets. Indeed, we can prove more strong result.

Theorem 3.2. *Let a family A satisfy the hypothesizes of Theorem 3.1. Then there exists a principal ideal of Rogers semilattice $\mathcal{R}_n^{-1}(A)$ isomorphic to upper semilattice $\mathbf{L}_\mathbf{m}^0$.*

Proof. We will show that the subset $\{\deg(\nu_R) \mid R \text{ is non-trivial c.e. set}\}$ of Rogers semilattice $\mathcal{R}_n^{-1}(A)$ forms a principal ideal. Let K be a creative set. Then $\nu_R \leqslant \nu_K$ for every non-trivial c.e. set R. Therefore, we have only to show that if β is Σ_n^{-1}–computable numbering of A reducible to ν_K then $\beta \equiv \nu_Q$ for some c.e. set Q. The latter is almost evident. Let f be a computable function such that $\beta(x) = \nu_K f(x)$ for all x. It is sufficient to take $Q = f^{-1}(K)$. $\qquad\square$

Corollary 3.1. *If a family A satisfies the hypothesizes of Theorem 3.1, and a family B is disjoint with A and has a minimal Σ_n^{-1}–computable numbering then Rogers semilattice $\mathcal{R}_n^{-1}(A \cup B)$ contains an ideal isomorphic to $\mathbf{L}_\mathbf{m}^0$.*

Proof. Let α be any minimal Σ_n^{-1}–computable numbering of B and consider the subset $\{\deg(\nu_R \oplus \alpha) \mid R \text{ is non-trivial c.e. set}\}$ of Rogers semilattice $\mathcal{R}_n^{-1}(A \cup B)$. It is obvious that, for every non-trivial c.e. sets R, Q,

$$R \leqslant_m Q \Leftrightarrow \nu_R \oplus \alpha \leqslant \nu_Q \oplus \alpha.$$

Therefore, we need only to show that, for every numbering β of $\mathcal{A} \cup \mathcal{B}$ if $\beta \leqslant \nu_K \oplus \alpha$ then there exists c.e. set R such that $\beta \equiv \nu_R \oplus \alpha$.

Since $\mathcal{A} \cap \mathcal{B} = \emptyset$ it follows that there exist a numbering γ of $\mathcal{A}$ and a numbering δ of $\mathcal{B}$ such that

$$\gamma \leqslant \nu_K, \ \delta \leqslant \alpha, \ \gamma \oplus \delta \equiv \beta.$$

As in the proof of Theorem 3.2, we can conclude that $\gamma \equiv \nu_R$ for some c.e. set R. $\delta \leqslant \alpha$ implies $\delta \equiv \alpha$ since α is minimal numbering. Hence, $\beta \equiv \nu_R \oplus \alpha$. $\qquad \square$

Corollary 3.2. *Let $\mathcal{A} = \{A, B\}$ be a family of Σ_n^{-1}-sets and let $n \geq 2$. Besides, let $B \in \Sigma_{n-1}^{-1}$ if n is odd. Then, for each of the following 6 cases, Rogers semilattice $\mathcal{R}_n^{-1}(\mathcal{A})$ contains an ideal isomorphic to $\mathbf{L_m^0}$:*

(1) A is computable set and there exist c.e. set B_0 and Σ_{n-1}^{-1}-set B_1 such that $B = B_0 \setminus B_1$ and $A \cap B_1 = \emptyset$;

(2) B is c.e. set and there exists c.e. set C such that $C \supseteq B \setminus A$ and $C \cap A = \emptyset$;

(3) $B \subset A$ and there exist c.e. set B_0 and Σ_{n-1}^{-1}-set B_1 such that $B = B_0 \setminus B_1$ and $B_1 \cap A = \emptyset$;

(4) $A \subset B$ and there exists c.e. set C such that $C \supseteq B \setminus A$ and $C \cap A = \emptyset$;

(5) B and $B \setminus A$ are c.e. sets;

(6) A and B are c.e. sets and $A \cap B = \emptyset$.

Proof. We will apply Theorem 3.2 and specify the sets A_0, A_1, B_0, B_1 and C to meet hypothesizes (i)–(iii) of Theorem 3.1 in each of cases (1)–(6).

(1) Since A is computable and hypothesis (i) holds it follows that we can take $C = \overline{A}$.

(2) Since B is c.e. set and hypothesizes (ii),(iii) hold for c.e. set C it follows that we can take $B_0 = B$ and $B_1 = \emptyset$.

(3) Hypothesis (i) holds. $B \subset A$ implies that we can take $C = \emptyset$.

(4) Hypothesizes (ii),(iii) hold for c.e. set C. Since $\mathcal{A} = \{A, B\}$ is Σ_n^{-1}-computable family it follows that there exist c.e. set B_0 and Σ_{n-1}^{-1}-set B_1 such that $B = B_0 \setminus B_1$. Inclusion $A \subset B$ implies $B_1 \cap A = \emptyset$ automatically.

(5) Let $C = B \setminus A$. Then hypothesizes (ii),(iii) hold evidently. To meet hypothesis (i) we can take $B_0 = B$ and $B_1 = \emptyset$.

(6) Since A, B are c.e. sets and $A \cap B = \emptyset$ it is sufficient to take $B_0 = B, B_1 = \emptyset, C = B$. $\qquad \square$

Corollary 3.3. *Let a family $\mathcal{A}$ meet one of cases (1)–(6) from Corollary 3.2. If $\mathcal{B}$ is a family disjoint with $\mathcal{A}$ and $\mathcal{B}$ has a minimal Σ_n^{-1}-com-*

putable numbering then Rogers semilattice $\mathcal{R}_n^{-1}(\mathcal{A} \cup \mathcal{B})$ *contains an ideal isomorphic to* $\mathbf{L}_m^0$.

Proof. This statement is a straightforward consequence of the preceding two corollaries. □

References

1. S.S. Goncharov, A. Sorbi, Generalized computable numberings and nontrivial Rogers semilattices. Algebra and Logic, 1997, v. 36, no. 4, 359–369.
2. S. Badaev, S. Goncharov, S. Podzorov, and A. Sorbi. Algebraic properties of Rogers semilattices of arithmetical numberings, in Computability and Models, eds. S.B. Cooper and S. Goncharov, Kluwer Academic/Plenum Publishers, New York, pp. 45–77, 2003.
3. S. Badaev, S. Goncharov, and A. Sorbi. Elementary properties of Rogers semilattices of arithmetical numberings, in: Proceedings of the 7th and 8th Asian Logic Conference, eds. Rod Downey and others, World Scientific, Singapore, pp. 1–10, 2003.
4. S. Badaev, S. Goncharov, and A. Sorbi. On elementary theories for Rogers semilattices, Algebra and Logic, 2005, v. 44, no. 3, 143–147.
5. S. Goncharov, S. Lempp, D.R. Solomon. Friedberg numberings of families of n-computably enumerable sets. Algebra and Logic, 2002, v. 41, no. 2, 81–86.
6. Zh. Talasbaeva. Positive Numberings of Families of Sets in the Ershov Hierarchy. Algebra and Logic, 2003, v. 42, no. 6, 413–418.
7. S. Badaev, S. Lempp. On decomposition of Rogers semilattices. In preparation.
8. H. Rogers, Jr. *Theory of Recursive Functions and Effective Computability.* McGraw-Hill, New York, 1967.
9. R.I. Soare *Recursively Enumerable Sets and Degrees.* Springer-Verlag, Berlin-Heidelberg, 1987.
10. Yu.L. Ershov, *Theory of Numberings.* Nauka, Moscow, 1977 (Russian).
11. Yu.L. Ershov, *Theory of Numberings.* Handbook of computability theory (E.R. Griffor, ed.), Noth-Holland, Amsterdam, 1999, pp. 473–503.
12. S.A. Badaev, S.S. Goncharov, *Theory of numberings: open problems.* In *Computability Theory and its Applications.* P. Cholak, S. Lempp, M. Lerman and R. Shore eds.—Contemporary Mathematics, American Mathematical Society, 2000, vol. 257, Providence, pp. 23-38.
13. Yu.L. Ershov *On some hierarchy of sets, I.* Algebra and Logic, 1968, v. 7, no. 1, 47–74 (Russian)
14. H. Putnam, *Trial and error predicates and the solution to a problem of Mostowski.* J. Symbolic Logic, 1965, v. 30, no. 1, 49–57.
15. A.B. Khutoretsky. On the cardinality of the upper semilattice of computable numberings. Algebra and Logic, 1971, v. 10, no. 5, 348–352.
16. K. Kuratowski and A. Mostowski, *Set Theory.* North-Holland Publishing Co., Amsterdam-Warszava, 1967.

ON BEHAVIOUR OF 2-FORMULAS IN WEAKLY o-MINIMAL THEORIES*

B. S. BAIZHANOV and B. Sh. KULPESHOV

*Institute for Problems of Informatics and Control,
050010, ul. Pushkina 125,
Almaty, Kazakhstan
E-mail: baizhanov@ipic.kz, kbsh@ipic.kz*

Here we investigate formulas with two free variables properly lying in the set of realizations of a non-algebraic one-type for weakly o-minimal theories. Our main result is a description of behaviour of such formulas that are ordered by the reverse ordering on the natural numbers in an one-type (Theorem 2.1). As a consequence, we obtain a description of the set of realizations of a non-algebraic one-type up to binarity in $\aleph_0$-categorical weakly o-minimal theories (Corollary 2.3).

1. Properties of 2-formulas

Let L be a countable first-order language. Everywhere in this paper we consider L-structures and assume that L contains a binary relation symbol $<$ that is interpreted as a linear ordering in these structures. This paper concerns the notion of *weak o-minimality* originally and deeply studied by D. Macpherson, D. Marker and Ch. Steinhorn in [1]. *A weakly o-minimal structure* is linearly ordered structure $M = \langle M, =, <, \ldots \rangle$ such that any definable (with parameters) subset of M is a finite union of convex sets in M.

Definition 1.1. A formula $F(x, y)$ is said to be *convex to the right (left)* if

$$M \models \forall y \forall x \left[F(x, y) \rightarrow (y \leq x \wedge \forall z(y \leq z \leq x \rightarrow F(z, y))) \right]$$
$$(M \models \forall y \forall x \left[F(x, y) \rightarrow (x \leq y \wedge \forall z(x \leq z \leq y \rightarrow F(z, y))) \right])$$

Given a formula like $F(x, y)$ we usually think of the second variable y as the variable for the parameter.

For arbitrary subsets A, B of a structure M we write $A < B$ if $a < b$ whenever $a \in A$ and $b \in B$. If $A \subset M$ and $x \in M$ then we write $A < x$ if

---

*This work is supported by Grant CRDF KMZ1-2620-AL-04

32

$A < \{x\}$. For any subset A of a structure M $A^+ := \{b \in M \mid A < b\}$ and $A^- := \{b \in M \mid b < A\}$. For an arbitrary complete type p we denote by $p(M)$ the set of realizations of the type p in M.

In the following definitions M is a weakly o-minimal structure, $A \subseteq M$, $p \in S_1(A)$ is non-algebraic.

Definition 1.2. An A-definable formula $F(x,y)$ is said to be p-*stable* if there are $\alpha, \gamma_1, \gamma_2 \in p(M)$ such that $\gamma_1 < F(M, \alpha) < \gamma_2$.

Definition 1.3. Let $F_1(x,y)$, $F_2(x,y)$ be p-stable convex to the right (left) formulas. We will say $F_2(x,y)$ is *greater than* $F_1(x,y)$ if there is $\alpha \in p(M)$ such that $F_1(M, \alpha) \subset F_2(M, \alpha)$.

Lemma 1.1. *Let M be a weakly o-minimal structure, $A \subseteq M$, $p \in S_1(A)$ be non-algebraic. Then the set of all p-stable convex to the right (left) formulas is linearly ordered.*

Definition 1.4. [2] A non-isolated type $p \in S_1(A)$ is said to be *quasirational to the right (left)* if there is an A-definable convex formula $U(x) \in p$ so that for any $\alpha \in p(M)$, $\beta \in U(M)$ such that $\alpha < \beta$ $(\alpha > \beta)$ we have $\beta \in p(M)$. Otherwise, it is said to be *irrational to the right (left)*.

Lemma 1.2. *Let M be a weakly o-minimal structure, $A \subseteq M$, M be $|A|^+$-saturated, $p \in S_1(A)$ be non-algebraic, $F(x,y)$ be a p-stable convex to the right (left) formula. Then the formula $F'(x,y) := \exists z[F(z,y) \wedge F(x,z)]$ also is p-stable convex to the right (left).*

Proof of Lemma 1.2.

Without loss of generality, suppose that $F(x,y)$ is convex to the right. First prove $F'(x,y)$ is convex to the right. Consider arbitrary elements $\alpha, \beta \in M$ such that $M \models F'(\alpha, \beta)$. There is $\gamma \in M$ such that $M \models F(\gamma, \beta) \wedge F(\alpha, \gamma)$. As $F(x,y)$ is convex to the right we have $\beta \leq \gamma \leq \alpha$. Understand for any $\gamma' \in M$ such that $\beta \leq \gamma' \leq \alpha$ we have $M \models F'(\gamma', \beta)$.

Let $\gamma' \leq \gamma$. As $F(x,y)$ is convex to the right and $M \models F(\gamma, \beta)$ we have $M \models F(\gamma', \beta) \wedge F(\beta, \beta)$ and consequently $M \models F'(\gamma', \beta)$.

Let $\gamma' > \gamma$. As $F(x,y)$ is convex to the right and $M \models F(\alpha, \gamma)$ we have $M \models F(\gamma', \gamma)$ and consequently $M \models F'(\gamma', \beta)$.

Thus, $F'(x,y)$ is convex to the right. Prove now $F'(x,y)$ is p-stable.

As $F(x,y)$ is p-stable then for any $\beta \in p(M)$ $F'(M, \beta) \subseteq p(M)$ and there are $\gamma_1, \gamma_2 \in p(M)$ such that $\gamma_1 < F(M, \beta) < \gamma_2$. It is obvious $\gamma_1 < F'(M, \beta)$. Prove that there is $\gamma \in p(M)$ such that $F'(M, \beta) < \gamma$.

Case 1. p is irrational to the right.

There is an infinite family of A-definable convex formulas $D_i(x), i < \lambda$ for some infinite cardinal λ such that for each $i < \lambda$ $p(M) < D_i(M)$. Consider the following set of formulas:

$$\{\forall x[D_i(x) \to y < x] | i < \lambda\} \cup \{\forall x[F'(x, \beta) \to y > x]\} \qquad (*)$$

It is locally consistent. Then by $|A|^+$-saturation of M there is $\gamma \in p(M)$ which satisfies $(*)$. Thus, $F'(x, y)$ is p-stable.

Case 2. p is quasirational to the right or isolated.

Then there is A-definable convex formula $U(x)$ such that $p(M)^+ = U(M)^+$.

For a contradiction, suppose there is $\beta \in p(M)$ such that $M \models \theta(\beta)$ where

$$\theta(y) := \forall x(x > y \wedge U(x) \to F'(x, y))$$

It is clear $\theta(y) \in p$. Consider $\beta, \gamma_1, \gamma_2 \in p(M)$ such that $\gamma_1 < F(M, \beta) < \gamma_2$. Let $f \in Aut_A(M)$ such that $f(\gamma_2) = \gamma_1$. As $\theta(y) \in p$ $M \models \theta(f^n(\beta))$ for each $n < \omega$, i.e. $M \models F'(\gamma_2, f^n(\beta))$. Then there is $\beta_n \in F(M, f^n(\beta))$ such that $M \models F(\gamma_2, \beta_n)$.

Understand that $f^n(\beta) < \beta_n < f^{n-1}(\beta)$ for each $n < \omega$.

As $f(\gamma_2) = \gamma_1 < \beta < \gamma_2$ we have $f^n(\gamma_2) < f^{n-1}(\beta) < f^{n-1}(\gamma_2) < \ldots < \gamma_2$ for each $n < \omega$. As $M \models \neg F(\gamma_2, \beta)$ then $M \models \neg F(f^n(\gamma_2), f^n(\beta))$ and consequently $M \models \neg F(\gamma_2, f^n(\beta))$ for each $n < \omega$.

As $M \models F(\gamma_2, \beta_n) \wedge F(\beta_n, f^n(\beta))$ we have $f^n(\beta) < \beta_n$.

As $M \models \neg F(f^n(\gamma_2), f^n(\beta))$ we have $\beta_n < f^n(\gamma_2)$ and so $\beta_n < f^{n-1}(\beta)$.

Thus, $f^n(\beta) \notin F(\gamma_2, M)$, $\beta_n \in F(\gamma_2, M)$ and $f^n(\beta) < \beta_n < f^{n-1}(\beta)$ for each $n < \omega$. It contradicts to weak o-minimality of M. $\square$

Definition 1.5. Let $F(x, y)$ be a p-stable convex to the right (left) formula. We will say $F(x, y)$ is *equivalence generating* in p if for any $\alpha, \beta \in p(M)$ such that $M \models F(\beta, \alpha)$ the following holds:

$$M \models \forall x \, [x \geq \beta \to [F(x, \alpha) \leftrightarrow F(x, \beta)]]$$
$$(M \models \forall x \, [x \leq \beta \to [F(x, \alpha) \leftrightarrow F(x, \beta)]])$$

Example 1.1. Let $M = \langle Q, =, <, R^2 \rangle$. M is a linearly ordered structure, Q is the ordering of rational numbers, for any $a, b \in M$ $M \models R(b, a) \Leftrightarrow a \leq b < a + \sqrt{2}$ and consequently $R(M, a) = \{b \in M | a \leq b < a + \sqrt{2}\}$ and $R(a, M) = \{b \in M | a - \sqrt{2} < b \leq a\}$.

Let $R^1(x, y) := \exists z[R(z, y) \wedge R(x, z)]$. It is obvious that for any $a, b \in M$ $R^1(b, a) \Leftrightarrow a \leq b < 2(a + \sqrt{2})$.

Let $R^n(x,y) := \exists z_1, \ldots, z_n[R(z_1, y) \wedge \wedge_{1 \le i < n} R(z_{i+1}, z_i) \wedge R(x, z_n)]$ for each $n < \omega$. One can see that for any $a, b \in M$

$$R^n(b, a) \Leftrightarrow a \le b < (n+1)(a + \sqrt{2})$$

Consequently, for any $a \in M$ we have

$$R(M, a) \subset R^1(M, a) \subset \ldots \subset R^n(M, a) \subset \ldots$$

i.e. we have infinitely many pairwise non-equivalent 2-formulas and thus $Th(M)$ is not $\aleph_0$-categorical. The formulas $R^n(x, y)$ for each $n < \omega$ including in addition atomic formulas we declare to be basic. It can now be shown by standard arguments that $Th(M)$ admits elimination of quantifiers relative to these basic formulas. Now it is not difficult to see that for any formula $\phi(x, \bar{y})$ for any model N of $Th(M)$ for all $\bar{a} \in N$, $\phi(N, \bar{a})$ is a union of finitely many convex subsets, that is, $Th(M)$ is a weakly o-minimal theory. Let $p(x) := \{x = x\}$. It is easy to see that $p(x) \in S_1(\emptyset)$. It is obvious that $R(x, y)$ is p-stable convex to the right and $R(x, y)$ is not equivalence generating.

Lemma 1.3. *Let M be a weakly o-minimal structure, $A \subseteq M$, M be $|A|^+$-saturated, $p \in S_1(A)$ be non-algebraic, $F(x, y)$ be a p-stable convex to the right (left) formula. Suppose that $F(x, y)$ is not equivalence generating. Then there are $\alpha, \beta \in p(M)$ such that*

$$M \models F(\beta, \alpha) \wedge \exists x[\neg F(x, \alpha) \wedge F(x, \beta)]$$

Proof of Lemma 1.3.

Without loss of generality, suppose that $F(x, y)$ is convex to the right. For a contradiction, suppose that there aren't $\alpha, \beta \in p(M)$ such that

$$M \models F(\beta, \alpha) \wedge \exists x[\neg F(x, \alpha) \wedge F(x, \beta)]$$

As $F(x, y)$ is not equivalence generating then there are $\alpha, \beta \in p(M)$ such that $M \models F(\beta, \alpha) \wedge \exists x[F(x, \alpha) \wedge \neg F(x, \beta)]$. Consequently there is $\gamma \in p(M)$ such that $\gamma \in F(M, \alpha) \setminus F(M, \beta)$.

Consider $f \in Aut_A(M)$ such that $f(\gamma) = \alpha$. As $\alpha < \beta < \gamma$ we have $f^n(\alpha) < f^n(\beta) < f^n(\gamma) = f^{n-1}(\alpha)$ for each $n < \omega$. As $M \models F(\gamma, \alpha)$ we have $M \models F(f^{n-1}(\alpha), f^n(\alpha))$ for each $n < \omega$. By our supposition we have $M \models F(\gamma, f^n(\alpha))$ for each $n < \omega$. As $M \models \neg F(\gamma, \beta)$ we have $M \models \neg F(f^n(\gamma), f^n(\beta))$ and consequently $M \models \neg F(\gamma, f^n(\beta))$ for each $n < \omega$.

Thus, $F(\gamma, M)$ is a union of infinitely many convex sets, contradicting to weak o-minimality of M. $\qquad\square$

Lemma 1.4. *Let M be a weakly o-minimal structure, $A \subseteq M$, $p \in S_1(A)$ be non-algebraic, M be $|A|^+$-saturated. Suppose that $F(x,y)$ is a p-stable convex to the right formula so that $F(x,y)$ is equivalence generating. Then*

1) $G(x,y) := F(y,x)$ is a p-stable convex to the left formula which is also equivalence generating.

2) $E(x,y) := F(x,y) \vee F(y,x)$ is an equivalence relation which partitions $p(M)$ into convex classes.

Proof of Lemma 1.4.

1) As $F(x,y)$ is convex to the right we have $M \models \forall y \forall x\, [F(x,y) \to y \leq x]$ and consequently $M \models \forall x \forall y\, [G(x,y) \to x \leq y]$.

Let $\alpha, \beta \in M$ such that $M \models G(\beta, \alpha)$. Then we have $\beta \leq \alpha$. Prove that for any β' such that $\beta \leq \beta' \leq \alpha$ we have $M \models G(\beta', \alpha)$.

As $M \models G(\beta, \alpha)$ then $M \models F(\alpha, \beta)$. As $F(x,y)$ is convex to the right we have $M \models F(\beta', \beta)$. As $F(x,y)$ is equivalence generating we have $M \models F(\alpha, \beta')$. Therefore $M \models G(\beta', \alpha)$ and $G(x,y)$ is convex to the left.

Prove that $G(x,y)$ is p-stable. Take an arbitrary $\alpha \in p(M)$ and consider $G(M, \alpha)$. It needs to find $\gamma_1, \gamma_2 \in p(M)$ such that $\gamma_1 < G(M, \alpha) < \gamma_2$. It is obvious for any $\gamma_2 \in p(M)$ such that $\gamma_2 > \alpha$ we have $G(M, \alpha) < \gamma_2$. Show an existence of such γ_1. For a contradiction, suppose that for any $\gamma_1 \in p(M)$ such that $\gamma_1 < \alpha$ we have $M \models G(\gamma_1, \alpha)$. Take an arbitrary $\beta \in p(M)$ such that $\beta < \alpha$ and consider $F(M, \beta)$. By our supposition $M \models G(\beta, \alpha)$ and consequently $M \models F(\alpha, \beta)$. As $F(x,y)$ is p-stable there is $\gamma_2 \in p(M)$ such that $F(M, \beta) < \gamma_2$. Consider $f \in Aut_A(M)$ such that $f(\gamma_2) = \beta$. As $\beta < \gamma_2$ then $f(\beta) < f(\gamma_2)$, i.e. $f(\beta) < \beta$. By the supposition we have $M \models G(f(\beta), \alpha)$. Consequently $M \models G(\beta, f^{-1}(\alpha))$ or equivalently $M \models F(f^{-1}(\alpha), \beta)$. As $\beta < \alpha$ then $f^{-1}(\beta) < f^{-1}(\alpha)$, i.e. $\gamma_2 < f^{-1}(\alpha)$. We have $M \models \neg F(\gamma_2, \beta) \wedge F(f^{-1}(\alpha), \beta) \wedge \gamma_2 < f^{-1}(\alpha)$ contradicting to $F(x,y)$ is convex to the right. Consequently $G(x,y)$ is p-stable.

Prove now that $G(x,y)$ is equivalence generating. For a contradiction, suppose that $G(x,y)$ is not equivalence generating. By Lemma 1.3 there are $\alpha, \beta \in p(M)$ such that $M \models G(\beta, \alpha) \wedge \exists x [\neg G(x, \alpha) \wedge G(x, \beta)]$.

There is $\gamma \in M$ such that $M \models \neg G(\gamma, \alpha) \wedge G(\gamma, \beta)$. Then we have

$$M \models F(\beta, \gamma) \wedge F(\alpha, \beta) \wedge \neg F(\alpha, \gamma) \wedge \gamma \leq \beta \leq \alpha$$

It contradicts to $F(x,y)$ is equivalence generating. Thus, $G(x,y)$ is equivalence generating.

2) It is obvious the relation $E(x,y)$ is reflexive and symmetrical. Prove transitivity of E. Let $M \models E(\alpha, \beta) \wedge E(\beta, \gamma)$. Without loss of generality suppose that $\alpha < \beta < \gamma$. Then we have $M \models F(\beta, \alpha) \wedge F(\gamma, \beta)$. As $F(x,y)$

is equivalence generating we have $M \models F(\gamma, \alpha)$ and consequently $M \models E(\gamma, \alpha)$. Thus, $E(x, y)$ is an equivalence relation.

Prove now that for any $\alpha \in M$ $E(\alpha, M)$ is convex. Let $\gamma_1, \gamma_2 \in E(\alpha, M)$. Without loss of generality we can assume $\alpha < \gamma_1 < \gamma_2$. Then we have $M \models F(\gamma_1, \alpha) \wedge F(\gamma_2, \alpha)$. As $F(x, y)$ is convex to the right $M \models F(\gamma, \alpha)$ and consequently $M \models E(\gamma, \alpha)$ for any γ such that $\gamma_1 \leq \gamma \leq \gamma_2$. $\qquad\square$

Definition 1.6. We will say p is *semiquasisolitary to the right (left)* if there is the greatest p-stable convex to the right (left) formula.

Definition 1.7. We will say p is *quasisolitary* if p is semiquasisolitary both to the right and to the left.

In Example 1.1 the type p is not quasisolitary.

Proposition 1.1. *Let M be a weakly o-minimal structure, $A \subseteq M$, M be $|A|^+$-saturated, $p \in S_1(A)$ be non-algebraic. Then*

1. If $F(x, y)$ is the greatest p-stable convex to the right (left) formula then $F(x, y)$ is equivalence generating.

2. Any semiquasisolitary one-type is quasisolitary.

Proof of Proposition 1.1.

1. Let $F(x, y)$ be the greatest p-stable convex to the right formula. Suppose that $F(x, y)$ is not equivalence generating. By Lemma 1.3 there are $\alpha, \beta \in p(M)$ such that $\quad M \models F(\beta, \alpha) \wedge \exists x[\neg F(x, \alpha) \wedge F(x, \beta)]$.

By Lemma 1.2 $F'(x, y) := \exists z[F(z, y) \wedge F(x, z)]$ is p-stable convex to the right. It is obvious $F(M, \alpha) \subset F'(M, \alpha)$, contradicting to $F(x, y)$ is the greatest.

2. Let $F(x, y)$ be the greatest p-stable convex to the right formula. By item 1 $F(x, y)$ is equivalence generating. By Lemma 1.4 $G(x, y) := F(y, x)$ is a p-stable convex to the left formula. For a contradiction, suppose $G(x, y)$ is not the greatest. Then there is a p-stable convex to the left formula $G'(x, y)$ which is greater than $G(x, y)$. Consequently there are $\alpha, \gamma \in p(M)$ such that $\gamma \in G'(M, \alpha) \backslash G(M, \alpha)$ $\quad(*)$. Let $G'_0(\gamma, x)$ be a convex subformula of $G'(\gamma, x)$ such that $M \models G'_0(\gamma, \alpha)$. As $G'(x, y)$ is p-stable there is $\gamma' \in p(M)$ such that $\gamma' < G'(M, \alpha)$. Consider $f \in Aut_A(M)$ such that $f(\gamma') = \alpha$. As $\gamma' < \gamma < \alpha$ then $\alpha = f(\gamma') < f(\gamma) < f(\alpha)$. As $M \models \neg G'(\gamma', \alpha)$ we have $M \models \neg G'(\alpha, f(\alpha))$ and consequently $M \models \neg G'(\gamma, f(\alpha))$.

We have $M \models \neg G'_0(\gamma, f(\alpha))$ and consequently $G'_0(\gamma, M) < f(\alpha)$ $\quad(**)$. Consequently $G'_0(\gamma, M) \subseteq p(M)$. Consider the following formula:

$$F'(x, y) := x \geq y \wedge [G'_0(y, x) \vee \exists z(G'_0(y, z) \wedge x \leq z)]$$

Prove $F'(x, y)$ is p-stable convex to the right and greater than $F(x, y)$.
Consider arbitrary $\gamma_1, \gamma_2 \in M$ such that $M \models F'(\gamma_2, \gamma_1)$. Then

$$M \models \gamma_2 \geq \gamma_1 \wedge [G_0'(\gamma_1, \gamma_2) \vee \exists z(G_0'(\gamma_1, z) \wedge \gamma_2 \leq z)]$$

Consider arbitrary $\beta \in M$ such that $\gamma_1 \leq \beta \leq \gamma_2$. If $M \models G_0'(\gamma_1, \beta)$ then $M \models F'(\beta, \gamma_1)$. If not we have $M \models \beta < \gamma_2 \wedge G_0'(\gamma_1, \gamma_2)$ and consequently $M \models F'(\beta, \gamma_1)$. Thus, $F'(x, y)$ is convex to the right.

Let $g \in Aut_A(M)$ such that $g(\gamma) = \gamma_1$. By $(**)$ we have $G_0'(\gamma_1, M) < g(f(\alpha))$ and consequently $F'(M, \gamma_1) < g(f(\alpha))$, i.e. $F'(x, y)$ is p-stable.

Understand that $F(M, \gamma) \subset F'(M, \gamma)$.

We have $M \models F'(\alpha, \gamma)$. If $M \models F(\alpha, \gamma)$ then $M \models G(\gamma, \alpha)$, contradicting to $(*)$. Consequently $\alpha \in F'(M, \gamma) \setminus F(M, \gamma)$ and thus, $F'(x, y)$ is greater than $F(x, y)$, contradicting to $F(x, y)$ is the greatest. $\qquad\square$

2. Main theorem

Lemma 2.1. *Let T be a weakly o-minimal theory, $M \models T$, $A \subseteq M$, M be $|A|^+$-saturated, $p \in S_1(A)$ be non-algebraic. Suppose that $E(x, y)$ is an A-definable non-trivial equivalence relation which partitions $p(M)$ into convex classes. Then E partitions $p(M)$ into infinitely many such classes, so that the induced ordering on classes is either a dense order without endpoints or a discrete order without endpoints.*

Proof of Lemma 2.1.

First show that there is no leftmost E-class which is contained in $p(M)$. By assumption there are elements $\alpha, \beta \in p(M)$ with $\alpha < \beta$ and $\neg E(\alpha, \beta)$. As all elements of $p(M)$ have the same type as β over A we have: for every element β' in $p(M)$ there exists $\alpha' \in p(M)$ such that $\alpha' < \beta'$ and $\neg E(\alpha', \beta')$. Therefore there is no smallest E-class in $p(M)$. We can also show that there is no the rightmost E-class. Thus, E partitions $p(M)$ into infinitely many classes. Now, consider the following formula:

$$\Phi(x) := \exists z[\neg E(z, x) \wedge x < z \wedge \forall t(x < t < z \rightarrow E(x, t) \vee E(z, t))]$$

If $\Phi(x) \in p$ then E-classes are discretely ordered. If not, then E-classes are densely ordered. $\qquad\square$

Corollary 2.1. *Let T be an $\aleph_0$-categorical weakly o-minimal theory, $M \models T$, $A \subseteq M$, M be $|A|^+$-saturated, $p \in S_1(A)$ be non-algebraic. Suppose that $E(x, y)$ is an A-definable non-trivial equivalence relation which partitions $p(M)$ into convex classes. Then the induced order on E-classes is a dense order without endpoints.*

Lemma 2.2. *Let T be an $\aleph_0$-categorical weakly o-minimal theory, $M \models T$, $A \subseteq M$, M be $|A|^+$-saturated, $p \in S_1(A)$ be non-algebraic. Suppose that $E_1(x, y)$, $E_2(x, y)$ are A-definable equivalence relations which partition $p(M)$ into convex classes so that there is an element $\alpha \in p(M)$ such that $E_1(M, \alpha) \subset E_2(M, \alpha)$. Then E_1 partitions each E_2-class into infinitely many E_1-classes.*

If $F(x, y)$ is a p-stable convex to the right (left) formula, we will say $F(x, y)$ is *trivial* if for any $\alpha \in p(M)$ $M \models \forall x\,[\,F(x, \alpha) \to x = \alpha\,]$. Otherwise, such a formula is said to be *non-trivial*. A quasisolitary type p is said to be *solitary* if the greatest p-stable convex to the right (left) formula is trivial.

If $A, B \subseteq M$, $n \in \omega$, we will say A is *n-indiscernible over B* in M if for any properly ordered n-tuples $\bar{a}$, $\bar{b} \in A^n$ $tp(\bar{a}/B)tp(\bar{b}/B)$, and we will say A is *indiscernible over B* in M if for any $n \in \omega$ A is n-indiscernible over B in M.

Lemma 2.3. *Let M be a linearly ordered structure, M be max $\{|A|^+, \omega\}$-saturated. Then $p(M)$ is $n + 1$-indiscernible over A if $p(M)$ is n-indiscernible over A and for every $\alpha_1, \ldots, \alpha_n \in p(M)$ such that $\alpha_1 < \cdots < \alpha_n$ the set $p(M) \cap \{\beta \in M | \alpha_n < \beta\}$ is 1-indiscernible over $A \cup \{\alpha_1, \ldots, \alpha_n\}$.*

Lemma 2.4. *Let M be a weakly o-minimal structure, $A \subseteq M$, $p \in S_1(A)$ be non-algebraic. Suppose that p is solitary. Then $p(M)$ is 2-indiscernible over A.*

Proof of Lemma 2.4.

By Fact 2.3 we have to prove that for $\alpha_1 \in p(M)$ all elements of $p(M)$ that are bigger then α_1 have the same type over $A \cup \{\alpha_1\}$. If this would not be the case, there would be a A-definable formula $F(x, \alpha_1)$ which does not hold for all elements of $p(M) \cap \{x | x > \alpha_1\}$. By changing $F(x, \alpha_1)$ either to the smallest connected component or to the formula

$$x \geq \alpha_1 \wedge \forall y[F(y, \alpha_1) \to x < y]$$

we can assume $F(x, y)$ to be convex to the right. Furthermore $F(x, y)$ is p-stable. This is a contradiction to the solitarity of p. $\qquad\square$

Theorem 2.1. *Let T be a weakly o-minimal theory, $M \models T$, $A \subseteq M$, M be $|A|^+$-saturated, $p \in S_1(A)$ be non-algebraic. Suppose that the set of all p-stable convex to the right formulas is ordered by ω^*, where ω^* is the reverse ordering on the natural numbers. Then any p-stable convex to the right (left) formula is equivalence generating.*

Proof of Theorem 2.1.

By the hypothesis there is the set of all p-stable convex to the right formulas ordered by ω^*: $F_1(x,y), F_2(x,y), \ldots, F_n(x,y), \ldots$ so that for any $\alpha \in p(M)$ we have $\quad F_1(M,\alpha) \supset F_2(M,\alpha) \supset \ldots \supset F_n(M,\alpha) \supset \ldots$.

Prove that for any $i \geq 1$ $\quad F_i(x,y)$ is equivalence generating.

Step i. Suppose that for any $j \in \{1, \ldots, i\}$ $F_j(x,y)$ is equivalence generating. Prove that $F_{i+1}(x,y)$ also is equivalence generating.

For a contradiction, suppose that $F_{i+1}(x,y)$ is not equivalence generating. By Lemma 1.3 there are $\alpha, \beta, \gamma \in p(M)$ such that

$$M \models F_{i+1}(\beta,\alpha) \wedge F_{i+1}(\gamma,\beta) \wedge \neg F_{i+1}(\gamma,\alpha) \qquad (*)$$

Consider the following formula: $F'(x,y) := \exists t\, [F_{i+1}(t,y) \wedge F_{i+1}(x,t)]$

By Lemma 1.2 $F'(x,y)$ is p-stable convex to the right.

By $(*)$ $\gamma \in F'(M,\alpha) \setminus F_{i+1}(M,\alpha)$, i.e. $F'(M,\alpha) \supset F_{i+1}(M,\alpha)$.

Consequently, there is $j \in \{1, \ldots, i\}$ $\quad F'(M,\alpha) = F_j(M,\alpha)$. Then the following holds:

$$M \models \forall x(F_j(x,\alpha) \to \exists t[F_{i+1}(t,\alpha) \wedge F_{i+1}(x,t)]) \qquad (1)$$

Consider an arbitrary element $\gamma_1 \in F_j(M,\alpha)\setminus F_{i+1}(M,\alpha)$. By (1) there is $\beta \in F_{i+1}(M,\alpha)$ such that $\gamma_1 \in F_{i+1}(M,\beta)$. Consider $f \in Aut_A(M)$ such that $f(\gamma_1)\alpha$. As $\alpha < \gamma_1$ we have $f^{n+1}(\alpha) < f^{n+1}(\gamma_1) = f^n(\alpha)$ for each $n < \omega$. As $M \models F_j(\gamma_1,\alpha)$ then $M \models F_j(f^n(\alpha), f^{n+1}(\alpha))$ for each $n < \omega$. As $F_j(x,y)$ is equivalence generating we have $M \models F_j(\gamma_1, f^n(\alpha))$ for each $n < \omega$. By (1) there is $\beta_n \in F_{i+1}(M, f^n(\alpha))$ such that $\gamma_1 \in F_{i+1}(M,\beta_n)$. As $M \models \neg F_{i+1}(\gamma_1,\alpha)$ then $M \models \neg F_{i+1}(f^n(\alpha), f^{n+1}(\alpha))$ and consequently $M \models \neg F_{i+1}(\gamma_1, f^n(\alpha))$ for each $n < \omega$.

As $M \models \neg F_{i+1}(\gamma_1, f^n(\alpha)) \wedge F_{i+1}(\beta_n, f^n(\alpha)) \wedge F_{i+1}(\gamma_1, \beta_n)$ we have $f^n(\alpha) < \beta_n$ for each $n < \omega$. As $M \models \neg F_{i+1}(f^{n-1}(\alpha), f^n(\alpha)) \wedge F_{i+1}(\beta_n, f^n(\alpha))$ we have $\beta_n < f^{n-1}(\alpha)$ for each $n < \omega$.

Thus, $f^n(\alpha) \notin F_{i+1}(\gamma_1, M), \beta_n \in F_{i+1}(\gamma_1, M)$ and $f^n(\alpha) < \beta_n < f^{n-1}(\alpha)$ for each $n < \omega$. Consequently $F_{i+1}(\gamma_1, M)$ is a union of infinitely many convex sets, contradicting to weak o-minimality of M. Step i is proved. $\square$

Corollary 2.2. *Let T be an $\aleph_0$-categorical weakly o-minimal theory, $M \models T$, $A \subseteq M$, $p \in S_1(A)$ be non-algebraic. Then any p-stable convex to the right (left) formula is equivalence generating.*

Proof of Corollary 2.2.

If A is finite it immediately follows from Theorem 2.1. Consider the case when A is infinite. Consider an arbitrary p-stable convex to the right formula $F(x, y)$. Let A_0 be a finite subset of A such that $F(x, y)$ is a formula over A_0. Let p_0 be $p|_{A_0}$. It is obvious that $F(x, y)$ is p_0-stable convex to the right. By Theorem 2.1 $F(x, y)$ is equivalence generating in p_0 and therefore also equivalence generating in p. $\qquad\square$

The following corollary is very close to results of Section 2 [3].

Corollary 2.3. *Let T be an $\aleph_0$-categorical weakly o-minimal theory, $M \models T$, $A \subseteq M$, A be finite, $p \in S_1(A)$ be non-algebraic. Suppose that $\{F_1(x, y), \ldots, F_m(x, y)\}$ is a complete list of all non-trivial p-stable convex to the right formulas so that for any $\alpha \in p(M)$ $F_1(M, \alpha) \subset \ldots \subset F_m(M, \alpha)$.*

Then the A-definable non-trivial equivalence relations with infinite classes on $p(M)$ are precisely E_i for $1 \leq i \leq m$ given by $E_i(x, y) := F_i(x, y) \vee F_i(y, x)$ so that the following holds:

(1) *E_m partitions $p(M)$ into infinitely many E_m-classes, every E_m-class is convex and open so that the induced ordering on classes is a dense order without endpoints*

(2) *For any $i \in \{1, \ldots, m - 1\}$ E_i partitions every E_{i+1}-class into infinitely many E_i-classes, every E_i-class is convex and open so that E_i-subclasses of every E_{i+1}-class are densely ordered without endpoints*

(3) *For any $\alpha \in p(M)$ $E_1(M, \alpha)$ is 2-indiscernible over A*

References

1. H.D. Macpherson, D. Marker, Ch. Steinhorn, *Weakly o-minimal structures and real closed fields*, **Trans. of Amer. Math. Soc.**, 352 (2000), pp. 5435–5483.
2. B.S. Baizhanov, *Expansion of a model of a weakly o-minimal theory by a family of unary predicates*, **JSL**, 66 (2001), pp. 1382–1414.
3. B. Herwig, H.D. Macpherson, G. Martin, A. Nurtazin, J.K. Truss, *On $\aleph_0$-categorical weakly o-minimal structures*, **APAL**, 101 (2000), pp. 65–93.

PROOFS ABOUT FOLKLORE:
WHY MODEL CHECKING = REACHABILITY?
(EXTENDED ABSTRACT)

K. CHOE, H. EO, S. O

Korea Advanced Institute of Science and Technology,
Kusong-dong Yusong-gu 373-1 Taejon 305-701, Korea,
E-mail: choe@cs.kaist.ac.kr, poisson@ropas.kaist.ac.kr, sho@pllab.kaist.ac.kr

N. V. SHILOV

Institute of Informatics Systems,
Lavren'ev av., 6, Novosibirsk 630090, Russia,
E-mail: shilov@iis.nsk.su

K. YI

Seoul National University,
San 56-1 Shilim-dong Kwanak-gu, Seoul 151-742, Korea,
E-mail: kwang@cse.snu.ac.kr

We demonstrate that Hintikka-like game-theoretic semantics for a so-called
Second-Order Elementary Propositional Dynamic Logic (SOEPDL) leads to a
principle opportunity to use solvers of simple reachability properties as engines
for model checking classical temporal and program logics like μ-Calculus (μC)
and Computation Tree Logic (CTL).

1. Introduction and Motivation

It is well-known that various propositional program logics like CTL [1]
(Computational Tree Logic) are easy to encode into the propositional μ-
Calculus (μC) of Kozen [4] due to its expressive power. It implies that any
model checking engine for μC can be used for checking CTL without any
model modification. But there is also interest [6, 7] to use of model checking
engines for simple temporal properties like safety or progress properties for
model checking other more complicated temporal properties but with aid of
some algebraic transformations of models. In particular, paper of Schuppan
and Biere [6] has admitted Cartesian products of models with finite sets
for an efficient reduction of model checking progress (liveness) properties
to model checking safety properties. It leads to a practically efficient model
checking progress properties via safety one.

Paper of Shilov and Yi [7] has demonstrated how (in principle) to use a model checker that can solve finite games for finite model checking of μC and second order propositional program logic 2M of Stirling [10]. For it, Shilov and Yi have admitted Cartesian products and power set construct on models, and introduced a very expressive Second-Order Elementary Propositional Logic (SOEPDL). The cited paper [7] has demonstrated that SOEPDL is more expressive than CTL, μC, and 2M, and that Second-order logic of several monadic Successors (**S(n)S**-Logic) of Rabin [5] can be interpreted in SOEPDL.

The reduction of SOEPDL model checking to μC model checking is based on Hintikka-like game theoretic semantics [3]. For a SOEPDL-formula we construct a game between two players Spoiler and Duplicator in a manner that validness of the formula is reduced to existence of a winning strategy for Duplicator. Because there exists a μC-formula WIN $\equiv \mu x.(p \vee \langle a \rangle \vee (\langle b \rangle true \wedge [b]x))$ that represents existence of a winning strategy of terminating games, model checking of given SOEPDL-formula is reduced to model checking of WIN in the model of the game.

Unfortunately, game-theoretic semantics suggested by Shilov and Yi [7] is extremely inefficient: the complexity of game construction is exponential to the size of model and length of formula. In the present paper we suggest more efficient game semantics whose complexity is exponential to the size of model and number of independent variables in the formula[a]. Then we study how to solve finite games by CTL model checkers, moreover, by checkers of $\forall$-reachability and $\exists$-reachability properties. It implies (in combination with reduction of SOEPDL model checking to existence of a winning strategy) a formal justification for a folklore opinion of software engineering community that finite-state model checking is basically a kind of reachability checking.

2. Preliminaries

Let *Prp* and *Act* be two disjoint alphabets of propositional variables and action symbols respectively. Let us assume that reader is familiar with basic concepts of CTL and μC. We would like to use a standard notation [1] for CTL, and adopt quite standard notation [7] for μC. In contrast, let us define below two less known second-order propositional program logics, namely 2M by Stirling [10] and SOEPDL by Shilov and Yi [7].

---

[a]Thus we try to restrict a set of 'valuable' variables in a manner that can improve upper complexity bound. This approach is very useful in complexity research, see for instance paper of Vardi [11].

Semantics of propositional program logics is defined in models, which are called labeled transition systems or Kripke structures.

Definition 2.1. A model M is a pair (D_M, I_M) where the domain D_M is a nonempty set, and the interpretation I_M is a pair of mappings (P_M, R_M). Elements of the domain D_M are called states. The interpretation maps propositional variables into sets of states and action symbols into binary relations on states: $P_M : Prp \to 2^{D_M}$ and $R_M : Act \to 2^{D_M \times D_M}$. (We use $I_M(a)$ and $I_M(p)$ instead of $R_M(a)$ and $P_M(p)$ frequently when it is explicit that $a \in Act$ and $p \in Prp$.)

Every model can be considered as labeled directed graph with nodes and edges marked by sets of propositional variables and action symbols respectively.

Definition 2.2. Assume we are given a set of formulae of any propositional program logic that is syntactically closed with respect to negation ($\neg$), conjunction ($\wedge$), and disjunction ($\vee$) (it can be a set of CTL-formulae, μC-formulae, etc.). A satisfiability relation $\models$ between models, states, and formulae is defined inductively with respect to the structure of formulae. For every model $M = (D_M, I_M)$, every state $s \in D_M$, and every formula ϕ let us write:

- "$s \models_M \phi$" iff $(M, s, \phi) \in \models$,
- "$s \not\models_M \phi$" iff $(M, s, \phi) \notin \models$.

For Boolean constants $s \models_M \mathit{true}$ and $s \not\models_M \mathit{false}$. For propositional variables we have: $s \models_M p$ iff $s \in I_M(p)$. For Boolean constructs $\models_M$ is defined in the standard manner: $s \models_M \neg\phi$ iff $s \not\models_M \phi$, $s \models_M \phi \wedge \psi$ iff $s \models_M \phi$ and $s \models_M \phi$, $s \models_M \phi \vee \psi$ iff $s \models_M \phi$ or $s \models_M \phi$. Formulae ϕ and ψ are said to be equivalent if for every $M = (D_M, I_M)$, every state $s \in D_M$: $s \models_M \phi \Leftrightarrow s \in D_M: s \models_M \psi$.

Definition 2.3. The simplest propositional program logic is EPDL (Elementary Propositional Dynamic Logic due to Harel [2]). Syntax of EPDL extends syntax of Classical Propositional Logic (CPL) by modalities associated with action symbols: if $a \in Act$ and ϕ is a formula then $([a]\phi)$ and $(\langle a \rangle \phi)$ are formulae[b]. Semantics of EPDL extends the above definition 2.2 of satisfiability relation $\models$ as follows:

---

[b]which are read as "box/diamond a ϕ" or "after a always/sometimes ϕ" respectively

- $s \models_M (\langle a \rangle \phi)$ iff for some state s': $(s, s') \in I_M(a)$ and $s' \models_M \phi$,
- $s \models_M ([a]\phi)$ iff for every state s': $(s, s') \in I_M(a)$ implies $s' \models_M \phi$.

Definition 2.4. Syntax of 2M and SOEPDL extends syntax of EPDL with quantifiers over propositional variables and special modalities: if $p \in Prp$ and ϕ is a formula then $(\forall p.\phi)$ and $(\exists p.\phi)$ are formulae[c] as well as $(\Box \phi)$ and $(\Diamond \phi)$[d].

Definition 2.5. Semantics of 2M-second-order quantifiers and SOEPDL-second-order quantifiers extends semantics of EPDL as follows:

- $s \models_M \forall p.\phi$ iff $s \models_{M_{S/p}} \phi$ for every $S \subseteq D_M$,
- $s \models_M \exists p.\phi$ iff $s \models_{M_{S/p}} \phi$ for some $S \subseteq D_M$,

where model $M_{S/p}$ agrees with M almost everywhere but $I_{M_{S/p}}(p) = S$.

Definition 2.6. Semantics of 2M- and SOEPDL-modalities $\Box/\Diamond$ extends semantics of EPDL as follows:

- **2M:** Think M as directed graph;
    - $s \models_M \Box \phi$ iff $s' \models_M \phi$
      for every s' in the strongly connected component of s.
    - $s \models_M \Diamond \phi$ iff $s' \models_M \phi$
      for some s' in the strongly connected component of s.

- **SOEPDL:**
    - $s \models_M \Box \phi$ iff $s' \models_M \phi$ for every state $s' \in D_M$,
    - $s \models_M \Diamond \phi$ iff $s' \models_M \phi$ for some state $s' \in D_M$.

Let us introduce several auxiliary definitions and concepts.

Definition 2.7.
A literal is a propositional variable or its negation. A formula of 2M or SOEPDL is said to be normal if negation $\neg$ occurs in the scope of literals in the formula.

E.g., $(p \wedge q) \vee \forall p.(\neg p \vee \neg q)$ is a normal formula, while $\neg(\neg p \vee \neg q)$ is not.

Proposition 2.1.
Every formula of 2M or SOEPDL is equivalent to some normal formula of the same logic that can be constructed in a linear time.

---

[c]which are read as "for every/some p ϕ" respectively
[d]which are read as "box/diamond ϕ" or "always/sometimes ϕ" respectively

Definition 2.8. Let ϕ be a formula of μC, 2M, or SOEPDL. Propositional variable p is said to be propositional constant[e] in the formula ϕ, if p has no bound instances in ϕ. Let $C(\phi)$ be set of propositional constants in ϕ. In contrast let $F(\phi)$ be set of propositional variables that have free instances in ϕ. Let independent variable index of ϕ be the maximal number of variables that are not constants in ϕ but have simultaneous free instances in some subformula of ϕ.

E.g., in $(p \land q) \lor \forall p.(\neg p \lor \neg q)$ the first instance of p is free, the second is bound, while both instances of q are free. It implies that q is a propositional constant in this formula. Propositional variables p and q have simultaneous free instances (in subformulae $(p \land q)$ and $(\neg p \lor \neg q)$). It means that the independent variable index of the above formula is 1 (since q is a constant).

3. Model Checking Game for SOEPDL

Definition 3.1. (of game semantics for SOEPDL)
Let $M = (D_M, I_M)$ be a model and ξ be a normal formula of SOEPDL. A game with winning positions between two players *Spoiler* and *Duplicator* is a tuple $G(M, \xi) = (P_s, P_d, M_s, M_d, W_s, W_d)$ where[f]:

- Positions are (s, ψ, S) where $s \in D_M$ is a state, ψ is a subformula of ξ, and $S : F(\psi) \to 2^{D_M}$ is a total function that maps each variable that is not a constant in ξ but has free instances in ψ to $S(x) \subseteq D$.
- Spoiler has moves of 4 kinds related to conjunctive subformulae, and wins in positions of 5 kinds.
- Duplicator has moves of 4 kinds related to disjunctive subformulae, and wins in positions of 5 kinds.

All moves and winning positions are represented in Fig. 1 and 2.

---

[e]Recall that *true* and *false* we call Boolean constants.

[f]Here we use the following notation for functions. First, the emptyset $\emptyset$ can be considered as a total function $\emptyset : \emptyset \to B$ for every set B. Then for every two elements a and b let $(a \mapsto b) : \{a\} \to \{b\}$ be a total function such that maps a into b. Next, let $F : A \to B$ be any total function, $C \subset A$, and $d \notin A$ and $b \in B$; then let $F|_C : C \to B$ be a restriction of F to C and $F_{b/d} : \{d\} \cup A \to B$ be an extension of F on d, i.e. the following functions:

$$(F|C)(x) = \begin{cases} F(x), & \text{if } x \in A \cap C, \\ \text{undefined otherwise}, \end{cases} \quad (F_{b/d})(x) = \begin{cases} F(x), & \text{if } x \in A, \\ b, & \text{if } x = d. \end{cases}$$

Spoiler	Duplicator
$i \in \{1, 2\}$:	
$(s, (\psi_1 \wedge \psi_2), S) \to (s, \psi_i, S|_{F(\psi_i)})$	$(s, (\psi_1 \vee \psi_2), S) \to (s, \psi_i, S|_{F(\psi_i)})$
$(s, t) \in I_M(a)$:	
$(s, ([a]\psi), S) \to (t, \psi, S)$	$(s, (\langle a \rangle \psi), S) \to (t, \psi, S)$
$t \in D_M$:	
$(s, (\Box \psi), S) \to (t, \psi, S)$	$(s, (\Diamond \psi), S) \to (t, \psi, S)$
$T \subseteq D$:	
$(s, (\forall y.\psi), S) \to (s, \psi, (S_{T/y})|_{F(\psi)})$	$(s, (\exists y.\psi), S) \to (s, \psi, (S_{T/y})|_{F(\psi)})$

Figure 1.   Moves of Spoiler and Duplicator

Spoiler	Duplicator	
$(s, false, \emptyset)$	$(s, true, \emptyset)$	
$(s, p, \emptyset)$ where $s \notin I_M(p)$	$(s, p, \emptyset)$ where $s \in I_M(p)$	$p \in C(\xi)$
$(s, \neg p, \emptyset)$ where $s \in I_M(p)$	$(s, \neg p, \emptyset)$ where $s \notin I_M(p)$	$p \in C(\xi)$
$(s, x, (x \mapsto T))$ where $s \notin T$	$(s, x, (x \mapsto T))$ where $s \in T$	$x \in Prp$
$(s, \neg x, (x \mapsto T))$, where $s \in T$	$(s, \neg x, (x \mapsto T))$, where $s \notin T$	$x \in Prp$.

Figure 2.   Winning positions of Spoiler and Duplicator

While the game in paper of Shilov and Yi [7] traces valuations of all variables for all positions, we only trace valuations of variables with free instances. Improvement of complexity comes from relative scarceness of free variables in each subformula.

Proposition 3.1.
For every finite model $M = (D_M, I_M)$ and every normal SOEPDL-formula ξ there exists a finite game $G(M, \xi)$ of two players Spoiler and Duplicator such that the following holds for every state $s \in D_M$ and every subformula ϕ of ξ: Duplicator has a winning strategy against Spoiler in a position $(s, \phi, (I_M)|_{F(\phi) \backslash C(\xi)})$ iff $s \models_M \phi$ The game can be constructed in time $O(d \times f \times 2^{d \times n})$, where d is a number of states in M, f is a size and n is independent variable index of ξ.

It implies the following theorem.

Theorem 3.1.
Let MC be a model checker for the following μC-formula WIN: $\mu x. (p \vee \langle a \rangle x \vee (\langle b \rangle true \wedge [b]x))$ that computes the satisfaction set of WIN in linear time in size of finite model. Then MC can be reused for checking all formulae of μC, 2M, and SOEPDL with upper time bound $f \times exp(d \times n)$ in

finite models (where d is number of states in a model, f is size and n is independent variable index of a formula).

4. Reduction to CTL

We study two opportunities how to use a model checker for CTL for solving μC, 2M, and SOEPDL formulas in finite models.

Definition 4.1. Let $G = (P_A, P_B, M_A, M_B, W_A, W_B)$ be a game with winning positions. Let D_G be $P_A \cup P_B$. A powerset model PT_G of this game is defined as follows. The domain of model is the powerset 2^{D_G}. A single propositional variable p is interpreted by the powerset 2^{W_A}. The interpretation of a single action symbols *next* comprises all pairs (S', S'') such that $S', S'' \subseteq D_G$ and

- for every $s' \in S'$, for some $(s', s'') \in (M_A \cup M_B)$:
  $s' \notin (W_A \cup W_B) \Rightarrow s'' \in S''$;
- for every $s' \in S'$, for every $(s', s'') \in M_B$:
  $s' \in (P_B \setminus (W_A \cup W_B)) \Rightarrow s'' \in S''$;
- for every $s'' \in S''$, for some $(s', s'') \in (M_A \cup M_B)$: $s' \in S'$.

Proposition 4.1.
For every game with winning positions G, for every set of positions S, if there is a finite upper bound on length of G sessions, then $S \models_{PT_G} \mathbf{EF}p$ iff a player A has winning strategies against the counterpart B in all positions within S.

In combination with theorem 3.1, it implies the following theorem.

Theorem 4.1.
Let MC be a model checker which can check CTL formula ($\mathbf{EF}p$) in finite models. Then MC can be reused for checking all formulae of SOEPDL, 2M, μC, and CTL in all finite models.

Unfortunately, this kind of reuse is double exponential in model size.

But there is another more efficient opportunity to use CTL model checker for solving finite games. This time we assume that we have an engine for solving $\mathbf{AF}$ and $\mathbf{EF}$ queries (formulas) and design a 'driver' that solves μC-formula WIN$\equiv \mu\ x.\ (p \vee \langle a \rangle x \vee (\langle b \rangle true \wedge [b]x))$ in finite games. These engines should be able to solve $\mathbf{AF}$ and $\mathbf{EF}$ queries in the following sense: for every finite model M, for every set of states P within this model

$X_0 := W_A$; %winning positions for A%
DO
$Y_i := \mathbf{AF}_{GB}X_i$; %$GB$ stays for G with B-moves only%
$X_{(i+1)} := Y_i \cup \mathbf{EF}_{GA}X_i$; % GA stays for G with A-moves only%
$i := i + 1$
$UNTIL\ X_i = X_{(i-1)}$;
$WIN := X_i$

Figure 3. A driver that solves finite games in terms of AF and EF

- let $\mathbf{AF}_M P$ be a set of states where $\mathbf{AG}p$ holds in M when p holds exactly at the states in P.
- let $\mathbf{EF}_M P$ be a set of states where $\mathbf{EF}p$ holds in M when p holds exactly at the states in P.

Proposition 4.2.
For every game with winning positions G with some upper bound on length of plays method depicted in Fig. 3 eventually terminates, and upon termination the set WIN consists of all positions where a player A has winning strategies against the counterpart B.

Proof. Termination of the method trivially follows from a simple observation: every turn in any game consists of one game step at least. Thus, if k is an upper bound for length of the game sessions, then $X_k = X_{(k+1)}$.

We show that for every $i \geq 1$, X_i consists of all positions within the game where player A has a strategy that leads to his/her win in all sessions that consist of $(i - 1)$ changes of turns at most (by induction on i). Proposition 4.2 follows from this claim since the loop condition is a fixpoint stabilization of X_i and the method terminates eventually.

Induction basis: $i = 1$. Then by construction $Y_i = \mathbf{AF}_{GB}W_A$, i.e. it consists of all positions where it is turn of B, but he/she loses every session that consists of his/her moves only. Hence $X_i = \mathbf{EF}_{GA}W_A \cup \mathbf{AF}_{GB}W_A$, i.e. it consists of all positions where it is turn of A, and he/she can run a session that consists of his/her moves only and leads to his/her win, or (alternatively) it is turn of B, but he/she loses every session that consists of his/her moves only.

Induction hypothesis: assume that for some $k \geq 1$ the claim holds.

Induction step: $i = (k + 1)$. Then by construction $Y_i = \mathbf{AF}_{GB}X_k$, i.e. it consists of all positions where it is turn of B, but he/she loses every session that starts with his/her turn and then consists of $(k - 1)$ turns at most, where A utilizes a strategy that guaranties win for A. Hence $X_i = Y_i \cup \mathbf{EF}_{GA} X_k$ consists of all positions where

- A can run a turn that leads to a position where he/she has a strategy that leads to A win in all sessions that consist of $(k-1)$ changes of turns at most,

or (alternatively)

- B loses every session that starts with turn of B and then consists of $(k-1)$ turns at most, where A utilizes a strategy that guaranties win for A,

i.e. where player A has a strategy that leads to his/her win in all sessions that consist of $k = (i-1)$ changes of turns at most. $\qquad\qquad\square$

Combining this proposition with theorem 3.1, we get another theorem.

Theorem 4.2.
*Let MC be a model checker that can solve **AF**- and **EF**-queries in linear time in size of finite model. Then MC can be reused for checking all formulae of μC, 2M, and SOEPDL with upper time bound $f^2 \times exp(d \times n)$ in all finite models (where d is number of states in a model, f is size and n is independent variable index of a formula).*

Observe that our above reuse theorems 1, 2, and 3 can be extended to the class of (not necessarily finite) models closed to Cartesian products and power-sets, because actual game model construction and CTL model construction uses only Cartesian products and power-sets.

References

1. Emerson E.A. *Temporal and Modal Logic. Handbook of Theoretical Computer Science*, v.B, Elsevier and The MIT Press, 1990, 995–1072.
2. Harel D. *First-Order Dynamic Logic.* Lecture Notes in Computer Science, v. 68, 1979.
3. Hintikka J., Sandu G. *Game-Theoretical Semantics.* Handbook of Logic and Language, 1997.
4. Kozen D. *Results on the Propositional Mu-Calculus.* Theoretical Computer Science, v. 27, n. 3, 1983, p. 333–354.
5. Rabin M.O. *Decidable Theories.* in Handbook of Mathematical Logic, ed. Barwise J. and Keisler H.J., Noth-Holland Pub. Co., 1977, 595–630.
6. Schuppan V. and Biere A. *Efficient reduction of finite state model checking to reachability analysis.* , International Journal on Software Tools for Technology Transfer (STTT), v.5 (2–3), p. 185–204, 2004.

7. Shilov N.V., Yi K. *On Expressive and Model Checking Power of Propositional Program Logics.* Lecture Notes In Computer Science, v. 2244, p. 39–46, 2001.

8. Shilov N.V., Yi K. *Engaging Students with Theory through ACM Collegiate Programming Contests.* Communications of ACM, v. 45, n. 9, 2002.

9. Shilov N.V., Yi K. *How to find a coin: propositional program logics made easy.* In Current Trends in Theoretical Computer Science, World Scientific, v. 2, 2004, p. 181–214.

10. Stirling C. *Games and Modal Mu-Calculus.* Lecture Notes in Computer Science, v. 1055, 1996, p. 298–312.

11. Vardi M.Y. On the complexity of bounded-variable queries. Proc. 14th ACM Symp. on Principles of Database Systems, 1995, p. 266–276.

A NOTE ON Δ_1 INDUCTION

C. DIMITRACOPOULOS*

*Department of History and Philosophy of Science,
University of Athens,
GR-157 71 Zografou, Greece
E-mail: cdimitr@phs.uoa.gr*

A. SIROKOFSKICH[†]

*Department of Mathematics,
University of Athens,
GR-157 84 Zografou, Greece
E-mail: asirokof@math.uoa.gr*

We give an alternative proof of a result of T. Slaman, concerning the strength of $I\Delta_1$, i.e. the theory of provably-Δ_1 induction.

1. Introduction

We work with subsystems of first-order Peano Arithmetic (PA). As usual, for $n \in \mathbf{N}$, $I\Sigma_n$ denotes the induction schema for Σ_n formulae (plus the well-known base theory PA^-), $L\Sigma_n$ the least number axiom schema for Σ_n formulae (plus PA^-), $B\Sigma_n$ denotes $I\Delta_0$ plus the collection schema for Σ_n formulae, exp denotes the axiom expressing "exponentiation is total" and Ω_1^k, $k \geq 1$, denotes the axiom expressing "$x^{log^{(k)}(x)}$ is total", where $log^{(k)}$ denotes the k-th iterate of the logarithmic function. Also, $I\Delta_n$ denotes the induction schema for provably-Δ_n formulae (plus PA^-) and $L\Delta_n$ the least number schema for provably-Δ_n formulae (plus PA^-). Finally, $\langle \, , \rangle$ denotes one of the usual pairing functions and $x \in y$ denotes the Δ_0 formula expressing "2^x appears in the binary expansion of y". For details, the reader can consult Hájek-Pudlák [6].

We will also deal with other fragments of PA, namely versions of the pigeonhole principle. As usual, $PHP\Sigma_n$ denotes PA^- plus the schema expressing this principle for Σ_n-definable functions:

---

*Research co-funded by the European Social Fund and National Resources – (EPEAEK II) PYTHAGORAS II.

[†]Research co-funded by the European Social Fund and National Resources – (EPEAEK II) HERAKLEITOS.

$\forall z$ "φ does not define a 1-1 function from $z + 1$ into z", for any $\varphi \in \Sigma_n$.

Actually, there is another version of the pigeonhole principle (see Dimitracopoulos-Paris [4]), but for our present purposes it suffices to consider the one above. Two apparently weaker fragments, that were studied widely in Paris et al. [8] for $n = 0$, are as follows:

$$\forall z \text{ "}(1 + \epsilon)z > z \rightarrow \varphi \text{ does not define a 1-1 function from } (1 + \epsilon)z \text{ into } z\text{",}$$
$$\text{for any } \varphi \in \Sigma_n \text{ and rational } \epsilon > 0$$
$$\forall z > 1 \text{"}\varphi \text{ does not define a 1-1 function from } z^2 \text{ into } z\text{", for any } \varphi \in \Sigma_n.$$

We will denote the corresponding fragments by $WPHP\Sigma_n$ and $wPHP\Sigma_n$ respectively.

Among these subsystems of PA, $I\Delta_n$ is one of the less well understood, the main problem concerning it being the following, which was set by J. Paris and appeared as Problem 34 in the list of open problems compiled by P. Clote and J. Krajiček [2].

Problem 1. *Does $I\Delta_n$ imply $B\Sigma_n$ $(n \geq 1)$?*

Motivated by this problem, several authors have recently studied the strength of $I\Delta_n$ and its parameter free counterpart, $I\Delta_n^-$, especially for the case $n = 1$ (see Beklemishev [1], Fernández-Margarit et al. [5], Slaman [10] and Thapen [11]). Before we refer briefly to some of their results, we mention a few early theorems concerning fragments of PA.

One of the first results concerning $B\Sigma_n$, proved by C. Parsons [9], is that it is not implied by the set of Π_{n+1} sentences true in the standard model.

Theorem 1. *For $n \geq 1$, $\Pi_{n+1}(\mathbf{N}) \not\Rightarrow B\Sigma_n$.*

Relationships among $B\Sigma_n$ and other fragments of PA were extensively studied in Paris-Kirby [7]; the ones that are relevant to the sequel are summarized as follows.

Theorem 2. *For all $n \in \mathbf{N}$, $I\Sigma_{n+1} \Rightarrow B\Sigma_{n+1} \Rightarrow I\Sigma_n \Leftrightarrow L\Sigma_n$ (and the implications are strict).*

By using (easy) modifications of some arguments in Paris-Kirby [7] and an argument due to R. Gandy (unpublished, see p. 66 of Hájek-Pudlák [6]), one obtains

Theorem 3. *For all $n \in \mathbf{N}$, $B\Sigma_{n+1} \Leftrightarrow L\Delta_{n+1} \Rightarrow I\Delta_{n+1}$.*

Turning now to recent work motivated by Problem 1, Fernández-Margarit et al. [5] contained a study of restrictions of $I\Delta_n$, $L\Delta_n$ etc. to $\Delta_n(T)$ formulae, i.e. formulae that are Δ_n provably in (a certain theory) T, while the other papers studied the original problem. In the rest of this section, we will refer briefly to results in Beklemishev [1], Slaman [10] and Thapen [11], in order to place our result in the appropriate background.

Motivated by the following picture,

$$
\begin{array}{ccc}
I\Sigma_1 & \Rightarrow & I\Delta_0 + exp \\
\Downarrow & & \Uparrow \\
B\Sigma_1 & \nLeftarrow & \Pi_2(\mathbf{N}) \\
\Downarrow & & \\
I\Delta_1 & &
\end{array}
$$

which emerged as a synthesis of earlier results for $n = 1$, Beklemishev considered the problem whether $\Pi_2(\mathbf{N})$ implies $I\Delta_1$ or not and he solved it by proving

Theorem 4. $\Pi_2(\mathbf{N}) \nRightarrow I\Delta_1$.

At the end of his paper, Beklemishev noted that Theorem 4 (and other results of his paper) can be easily generalized for any $n > 1$ and stated some problems motivated by Problem 1. Two of his problems were the following.

Problem 2. *Does $I\Delta_1^-$ follow from EA? From any r.e. set of true Π_1 sentences?*

($I\Delta_1^-$ denotes parameter free $I\Delta_1$, while EA is a theory he called "Elementary Arithmetic" and is easily seen to be equivalent to $I\Delta_0 + exp$ - see, e.g., section 1(b) in chapter I of Hájek-Pudlák [6]).

Problem 3. *Does $B\Sigma_1$ follow from $I\Delta_1$ together with all true Π_2 sentences?*

Slaman considered a variant of Problem 1, namely the same question but taking as base theory $I\Delta_0 + exp$ instead of $I\Delta_0$, and showed that the answer is "yes", i.e. he proved

Theorem 5. *For all $n \geq 1$, $I\Delta_n + exp \Rightarrow B\Sigma_n$.*

As a consequence, he obtained positive solutions to

(i) Problem 1 for $n > 1$, since $I\Delta_n \Rightarrow I\Sigma_{n-1} \Rightarrow exp$, for such n
(ii) Problem 3, since exp is a particular case of a true Π_2 sentence.

Thapen improved Slaman's result for $n = 1$, by showing

Theorem 6. $I\Delta_1 + t \Rightarrow B\Sigma_1$, *where* t *is the axiom* $\forall x \exists y \exists z (x < p(y) \wedge z = x^y)$, *with* p *being any primitive recursive function.*
In particular, it follows that $I\Delta_1 + \Omega_1^k \Rightarrow B\Sigma_1$.

Cordón-Franco et al. [3] produced a negative solution to Problem 2. Actually, they proved the following more general result:

Theorem 7. *For any* $n \in \mathbf{N}$, *there is no r.e. set of true* Π_{n+2}*-sentences which implies* $I\Delta_{n+1}^-$.

In the next section, we give an alternative argument for Slaman's result and discuss the plausibility of the conjecture that a modification of this argument can lead to a proof, in the same spirit, of Thapen's result. The main difference between our approach and those of Slaman and Thapen is that we exploit results of Dimitracopoulos-Paris [4] and Paris et al. [8], concerning the provability of versions of the pigeonhole principle in fragments of PA.

2. Δ_1-induction vs. Σ_1-pigeonhole principle

Before giving proofs we first recall some results of Dimitracopoulos-Paris [4].

Proposition 8. $I\Delta_0 + exp \Rightarrow PHP\Delta_0$.

Idea of proof. Working in $M \models I\Delta_0 + exp$, assume $PHP\Delta_0$ fails, i.e. for some $\theta \in \Delta_0, a \in M$ we have $M \models ``\theta$ defines a 1-1 function $f : a + 1 \to a"$. Since $M \models exp$, f can be coded and hence

$$M \models \exists z \leq a \exists t < 2^{\langle a, a-1 \rangle + 1} ``t \text{ codes a 1-1 function from } z + 1 \text{ into } z".$$

But now we can use $L\Delta_0$ to find the least such z and then derive a contradiction. $\square$

Proposition 9. *For all* $n > 0$, $PHP(\Sigma_n \vee \Pi_n) \Rightarrow I\Sigma_n$.

Theorem 10. *For all* $n > 0$, $PHP\Sigma_n \Rightarrow B\Sigma_n$.

Idea of proof. Working in $M \models PHP\Sigma_n$, assume $B\Sigma_n$ fails, i.e. for some $\theta \in \Pi_{n-1}, a \in M$ we have

$$M \models \forall x \leq a \exists y \theta(x, y) \wedge \neg \exists t \forall x \leq a \exists y < t \theta(x, y).$$

By Proposition 9, $M \models L\Sigma_{n-1}$ and hence the formula $\theta(x, y) \wedge \forall u < y \neg\theta(x, u)$ defines a function $f : a \to M$ with unbounded range. By considering the elements in the range of f in increasing order, we can now produce a Σ_n formula defining a 1-1 function g from $a + 1$ onto a (the idea is that $g(x) = y \Leftrightarrow f(y)$ is the immediate successor of $f(x)$ in the range of f). But such a function cannot exist, since $PHP\Sigma_n$ holds. $\square$

In view of Theorem 10, to prove that $I\Delta_1 + exp \Rightarrow B\Sigma_1$, it suffices to prove an apparently stronger result, i.e.

Theorem 11. $I\Delta_1 + exp \Rightarrow PHP\Sigma_1$.

Proof. Let $M \models I\Delta_1 + exp$, $a \in M$ and $\theta(x, y) \in \Sigma_1$ such that θ is of the form $\exists z \varphi(x, y, z)$, $\varphi \in \Delta_0$, and it defines a 1-1 function from $a + 1$ into a.
Case 1. $M \models \exists w \forall x < a + 1 \exists y < a \exists z < w \varphi(x, y, z)$, say $d \in M$ satisfies $M \models \forall x < a + 1 \exists y < a \exists z < d \varphi(x, y, z)$.
Then the Δ_0 formula $\exists z < d \varphi(x, y, z)$ defines a 1-1 function from $a + 1$ into a, which contradicts Proposition 8.
Case 2. $M \not\models \exists w \forall x < a + 1 \exists y < a \exists z < w \varphi(x, y, z)$.
Now consider the set defined as follows

$$A = \{b : \exists y < a \exists w [\varphi(b, y, w) \wedge \forall u < w \neg\varphi(b, y, u) \wedge$$
$$\forall x < b \exists y < a \exists z < w \varphi(x, y, z)]\}.$$

Clearly, A is a Σ_1-definable proper subset of $a + 1$. If we show that A is closed under successor and Π_1-definable, then we will contradict the fact that M satisfies $I\Delta_1$.

To prove A is closed under successor in M, let $b \in A$. By the definition of A, there exist $c < a, d \in M$ such that

$$M \models \varphi(b, c, d) \wedge \forall u < d \neg\varphi(b, c, u) \wedge \forall x < b \exists y < a \exists z < d \varphi(x, y, z).$$

By hypothesis about φ, there exist unique $c^* < a, d^*$ such that

$$M \models \varphi(b + 1, c^*, d^*) \wedge \forall u < d^* \neg\varphi(b, c^*, u).$$

Setting now $\hat{d} = max(d, d^*)$, we can easily see that

$$M \models \forall x < b + 1 \exists y < a \exists z < \hat{d} \varphi(x, y, z)$$

and hence $b + 1 \in A$, as required.
Concerning now the Π_1-definability of A, we claim that for any $b \in M$, $b < a + 1$:

$$b \in A \Leftrightarrow M \models \forall y < a \forall w [\varphi(b, y, w) \to \forall x < b \exists y < a \exists z < w \varphi(x, y, z)].$$

($\Rightarrow$) Assume $b \in A, c < a, d$ satisfy

$$M \models \varphi(b, c, d) \wedge \forall u < d \neg \varphi(b, c, u) \wedge \forall x < b \exists y < a \exists z < d \varphi(x, y, z)$$

and c^*, d^* are arbitrary elements of M such that $M \models \varphi(b, c^*, d^*)$. Then $c = c^*$ and $d \leq d^*$ and hence $M \models \forall x < b \exists y < a \exists z < d^* \varphi(x, y, z)$, as required.

($\Leftarrow$) Assume b satisfies

$$M \models \forall y < a \forall w [\varphi(b, y, w) \to \forall x < b \exists y < a \exists z < w \varphi(x, y, z)] \quad (*).$$

By hypothesis about θ, there exist $c < a$ and d such that $M \models \varphi(b, c, d) \wedge \forall u < d \neg \varphi(b, c, u)$. From (*) we can now deduce that $M \models \forall x < b \exists y < a \exists z < d \varphi(x, y, z)$, as required. $\square$

Although we believe the above approach can lead to an alternative proof of Thapen's result, we have not been able to obtain such a proof. As a small hint in this direction, we mention the following result.

Theorem 12. $I\Delta_1 + \Omega_1 \vdash WPHP\Sigma_1$.

For the proof, we need a result of Paris et al. [8], namely Corollary 2 in that paper.

Lemma 13. $I\Delta_0 + \Omega_1 \vdash WPHP\Delta_0$.

Proof of Theorem 12. Let $M \models I\Delta_1 + \Omega_1$, $a \in M$ and $\theta(x, y) \in \Sigma_1$ such that θ is of the form $\exists z \varphi(x, y, z)$, $\varphi \in \Delta_0$, and it defines a 1-1 function from $(1 + \epsilon)a$ into a.

Case 1. $M \models \exists w \forall x < (1 + \epsilon)a \exists y < a \exists z < w \varphi(x, y, z)$, say $d \in M$ satisfies $M \models \forall x < (1 + \epsilon)a \exists y < a \exists z < d \varphi(x, y, z)$. But then the Δ_0 formula $\exists z < d \varphi(x, y, z)$ defines a 1-1 function from $(1 + \epsilon)a$ into a, which is impossible, by Lemma 13.

Case 2. $M \not\models \exists w \forall x < (1 + \epsilon)a \exists y < a \exists z < w \varphi(x, y, z)$. Then we can repeat the argument used for Case 2 in the proof of Theorem 11 to obtain a contradiction. $\square$

Remark. According to Theorem 1 of Paris et al. [8], for any $k \geq 1$,

$$I\Delta_0 + \Omega_1^k \vdash wPHP\Delta_0.$$

Using this result instead of Lemma 13, one can immediately modify the proof above to show that $I\Delta_1 + \Omega_1^k \vdash wPHP\Sigma_1$, for any $k \geq 1$.

As mentioned by Paris et al. [8], it is not known whether Ω_1 in Lemma 13 can be weakened to Ω_1^k, for any/some $k > 1$; if this were proved, Theorem 12 could be strengthened accordingly.

Acknowledgement. The authors would like to thank Jeff Paris for comments and corrections which helped to improve this paper.

References

1. L. D. Beklemishev: *On the induction schema for decidable predicates*, J. Symbolic Logic 68 (2003), 17–34.
2. P. Clote and J. Krajiček: *Open problems*, Oxford Logic Guides, 23, Arithmetic, proof theory, and computational complexity (Prague, 1991), 1–19, Oxford Univ. Press, New York, 1993.
3. A. Cordón-Franco, A. Fernández-Margarit and F. F. Lara-Martín: *Fragments of Arithmetic and True Sentences*, Mathematical Logic Quarterly. Vol. 51. Num. 3. 2005. Pag. 313-328.
4. C. Dimitracopoulos and J. Paris: *The pigeonhole principle and fragments of arithmetic*, Z. Math. Logik Grundlag. Math. 32 (1986), 73–80.
5. A. Fernández-Margarit and F. F. Lara-Martín: *Induction, minimization and collection for Δ_{n+1}-formulas*, Arch. Math. Logic 43 (2004), 505–541.
6. P. Hájek and P. Pudlák: *Metamathematics of first-order arithmetic*, Springer-Verlag, Berlin, 1993.
7. J. B. Paris and L. A. S. Kirby: *Σ_n-collection schemas in arithmetic*, Logic Colloquium '77, North-Holland, Amsterdam, 1978, 199–209.
8. J. B. Paris, A. J. Wilkie and A. R. Woods: *Provability of the pigeonhole principle and the existence of infinitely many primes*, J. Symbolic Logic 53 (1988), no. 4, 1235–1244.
9. C. Parsons: *On a number theoretic choice schema and its relation to induction*, 1970, Intuitionism and Proof Theory (Proc. Conf., Buffalo, N.Y., 1968), 459–473 North-Holland, Amsterdam.
10. T. A. Slaman: *Σ_n-bounding and Δ_n-induction*, Proc. Amer. Math. Soc. 132 (2004), 2449–2456.
11. N. Thapen: *A note on Δ_1 induction and Σ_1 collection*, Fund. Math. 186 (2005), no. 1, 79–84.

ARITHMETIC TURING DEGREES AND CATEGORICAL THEORIES OF COMPUTABLE MODELS*

E. FOKINA

Sobolev Institute of Mathematics
Siberian Branch of the Russian Academy of Sciences
4 Acad. Koptyug avenue, Novosibirsk, 630090, Russia
E-mail: e_fokina@math.nsc.ru

In this paper we study the complexity of uncountably categorical and of countably categorical theories with computable models. We also study the complexity of index sets of countably categorical computable d-decidable models.

1. Introduction

One of the approaches of the computable model theory deals with the existence of computable models for the first order theories. Every consistent decidable, that is with computable set of theorems, theory T has a decidable model. For the uncountably categorical first order theories Harrington and Khissamiev in [7, 8], showed that indeed all countable models of such theory T are decidable. If T is uncountably categorical but not decidable then some of its models can be computable while the others are not. In the paper of Baldwin and Lachlan [1] it was proved that all countable models of an uncountably categorical theory can be listed into a chain of the elementary embeddings $\mathbf{A}_0 \preceq \mathbf{A}_1 \preceq \mathbf{A}_2 \preceq \ldots \mathbf{A}_\omega$, where $\mathbf{A}_0$ is a prime model, $\mathbf{A}_\omega$ is a saturated model and every $\mathbf{A}_{i+1}$ is a minimal proper elementary extension of $\mathbf{A}_i$. Then the spectrum of computable models of the theory T is the set $SCM(T) = \{i \mid \mathbf{A}_i$ has a computable presentation$\}$. Thus, the result of Harrington and Khissamiev can be presented as $SCM(T) = \omega \bigcup \{\omega\}$. This result led to the investigation of computable models of $\aleph_1$-categorical undecidable theories. In [4] Goncharov showed that there existed a $\aleph_1$-categorical, $\mathbf{0}'$-computable theory with $SCM(T) = \{0\}$. This result was generalized by Kudaiberguenov in [11], where he presented a $\aleph_1$-categorical, $\mathbf{0}'$-computable theory with $SCM(T) = \{0, 1 \ldots, n\}$. In [9] Khoussainov, Nies, Shore built examples of $\aleph_1$-categorical, $\mathbf{0}''$-computable theories T_1

---

*This work was partially supported by RFBR grant 05-01-00819 and grant of Scientific schools of Russia 4413.2006.1

and T_2 with $SCM(T_1) = \omega$ and $SCM(T_2) = \omega \bigcup \{\omega\} \setminus \{0\}$. Moreover Nies in [14] built an example of $\aleph_1$-categorical theory T with $SCM(T) = \{1\}$ and proved that for an arbitrary $\aleph_1$-categorical theory T $SCM(T) \in \Sigma_3^0(\emptyset^\omega)$. All given examples of $\aleph_1$-categorical theories are $\mathbf{0}''$-computable. In [5] Goncharov and Khoussainov for any $n \geq 1$ built an example of $\aleph_1$-categorical theory which is Turing equivalent to 0^n and has a computable model. Using a modification of the construction from [5] for any arithmetic degree a we built a $\aleph_1$-categorical theory which is Turing equivalent to a and has a computable model. Moreover we show that every countable model of this theory has a computable presentation, that is $SCM(T) = \omega \bigcup \{\omega\}$.

Lerman and Schmerl in [12] gave some sufficient conditions for the countably categorical theory to have a computable model. They showed that any countably categorical arithmetic theory T for which the set of all sentences beginning with $\exists$ and having $n + 1$ changes of quantifiers is a Σ_{n+1}^0–set for all n, has a computable model.

In [10] Knight improved this result omitting the condition that T is arithmetic but requiring certain uniformity. However all known examples of $\aleph_0$-categorical theories with computable models were of low complexity. In [5] Goncharov and Khoussainov for any n built an example of countably categorical theory, Turing equivalent to 0^n, with computable model. Using the technics from [5] for any given arithmetic degree a we built a countably categorical theory which is Turing equivalent to a and has a computable model.

Let's introduce some basic definitions. We fix some computable Godel numbering of a language L. An algebraic structure $\mathbf{A}$ of the language L is **computable** if its domain is computable and all basic operations and predicates are uniformly computable. This definition is equivalent to the condition that the atomic diagram of $\mathbf{A}$ is computable. The algebraic structure $\mathbf{B}$ is **computably presentable** if it's isomorphic to a computable structure. In this case any isomorphism of $\mathbf{B}$ onto $\mathbf{A}$ is called the **computable presentation** of $\mathbf{B}$. A complete theory T is α-**categorical** if any two models of T of the power α are isomorphic. It's well-known that if theory is β-categorical for some uncountable β then it is α-categorical for any uncountable α.

To prove the basic results of the paper we need, like in [5], to define two operators. The construction of the operators follows the ideas of Marker from [13]. Their definition and properties are formulated in the next section. The detailed information can be found in [5]. In the section 3 we give the definition of 1-to-1-representation of Σ_2^0–sets and state 1-to-1-representation lemmas and corollaries. The proof of the lemmas is in [5, 6]. In the next two sections the following theorems are proved:

Theorem 4.1. *For any arithmetic Turing degree there exists a $\aleph_1$-categorical theory T of a finite signature which is Turing equivalent to this degree and has a computable model. Moreover, all countable models of T have a computable presentation.*

Theorem 5.1. *For any arithmetic Turing degree there exists a $\aleph_0$-categorical theory T of a finite signature which is Turing equivalent to this degree and has a computable model.*

In the last section we are interested in the complexity of the index sets of d-decidable $\aleph_0$-categorical models, where d is arithmetic. More precisely we prove the following:

Theorem 6.1. *For every arithmetic Turing degree d the index set of all d-decidable models has the Turing degree $d^{(3)}$ in the universal computable numbering of all computable $\aleph_0$-categorical models in the signature with one binary predicate.*

2. Marker's construction

Let L be a finite language with no functional symbols, and let $\mathcal{A} = (A, P_0^{n_0}, \ldots, P_m^{n_m})$ be a structure of L. We assume that for every P of this structure the sets P and $A^k \setminus P$ are infinite where k is the arity of P. For each k-ary predicate P of this structure we define $\exists$- and $\forall$-extensions of P.

Marker's $\exists$-extension of P is a $(k+1)$-ary predicate denoted by $P_\exists$ with the following properties. Let X be an infinite set disjoint with A. Then $P_\exists$ satisfies the following conditions:

(1) If $P_\exists(a_1, a_2, \ldots, a_k, a_{k+1})$ then $P(a_1, \ldots, a_k)$ and $a_{k+1} \in X$.
(2) For all $a_{k+1} \in X$ there exists a unique tuple $(a_1, \ldots, a_k)$ such that $P_\exists(a_1, a_2, \ldots, a_k, a_{k+1})$.
(3) If $P(a_1, \ldots, a_k)$ then there exists a unique a such that $P_\exists(a_1, a_2, \ldots, a_k, a)$.

Marker's $\forall$-extension of the predicate P is a $(k+1)$-ary predicate $P_\forall$ with the following properties. Let X be an infinite set disjoint with A. Then $P_\forall$ satisfies the following conditions:

(1) If $P_\forall(a_1, a_2, \ldots, a_k, a_{k+1})$ then $a_1, \ldots, a_k \in A$ and $a_{k+1} \in X$.

(2) For all $(a_1, \ldots, a_k) \in A$ there exists at most one $a_{k+1} \in X$ such that $\neg P_\forall(a_1, a_2, \ldots, a_k, a_{k+1})$.

(3) If $P_\forall(a_1, a_2, \ldots, a_k, a_{k+1})$ for all $a_{k+1} \in X$ then $P(a_1, \ldots, a_k)$.

(4) For all $a_{k+1} \in X$ there exists a unique tuple $(a_1, \ldots, a_k)$ such that $\neg P_\forall(a_1, a_2, \ldots, a_k, a_{k+1})$.

The set X in $\exists$- or $\forall$-extension is called a **fellow of P**.

Definition 2.1. Let $\mathcal{A} = (A, P_0^{n_0}, \ldots, P_m^{n_m})$ be a model.

(1) The model $\mathcal{A}_\exists$ is a model $(A \cup X_0 .. \cup X_m, P_0^{n_0+1}, .., P_m^{n_m+1}, X_0, .., X_m)$, where each $P_i^{n_i+1}$, $i = 0, \ldots, m$, is a Marker's $\exists$-extension of $P_i^{n_i}$ such that fellows X_i of distinct predicates are pairwise disjoint sets.

(2) The model $\mathcal{A}_\forall$ is a model $(A \cup X_0 .. \cup X_m, P_0^{n_0+1}, .., P_m^{n_m+1}, X_0, .., X_m)$, where each $P_i^{n_i+1}$, $i = 0, \ldots, m$, is a Marker's $\forall$-extension of $P_i^{n_i}$ such that fellows X_i of distinct predicates are pairwise disjoint sets.

Theorem 2.1. *Let $\mathcal{A}_\exists$ and $\mathcal{A}_\forall$ be the Marker's extensions of the model $\mathcal{A}$. Then they satisfy the following properties:*

(1) *The model $\mathcal{A}$ is definable in each of the extensions.*

(2) *If the theory of the model $\mathcal{A}$ is $\aleph_0$-categorical then so is the theory of each of the extensions.*

(3) *If the theory of the model $\mathcal{A}$ is $\aleph_1$-categorical then so is the theory of each of the extensions.*

(4) *If the theory of $\mathcal{A}$ is almost strongly minimal then so is the theory of each of the extensions.*

(5) *Any automorphism of $\mathcal{A}$ can be extended to automorphisms of each of the extensions.*

Let $\mathcal{A}$ be a structure and w be a word over the alphabet $\{\exists, \forall\}$. We define $\mathcal{A}_w$ by induction. If w is an empty string then $\mathcal{A}_w = \mathcal{A}$. If $w = w'\exists$ or $w = w'\forall$ and $\mathcal{B} = \mathcal{A}_{w'}$ then $\mathcal{A}_{w'\exists} = \mathcal{B}_\exists$ and $\mathcal{A}_{w'\forall} = \mathcal{B}_\forall$. Therefore we have the following corollary:

Corollary 2.1. *Let $\mathcal{A}$ be a structure and w be a word over the alphabet $\{\exists, \forall\}$. Then*

(1) *The model $\mathcal{A}$ is definable in $\mathcal{A}_w$.*

(2) *If the theory of the model $\mathcal{A}$ is $\aleph_0$-categorical ($\aleph_1$-categorical) then so is the theory of $\mathcal{A}_w$.*

(3) *Any automorphism of $\mathcal{A}$ can be extended to an automorphism of $\mathcal{A}_w$.*

Our next goal is to show that $\mathcal{A}_{\exists\forall}$ is less complex than $\mathcal{A}$ itself from a computability theoretic point of view.

3. On one-to-one representation of Σ_2^0-sets

The following definition and lemmas can be found in [5]. We will need them for the proof of the main results of the paper.

Definition 3.1. A Σ_2^0-set A is **one-to-one representable** if for some computable predicate $Q \subset \omega^3$ the following is true:

(1) For every $n \in \omega$, $\exists a \forall b\, Q(n, a, b)$ if and only if $n \in A$.
(2) For every $n \in \omega$, $\exists a \forall b\, Q(n, a, b)$ if and only if $\exists^{=1} a \forall b\, Q(n, a, b)$[a].
(3) For every b there exists a unique pair $\{n, a\}$ such that $\neg Q(n, a, b)$.
(4) For every pair $\{n, a\}$ either $\exists^{=1} b \neg Q(n, a, b)$ or $\forall b\, Q(n, a, b)$.
(5) For every a there exists a unique n such that $\forall b\, Q(n, a, b)$.

Lemma 3.1. *Let A be a coinfinite Σ_2^0-set that possesses an infinite computable subset S such that $A \setminus S$ is infinite. Then A has a one-to-one-representation.*

The definition of a one-to-one-representation of a Σ_2^0-set can be relativized with respect to any oracle X. The relativized version of the lemma will be used in the proofs in the next sections.

Lemma 3.2. *Let A be a coinfinite $\Sigma_2^{0,X}$-set that possesses an infinite X-computable subset S such that $A \setminus S$ is infinite. Then there exists a X-computable set Q such that Q is a one-to-one-representation of A.*

The following two theorems are the corollaries of the lemma 3.2 and the corollary 2.1.

Theorem 3.1. *For any Turing degree d and for every computable sequence of d'-computable models $\mathcal{M}_0, \ldots \mathcal{M}_n, \ldots$ of a finite signature there exists a computable sequence $(\mathcal{M}_0)_{\forall\exists}, \ldots, (\mathcal{M}_n)_{\forall\exists}, \ldots$ of d-computable models.*

Proof of the Theorem 3.1. The proof of the lemma 3.1 in [5] shows that the construction of one-to-one-representations may be arranged uniformly for all n. Using the uniform version of the lemma 3.1 one can construct the sequence $(\mathcal{M}_0)_{\forall\exists}, \ldots, (\mathcal{M}_n)_{\forall\exists}, \ldots$ and show that every $(\mathcal{M}_n)_{\forall\exists}$ is d-computable.

---

[a] $\exists^{=1} x P(x)$ means that there exists a unique x satisfying P.

Theorem 3.2. *For every Turing degree d a model $\mathcal{M}$ is d-decidable if and only if $\mathcal{M}_\forall$ and $\mathcal{M}_\exists$ are d-decidable.*

Proof of the Theorem 3.2. According to the theorem 2.1 the model $\mathcal{M}$ is definable in each of the extensions $\mathcal{M}_\forall$ and $\mathcal{M}_\exists$. Therefore, if $\mathcal{M}_\forall$ or $\mathcal{M}_\exists$ is d-decidable then so is $\mathcal{M}$. On the other hand, the properties of $\mathcal{M}_\forall$ or $\mathcal{M}_\exists$ are completely determined by $\mathcal{M}$. Thus, if $\mathcal{M}$ is d-decidable then $\mathcal{M}_\forall$ and $\mathcal{M}_\exists$ are d-decidable.

4. $\aleph_1$-categorical theory with computable models

Theorem 4.1. *For any arithmetic Turing degree there exists a $\aleph_1$-categorical theory T of a finite signature which is Turing equivalent to this degree and has a computable model. Moreover, all countable models of T have a computable presentation.*

Proof. In the proof we follow the ideas from [5]. Let X be a Σ_n^0-set. We consider the structure $\mathcal{M} = (M, P)$, where P is a binary predicate on M with the following properties:

(1) Predicate P is antireflexive that is $\neg P(x, x)$ for all x.

(2) For any x there exists a unique y such that $P(x, y)$. For any y there exists a unique x such that $P(x, y)$.

(3) $m \in X$ if and only if there exists a unique P-cycle of the length $3m + 1$ and there is no P-cycle of the length $3m + 2$.

$m \notin X$ if and only if there exists a unique P-cycle of the length $3m + 2$ and there is no P-cycle of the length $3m + 1$.

For any m there exists a P-cycle of the length $3m$.

(4) Any $x \in M$ is in some P-cycle.

We denote $T_X = Th(\mathcal{M})$. The properties of the theory T_X are:

(1) Theory T_X is $\aleph_1$-categorical.

(2) Theory T_X is complete.

(3) $T_X \equiv_T X$.

It's easily seen that up to isomorphism $\mathcal{M}$ has a presentation $\mathcal{M} = (\omega, P)$, where $P \in \Sigma_n^0$ and P satisfies the conditions of lemma 3.2. P is coinfinite and possesses an infinite $0^{(n-1)}$-computable subset S of $3m$-cycles such that $P \setminus S$ is infinite. According to the lemma 3.2 P has a 1-to-1-representation, that is, there exists $Q_1 \subset \omega^4$ such that $Q_1 \in \Sigma_{n-2}^0$ and

(1) For all (x, y) $\exists a \forall b\, Q_1(x, y, a, b)$ if and only if $(x, y) \in P$.

(2) For all (x, y) $\exists a \forall b\, Q_1(x, y, a, b)$ if and only if $\exists^{=1} a \forall b\, Q_1(x, y, a, b)$.

(3) For every b there exists a unique tuple $\{x, y, a\}$ such that $\neg Q_1(x, y, a, b)$.

(4) For every $\{x, y, a\}$ either $\exists^{=1} b \neg Q_1(x, y, a, b)$ or $\forall b\, Q_1(x, y, a, b)$.

(5) For every a there exists a unique (x, y) such that $\forall b\, Q_1(x, y, a, b)$.

Let's consider models $\mathcal{M}_1 = \mathcal{M}_{\forall \exists}$ and $\mathcal{N}_1 = (M \cup X_1 \cup X_2 \cup Y_1, P_1^4, A_1^1, A_2^1, B_1^2)$, where X_1, X_2, Y_1 are infinite pairwise disjoint sets that are disjoint with M, $P_1(x, y, u_1, v_1) \iff (x, y) \in M$, $u_1 \in X_1$, $v_1 \in Y_1$ and $Q_1(x, y, u_1, v_1)$, A_1 and A_2 are for X_1 and X_2 correspondingly; $v_1 \in Y_1 \iff (\exists u_2 \in X_2) B_1^2(v_1, u_2)$ and the predicates satisfy the conditions from the definition of the $\exists$-extension.

Then P_1 is $0^{(n-1)}$-computable, $\mathcal{N}_1$ is $0^{(n-1)}$-computable and the following holds:

(1) From the definition of P_1 and the properties of P it follows that P_1 satisfies the conditions of the lemma 3.2.

(2) The mapping $F: X_1 \to P$ such that $F(u_1) = (x, y) \iff (\forall v_1)\, P_1(x, y, u_1, v_1)$ is a bijection. Proof. From the properties of Q_1 for every $u_1 \in X_1$ there exists a unique pair $(x, y) \in P$ such that $(\forall v_1) Q_1(x, y, u_1, v_1)$, i.e. $(\forall v_1) P_1(x, y, u_1, v_1)$; for every pair $(x, y) \in P$ there exists a unique element $u_1 \in X_1$ such that $(\forall v_1) Q_1(x, y, u_1, v_1)$, i.e. $(\forall v_1) P_1(x, y, u_1, v_1)$.

(3) The mapping G such that $G(x, y, u_1) = v_1 \iff \neg P_1(x, y, u_1, v_1)$ is a bijection. Proof. From the properties of Q_1 for every tuple (x, y, u_1) either there exists a unique element $v_1 \in Y_1$ such that $\neg Q_1(x, y, u_1, v_1)$ (i.e. $\neg P_1(x, y, u_1, v_1)$) or $(\forall v_1) Q_1(x, y, u_1, v_1)$, i.e. $(\forall v_1) P_1(x, y, u_1, v_1)$; for every $v_1 \in Y_1$ there exists a unique (x, y, u_1) such that $\neg Q_1(x, y, u_1, v_1)$, i.e. $\neg P_1(x, y, u_1, v_1)$.

Thus, from the properties 2 and 3 it follows that $\mathcal{M}_1 \cong \mathcal{N}_1$. The properties of $\mathcal{M}_1$ guarantee that $T_1 = Th(\mathcal{M}_1) = Th(\mathcal{N}_1)$ is $\aleph_1$-categorical and $T_1 \equiv_T T_X \equiv_T X$.

By induction we build $\mathcal{M}_0 = M$, $\mathcal{M}_1, \ldots, \mathcal{M}_{n-1}$, where $\mathcal{M}_k = (\mathcal{M}_{k-1})_{\forall \exists}$. We also build $\mathcal{N}_1, \ldots, \mathcal{N}_{n-1}$. Using the lemma we find Q_k which is one-to-one-representation of P_{k-1}. We define $\mathcal{N}_k = (M \cup X_{1,1} \cup X_{2,1} \cup X_{1,2} \cup \ldots \cup X_{1k} \cup \ldots \cup X_{2^{2k-1}k} \cup Y_{1,1} \cup Y_{1,2} \ldots \cup Y_{2^{2k-2}k}, P_k^{2k+2}, A_{1,1}^{2k-1} \ldots A_{1k}^1 \ldots A_{2^{2k-1}k}^1, B_{1,1}^{2k} \ldots B_{2^{2k-2}k}^2)$ where the sets $X_{i,j}$, $Y_{i,j}$ and M are pairwise disjoint. $P_k(x, y, u_{1,1}, u_{1,2}, .. u_{1,k}, v_{1,1}, .., v_{1,k}) \iff (x, y) \in M$, $u_{i,j} \in X_{i,j}$, $v_{i,j} \in Y_{i,j}$ and $Q_k(x, y, u_{1,1}, u_{1,2}, .. u_{1,k}, v_{1,1}, .., v_{1,k})$, and $u_{1,1} \in X_{1,1} \Leftrightarrow \exists u_{2,k} \forall v_{2,k} \ldots \forall v_{2,2} A_{1,1}(u_1, \ldots, v_{2,k})$ etc.; $A_{1,k}^1, \ldots A_{2^{2k-1},k}^1$ are for $X_{1,k} \ldots X_{2^{2k-1},k}$.

Similar to the case $k = 1$ P_k satisfies the conditions of the lemma, Q_k, P_k are $O^{(n-k-1)}$-computable, $\mathcal{N}_k$ is $0^{(n-k-1)}$-computable. From the properties of $\mathcal{M}_k$ and $\mathcal{N}_k$: $\mathcal{M}_k \cong \mathcal{N}_k$, $T_k = Th(\mathcal{N}_k)$ is $\aleph_1$-categorical, $T_k \equiv_T T_{k-1} \equiv_T \cdots \equiv_T T_X \equiv_T X$. In particular $\mathcal{N}_{n-1}$ is computable and its theory $T_{n-1} = Th(\mathcal{N}_{n-1})$ is $\aleph_1$-categorical and $T_{n-1} \equiv_T X$.

We prove now that all models of $T_{n-1} = Th(\mathcal{N}_{n-1})$ have a computable presentation. Let's note the following. Let $\mathcal{A} = (A, P)$ be some algebraic structure and $\mathcal{A}_{\forall \exists} = (A \cup X_1 \cup X_2 \cup Y_1, P_{\forall \exists}, U_1^1, U_2^1, V_1^2)$ is its Marker extension. We define $A_1 = (A, P_1)$ as a structure in which $\mathcal{A}_1 \models P_1(x, y) \iff \mathcal{A}_{\forall \exists} \models \exists u_1 \forall v_1 P_{\forall \exists}(x, y, u_1, v_1)$. The properties of the Marker extensions guarantee that $P_{\forall \exists}$ is 1-to-1-representation P_1 and $P = P_1$, $\mathcal{A} = \mathcal{A}_1$.

As it was proved in [1], all models $\mathcal{M}_i^X, i = 1, 2 \ldots$ of the theory T_X can be listed into a chain of elementary embeddings from the prime model to the saturated model. The prime model consists only of finite P-cycles and every next model contains one more infinite chain than the previous. Thus, every model $\mathcal{M}_i^X$ of T_X is X-computably presentable. We apply to M_i^X the operator $\forall \exists$ $n-1$ times and we get a computable model $(\mathcal{M}_i^X)^{n-1}$ of the theory T_{n-1}. Moreover, if $m \in X$ then $\varphi_m \in T_X$, where φ_m is a statement saying that there exists a unique cycle of the length $3m + 1$ and there is no cycles of the length $3m + 2$. Then, $\mathcal{M}_i^X \models \varphi_m$ and the corresponding statement φ_m^{n-1} for $\exists u_{1,n} \forall v_{1,n} \ldots \exists u_{1,1} \forall v_{1,1} P_{n-1}(x, y, u_{1,1} \ldots u_{1,n}, v_{1,1} \ldots v_{1,n})$ is true in $(\mathcal{M}_i^X)^{n-1}$. Similarly, if $m \notin X$ then $\neg \varphi_m^{n-1} \in T_{n-1}$. Let $\mathcal{M}^{n-1} = (M \cup N_{11}^1 \ldots N_{2^{2n-1}n}^1 \cup N_{11}^2 \ldots N_{2^{2n-2}n}^2, P_{n-1}, U_{11} \ldots U_{2^{2n-1},n}, V_{11} \ldots V_{2^{2n-2},n})$ be a model of T_{n-1}. We consider a model $\mathcal{N} = (M, P)$ where $\mathcal{N} \models P(x, y) \iff \mathcal{M}^{n-1} \models \exists u_{1,n} \forall v_{1,n} \ldots \exists u_{1,1} \forall v_{1,1} P_{n-1}(x, y, u_{1,1} \ldots, v_{1,n})$. Then $\mathcal{N}$ is a model of T_X and it has a X-computable presentation. We apply the operator $\forall \exists$ to $\mathcal{N}$ $n - 1$ times and get a computable model $\mathcal{N}_{n-1}$ which is isomorphic to $\mathcal{M}_{n-1}$. Thus, $\mathcal{M}_{n-1}$ has a computable presentation, i.e. $SCM(T_{n-1}) = \omega \bigcup \{\omega\}$.

5. $\aleph_0$-categorical theory with computable model

In this section we prove the following theorem.

Theorem 5.1. *For any arithmetic Turing degree there exists a $\aleph_0$-categorical theory T of a finite signature which is Turing equivalent to this degree and has a computable model.*

Proof. We code a Σ^0_{n+1}-set Y into a $\aleph_0$-categorical theory T_Y so that Y and T_Y have the same Turing degree.

The construction of T_Y is similar to the construction in [5]. The language of T_Y consists of one binary predicate R. For every $n \in \omega$ we define a **cycle** $C_n = (\{0, 1, \ldots, n+2\}, R)$ of the length $n+3$, where $R(x, y)$ is true if and only if $\{x, y\} = \{i, i+1\}$ or $\{x, y\} = \{n+2, 0\}$.

Let $\mathcal{K}_Y$ be a class of all finite graphs $\mathcal{G}$ such that $m \in Y$ if and only if $C_{3m+1} \in \mathcal{K}_Y$ and $C_{3m+2} \notin \mathcal{K}_Y$, $m \notin Y$ if and only if $C_{3m+2} \in \mathcal{K}_Y$ and $C_{3m+1} \notin \mathcal{K}_Y$

As in [5] the class $\mathcal{K}_Y$ satisfies the following properties.

If $\mathcal{A}, \mathcal{B}_1, \mathcal{B}_2$ are in $\mathcal{K}_Y$ and there are embeddings $e : \mathcal{A} \to \mathcal{B}_1$, $f : \mathcal{A} \to \mathcal{B}_2$ then there exists $\mathcal{C}$ from $\mathcal{K}_Y$ and embeddings $g : \mathcal{B}_1 \to \mathcal{C}$ and $h : \mathcal{B}_2 \to \mathcal{C}$ such that $ge = hf$.

If $\mathcal{A} \in \mathcal{K}_Y$ and $\mathcal{B}$ is a subgraph of $\mathcal{A}$ then $\mathcal{B} \in \mathcal{K}_Y$.

For all $\mathcal{A}$ and $\mathcal{B}$ from $\mathcal{K}_Y$ there exists $\mathcal{C} \in \mathcal{K}_Y$ such that there are embeddings of $\mathcal{A}$ and $\mathcal{B}$ into $\mathcal{C}$.

The axioms of T_Y are the following. The first of them states that R is antireflexive. The infinite list of universal sentences says that any $\mathcal{B} \notin \mathcal{K}_Y$ can not be embedded into models of T_Y. The infinite list of $\forall\exists$-sentences guarantees the following property. If $\mathcal{A}, \mathcal{B} \in \mathcal{K}_Y$ and there is an embedding $f : \mathcal{A} \to \mathcal{B}$ then there is an extension $\mathcal{A}'$ of $\mathcal{A}$ and an isomorphic mapping $f' : \mathcal{A}' \to \mathcal{B}$ that extends f.

The sentence that guarantees the existence of $3n + 1$-cycle belongs to T_Y if and only if $n \in Y$. At the same time $n \notin Y$ if and only if the sentence that guarantees the existence of $3n + 2$-cycle belongs to T_Y. Therefore, Y and T_Y have the same Turing degree.

Using the properties of $\mathcal{K}_Y$, for every $\mathcal{A} \in \mathcal{K}_Y$ we find $\mathcal{A}^*$ that satisfies the following property. For all $\mathcal{B}, \mathcal{C} \in \mathcal{K}_Y$ and an embedding f of $\mathcal{B}$ into $\mathcal{A}$ such that $\mathcal{B}$ is a subgraph of $\mathcal{C}$, and $card(\mathcal{C}) = card(\mathcal{B}) + 1$ there is an embedding $g : \mathcal{C} \to \mathcal{A}^*$ that extends f.

Then the model $\mathcal{A}$ of T_Y is constructed as follows. Let $\mathcal{A}_0$ be a model from $\mathcal{K}_Y$. We consider the chain

$$\mathcal{A}_0 \subset \mathcal{A}_1 \subset \mathcal{A}_2 \ldots$$

of models from $\mathcal{K}_Y$ such that $\mathcal{A}_{n+1} = \mathcal{A}_n^*$. Let $\mathcal{A}$ be the union of the chain. Then $\mathcal{A}$ is a model of T_Y.

Using the back-and-forth method we can show that any two countable models $\mathcal{A}$ and $\mathcal{B}$ of T_Y are isomorphic. That is, T_Y is $\aleph_0$-categorical.

As $T_Y \equiv_T Y$ then T_Y has a model $\mathcal{A} = (\omega, R)$, $R \in \Sigma^0_{n+1}$ and R satisfies the conditions of the lemma 3.2. We construct the sequence of models $\{\mathcal{A}_i\}_{i \leq n}$ such that:

(1) $\mathcal{A}_0$ is $\mathcal{A}$.
(2) $\mathcal{A}_i$, $1 \leq i \leq n$, is a Marker's $\forall\exists$-extension of $\mathcal{A}_{i-1}$.
(3) $\mathcal{A}_i$ is $\mathbf{0}^{n-i}$-computable.

In particular the structure $\mathcal{A}_n$ is computable. The corollary 2.1 guarantees that its theory T is $\aleph_0$-categorical and $T \equiv_T Y$. The theorem is proved.

6. Complexity of index sets

In the previous section it was proved that for every arithmetic Turing degree d there exists a countably categorical theory of Turing degree d with a computable model. Now we are interested in the complexity of the set of such countably categorical computable models with d-decidable theories.

More precisely, let $\mathcal{M}_0, \mathcal{M}_1, \ldots, \mathcal{M}_n, \ldots$ be a universal numbering of all computable models of a fixed signature. For every arithmetic degree d we consider the index set

$$CK^d = \{n | \mathcal{M}_n \text{ is a } d\text{-decidable model}\}.$$

Theorem 6.1. *For every arithmetic Turing degree d the index set CK^d of all d-decidable models has the Turing degree $d^{(3)}$ in the universal computable numbering of all computable $\aleph_0$-categorical models in the signature with one binary predicate.*

Lemma 6.1. $CK^d \in \Sigma^{0,d}_3$.

Proof of the Lemma 6.1. We need to write that there exists a d-computable function which is the characteristic function of the full diagram of $\mathcal{M}_n$. This statement is a $\Sigma^{0,d}_3$-sentence.

Lemma 6.2. *For every d there exists an ordering $L_d = \langle N, \preceq_d \rangle$ that is d-computable, has the type $\omega + \omega*$ but ω is not d-c.e.*

Proof of the Lemma 6.2. The lemma is the relativized modification of the lemma from [3]. The original version states the existence of a computable ordering of the type $\omega + \omega*$ such that its initial segment of the type ω is not computable. It's not hard to see that in this case ω or $\omega*$ is not c.e. because otherwise we can enumerate both of them. Therefore, they both are computable.

Lemma 6.3. *If $A \in \Sigma_3^{0,d}$ then there exists a d-computable predicate $Q(n,x,y)$ such that*

$$n \in A \Longleftrightarrow (\exists x)(\exists^\infty y)Q(n,x,y)$$

and for all n $Q(n,0,0)$ and $Q(n,0,1)$.

Proof of the Lemma 6.3 can be found in [15].

Proof of the Theorem 6.1. For all $x \in L_d$ and for all n we consider the set $L_{(n,x)} = \{\langle x', y'\rangle | x' \leq x \text{ and } Q(n,x',y')\}$ that is uniformly d-computable. Let $R_{(n,x)}$ be a linear ordering of $L_{(n,x)}$ such that

if $L_{(n,x)}$ is infinite then $(L_{(n,x)}, R_{(n,x)})$ has the type η and $\langle 0,0\rangle < \langle 0,1\rangle$;

if $L_{(n,x)}$ is finite then $(L_{(n,x)}, R_{(n,x)})$ is a linear ordering.

We define now

$$L_n \leftrightharpoons \sum_{x \in L_a} L_{(n,x)}.$$

If $n \in A$ then according to the lemma 6.3 $(\exists x_0)(\exists^\infty y)Q(n,x_0,y)$. Thus, for all x such that $x_0 \leq x$ the set $L_{(n,x)}$ is infinite. By the definition of $R_{(n,x)}$

$$L_n \cong \sum_{k=0}^{r(n)} S_k + P_k,$$

where S_k is finite and $P_k \cong \eta$ for every $k \leq r(n)$, $r(n)$ is finite and depends on n. Thus, L_n is d-decidable. If $n \notin A$ then for all x the set $L_{(n,x)}$ is finite and $L_n \cong \omega + \omega*$. By the lemma 6.2 L_n is d-computable as L_d is d-computable. At the same time L_n is not d-decidable. If L_n were d-decidable then we could enumerate ω with the oracle d and ω would not be c.e. over d.

For every n the structure L_n satisfies the conditions of the lemma 3.2. Trere exists m such that each R_n is a coinfinite Σ_0^m-set. The set S_n of all pairs $\langle 0,0\rangle$ and $\langle 0,1\rangle$ from all $L_{(n,x)}$ form an infinite subset which is $\mathbf{0}^{(m-1)}$-computable and the lemma 3.2 gives a 1-to-1-representation R_n^1 of R_n.

Again as in the previous two sections for all n we build the sequence of models $L_n^i, 1 \leq i \leq m$, such that L_n^0 is L_n and each L_n^i is a $\forall\exists$-extension of L_n^{i-1}. Now we use the theorems 3.1 and 3.2. Each L_n^i is $\mathbf{0}^{(m-i)}$-computable. According to the corollary 2.1 for all n and for all $i \leq m$ L_n^i is $\aleph_0$-categorical. In particular, L_n^m is computable and $\aleph_0$-categorical. If $n \in A$ then L_n^m is d-decidable and if $n \notin A$ then L_n^m is not d-decidable.

Corollary *The index set of all decidable models has the Turing degree $0^{(3)}$ in the universal computable numbering of all computable $\aleph_0$-categorical models in the signature with one binary predicate.*

References

1. J. Baldwin, A. Lachlan, *On Strongly Minimal Sets*, Journal of Symbolic Logic, 36, 1971, 79–96.
2. C. C. Chang and H. J. Keisler, Model Theory, 3rd ed., Stud. Logic Found. Math., 73, 1990.
3. Yu. Ershov, Theory of Numberings 3, Novosibirsk, Novosibirsk State University, 1974.
4. S. Goncharov,*Constructive Models of ω_1-categorical Theories*, Matematicheskie Zametki, 23, 1978, 885–888.
5. S. Goncharov and B. Khoussainov, *Complexity of categorical theories with computable models*, Dokl. Russian Academy of Science, 2002, 385, N 3, 299–301.
6. S. Goncharov and B. Khoussainov, *On comlexity of theories of computable $\aleph_1$-categorical models*, Vestnik of Novosibirsk State University, series: mathematics, mechanics and informatics , 2001, 1, N 2, 63–76.
7. L. Harrington, *Recursively Presentable Prime Models*, J. of Symbolic Logic, 39, 1974, 305–309.
8. N. Khisamiev, *Strongly Constructive Models of a Decidable Theory*, Izv. Akad. Nauk Kazakh. SSR, Ser. Fiz.-Mat., 1, 1974, 83–84.
9. B. Khoussainov, A. Nies, R. Shore, *On Recursive Models of Theories*, Notre Dame Journal of Formal Logic 38, 2, 1997, 165–178.
10. J. Knight, *Nonarithmetical $\aleph_0$-categorical Theories with Recursive Models*, J. Symbolic Logic, 59, N 1, 1994, 106–112.
11. K. Kudaibergenov, *On Constructive Models of Undecidable Theories*, Siberian Mathematical Journal, v. 21, no 5, 1980, 155–158.
12. M. Lerman, J. Schmerl, *Theories With Recursive Models*, J. Symbolic Logic 44, N 1, 1979, 59–76.
13. D. Marker, *Non-Σ_n-axiomatizable almost strongly minimal theories*, J. Symbolic Logic 54, 1989, 921–927.
14. A. Nies, *A new spectrum of recursive models*, Notre Dame Journal of Formal Logic, 40, 1999, 307–314
15. H. Rogers, Theory of recursive functions and effective computability, McGraw-Hill, 1967.

EQUIVALENCE RELATIONS AND CLASSICAL BANACH SPACES*

SU GAO

*Department of Mathematics, PO Box 311430,
University of North Texas, Denton, TX 76210, U.S.A.
E-mail: sgao@unt.edu*

We give a survey of results on Borel reducibility among equivalence relations induced by classical Banach spaces. We present an application of this study to a classification problem related to the big O notation. Finally we study a question of Kanovei and provide some information on the complexity of the equivalence relations involved.

1. Preamble

Classical Banach spaces and their actions give important examples of equivalence relations and are intensively studied in the descriptive set theory of Borel reducibility. In this article we give a survey of the results related to these equivalence relations and discuss some intriguing open problems. Many folklore results in the area have simple proofs which are hard to find in the literature; for some of them we give the proofs here. Our selection of results is not complete and inevitably reflects personal taste. At times attributions of the results are hard to determine and might not be accurate. The main objective of the paper is to provide an overview of the area for the reader and to motivate further research.

We recall the definition of Borel reducibility. Let E, F be equivalence relations on Polish spaces X, Y, respectively. We say that E is *Borel reducible to F*, and denote $E \leq_B F$, if there is a Borel function $\theta : X \to Y$ such that, for all $x_1, x_2 \in X$,

$$x_1 E x_2 \iff \theta(x_1) F \theta(x_2).$$

If $E \leq_B F$ and $F \leq_B E$ then we say that E and F are *Borel bireducible* and denote $E \sim_B F$. If $E \leq_B F$ but $F \not\leq_B E$ then we write $E <_B F$.

In this article we focus on classical Banach spaces $\ell^p (p \geq 1), c_0, \ell^\infty$

---

*Research partially supported by the U.S. NSF grant DMS-0501039. I would like to thank the Chinese Academy of Sciences for a partial travel grant.

and briefly on $C_0(\mathbb{R}^+)$. There is in fact very interesting work done on non-classical Banach spaces, but it is not covered in this paper.

2. ℓ^1

This notation is now overloaded with several meanings. As a Banach space it denotes the linear subspace of $\mathbb{R}^\omega$ given by

$$\ell^1 = \left\{ (x_n) \in \mathbb{R}^\omega \ : \ \sum_{n=0}^{\infty} |x_n| < \infty \right\}$$

endowed with the complete norm

$$\|(x_n)\|_1 = \sum_{n=0}^{\infty} |x_n|.$$

As an equivalence relation its underlying space is $\mathbb{R}^\omega$ and it is defined as

$$(x_n)\ell^1(y_n) \iff (x_n - y_n) \in \ell^1,$$

where the ℓ^1 on the right hand side is the above space. Sometimes the equivalence relation is also represented by the quotient space and is denoted by $\mathbb{R}^\omega/\ell^1$. This last notation emphasizes the fact that the equivalence relation is induced by the additive action of ℓ^1 as an additive group. Thus the notation ℓ^1 is also used to denote the Polish group under addition.

Kechris noticed early on that the equivalence relation ℓ^1 is related to ideals. Recall that the summable ideal on $\mathbb{N}$ is defined by

$$I = \left\{ A \subseteq \mathbb{N} : \sum_{n \in A} \frac{1}{n+1} < \infty \right\}.$$

For any ideal I on $\mathbb{N}$ define the equivalence relation E_I on $2^\mathbb{N}$ by

$$x E_I y \iff x \triangle y \in I,$$

where $x, y \in 2^\mathbb{N}$ are understood as subsets of $\mathbb{N}$ in a natural way and $x \triangle y = (x \setminus y) \cup (y \setminus x)$ is the symmetric difference of x and y. The following simple fact was the starting point of the study of the equivalence relation ℓ^1.

Lemma 2.1 (Kechris). *Let I be the summable ideal on $\mathbb{N}$. Then $\ell^1 \sim_B E_I$.*

Hjorth discovered the following important dichotomy regarding equivalence relations below ℓ^1 in the Borel reducibility hierarchy. Recall that an equivalence relation F is *countable* if every F-equivalence class is countable, and an equivalence relation E is *essentially countable* if there is a countable equivalence relation F such that $E \leq_B F$. Among the countable equivalence relations there is a universal one E_∞, that is, for any countable equivalence relation E we have $E \leq_B E_\infty$ (see [4]).

Theorem 2.2 (Hjorth [10]). *If $E \leq_B \ell^1$ then either $E \sim_B \ell^1$ or E is essentially countable.*

It is still open whether the second possibility can be replaced by $E \leq_B E_0$, where E_0 is the equivalence relation on $2^{\mathbb{N}}$ defined by eventual agreement:

$$x E_0 y \iff \exists m \forall n \geq m \, (\, x(m) = y(m) \,).$$

In the same spirit Kanovei asked the following concrete question.

Question 2.3 (Kanovei [12]). *Is $E_\infty \leq_B \ell^1$?*

In any case the equivalence relation ℓ^1 has a rather modest complexity in the Borel reducibility hierarchy. This is in contrast with the other equivalence relations the actions of the group ℓ^1 is capable of producing. Recall that if a Polish group G acts continuously on a Polish space X we say that X is a *Polish G-space* and denote the induced orbit equivalence relation by E_G^X, thus for $x_1, x_2 \in X$,

$$x_1 E_G^X x_2 \iff \exists g \in G \, (\, g \cdot x_1 = x_2 \,).$$

Theorem 2.4 (Gao-Pestov [9]). *Let G be any abelian Polish group and X be a Polish G-space. Then there is a Polish ℓ^1-space Y such that $E_G^X \leq_B E_{\ell^1}^Y$. Moreover, if E_G^X is Borel, then Y can be chosen so that $E_{\ell^1}^Y$ is Borel.*

In fact ℓ^1 is capable of producing a universal equivalence relation among the orbit equivalence relations given by actions of abelian Polish groups. We give an example of this universal equivalence relation. Given a Polish space X let $F(X)$ be the set of all closed subsets of X. It is possible to equip $F(X)$ with a Polish topology so that it becomes a Polish space (see e.g. [13]). Most of the time the particular Polish topology is not as important as the induced Borel structure, and the space $F(X)$ with this Borel structure is called the *Effros Borel space*. Consider $F(\ell^1)$ with the action by ℓ^1:

$$g \cdot F = \{g + f : f \in F\}.$$

Then results of [9] show that the orbit equivalence relation $E_{\ell^1}^{F(\ell^1)}$ is a universal orbit equivalence relation for abelian Polish group actions.

This last result was proved indirectly. It is still of interest to produce a meaningful reduction from ℓ^1 to $E_{\ell^1}^{F(\ell^1)}$. In other words, which closed subsets of ℓ^1 can code elements of $\mathbb{R}^\omega$ in an ℓ^1-invariant way?

A more intriguing question is whether it is possible to Borel reduce E_∞ to any orbit equivalence relation by abelian Polish group action. This is weaker than Kanovei's question above, but a negative answer would be more striking.

Question 2.5. Is $E_\infty \leq_B E_{\ell^1}^{F(\ell^1)}$?

3. ℓ^p $(p \geq 1)$

Similar to the situation for ℓ^1, ℓ^p for all $p \geq 1$ have been investigated as equivalence relations and Polish groups in action. One naturally wonders about their Borel reducibility, and Dougherty and Hjorth gave the full answer.

Theorem 3.1 (Dougherty-Hjorth [3]). *For any* $1 \leq p < q < \infty$, $\ell^p <_B \ell^q$.

Thus they form a chain of order type $\mathbb{R}^{\geq 0}$ (the nonnegative real numbers). A natural question is whether they exhaust all equivalence relations on a chain in the Borel reducibility hierarchy. We note next that this is not the case. For this we need a definition.

Let X_n, $n \in \omega$, be Polish spaces and E_n be equivalence relations on X_n respectively. The *direct sum* of X_n, denoted $\bigoplus_{n \in \omega} X_n$ or simply X_ω, is the disjoint union of all X_n with the topology naturally induced by the topologies of all X_n, with each X_n clopen. The *direct sum* of E_n, denoted $\bigoplus_{n \in \omega} E_n$ or simply E_ω, is defined by

$$x E_\omega y \iff \exists n \in \omega\,(\,x, y \in X_n \wedge x E_n y\,).$$

In particular, if $X_n = X$ and $E_n = E$ for all $n \in \omega$, E_ω is an equivalence relation on X_ω. An equivalence relation E on X is called *splitting* if there is a Borel isomorphism $\varphi : X_\omega \to X$ such that

$$x E_\omega y \iff \varphi(x) E \varphi(y).$$

By the Shröder-Bernstein theorem E is splitting iff there is a Borel isomorphic embedding $\varphi : X_\omega \to X$ as above, so that $\varphi(X_\omega)$ is an E-invariant subset of X. We note the following simple fact.

Lemma 3.2. *For any* $p \geq 1$, ℓ^p *is splitting.*

Proof. Let $Y_n \subseteq \mathbb{R}^\omega$ be the set of all (x_k) such that $x_k = n$ if k is even. Let E_n be the $\ell^p \upharpoonright Y_n$. It is clear that E_n is Borel isomorphic to ℓ^p. However, $E_\omega = \ell^p \upharpoonright \bigcup_{n \in \omega} Y_n$. In particular, for $(x_k) \in Y_n$, $(x_k') \in Y_m$ and $n \neq m$, (x_k) is not ℓ^p-equivalent to (x_k'). Finally, notice that the saturation of Y_ω in $\mathbb{R}^\omega$ can be obtained by an action of ℓ^p on Y_ω. This action actually gives a Borel isomorphism of Y_ω with an invariant subset of $\mathbb{R}^\omega$. $\square$

For each $p > 1$, define an equivalence relation $\ell^{<p}$ by the following. Fix any increasing sequence $1 \leq p_0 < p_1 < \cdots < p_n < \cdots$ converging to p, let E_n be the ℓ^{p_n}-equivalence relation on X_n, a copy of $\mathbb{R}^\omega$. Then define $\ell^{<p}$ to be E_ω on X_ω. First note that the Borel reducibility class of $\ell^{<p}$ does not depend on the choice of the sequence (p_n). In fact, if (p_n') is another sequence and E_ω' be obtained using (p_n') in the same fashion. Without loss of generality we may assume $p_n' < p_n$ for all $n \in \omega$. Then the Dougherty-Hjorth theorem implies that $E_\omega' \leq_B E_\omega$. Since this is symmetric we have that in fact $E_\omega' \sim_B E_\omega$. We also note the following fact.

Lemma 3.3. *For each* $1 \leq r < p$, $\ell^r <_B \ell^{<p} \leq_B \ell^p$.

Proof. The first reduction is trivial from the definition of $\ell^{<p}$. The nonreduction follows from the Dougherty-Hjorth theorem. The second reduction follows from the fact that ℓ^p is splitting. $\square$

Lemma 3.4. *For each* $p > 1$, $\ell^{<p} <_B \ell^p$.

Proof. Toward a contradiction assume that $\theta : \mathbb{R}^\omega \to X_\omega$ is a Borel reduction of ℓ^p to $\ell^{<p}$. Let $\theta' : \mathbb{R}^\omega \to \mathbb{N}$ be defined by $\theta'(x) = n$ iff $\theta(x) \in X_n$. Then θ' is an invariant map. Since every orbit of ℓ^p is dense and meager, it follows (see, e.g., [11, Theorem 3.2]) that there is a comeager set $C \subseteq \mathbb{R}^\omega$ such that θ' is constant on C. Thus there is $n \in \omega$ such that θ is a reduction of ℓ^p on C to ℓ^{p_n} for $p_n < p$. The proof of the Dougherty-Hjorth theorem [3] shows that this is impossible. $\square$

We do not know if $\ell^{<p}$ is the least upper bound for all ℓ^r, $r < p$. However, it is easy to see that, if E is splitting and $\ell^r \leq_B E$ for all $r < p$, then $\ell^{<p} \leq_B E$. Also, $\ell^{<p}$ is also splitting. On the other hand, one could ask if ℓ^p is the greatest lower bound l^q, $q > p$. We do not know the answer to this question.

In a similar fashion one can define an upper bound for all ℓ^p, $p \geq 1$, as follows. The equivalence relation ℓ^ω is defined on a direct sum $X = \bigoplus_{i \in \mathbb{N}^+} X_i$ where each X_i is a copy of $\mathbb{R}^\omega$. For $x, y \in X$,

$$x \ell^\omega y \Leftrightarrow \exists i (x, y \in X_i \wedge \|x - y\|_i < \infty).$$

It is easy to see that for any $1 \leq p < \infty$, $\ell^p <_B \ell^\omega$.

Also observe that if each E_n on X_n is an orbit equivalence relation induced by an action of a Polish group G_n, then E_ω is an orbit equivalence relation induced by an action of the product group $\prod_{n \in \omega} G_n$. Thus all the equivalence relations considered so far are orbit equivalence relations. By results of the preceding section all of them are Borel reducible to some orbit equivalence relation of ℓ^1 action. It is again of interest to produce concrete such examples of ℓ^1 actions.

Next we turn to products of the equivalence relations ℓ^p. Recall that if Λ is an index set and E_λ are equivalence relations on X_λ respectively. The product equivalence relation $E^\Lambda = \prod_{\lambda \in \Lambda} E_\lambda$ is the equivalence relation on $X^\Lambda = \prod_{\lambda \in \Lambda} X_\lambda$ defined by

$$xE^\Lambda y \iff \forall \lambda \in \Lambda \, (\, x(\lambda)E_\lambda y(\lambda)\,).$$

We note two simple facts.

Lemma 3.5. *For any $0 < n < \omega$, $(\ell_p)^n \sim_B \ell^p$.*

Proof. The obvious homeomorphism between $\mathbb{R}^\omega$ and $(\mathbb{R}^\omega)^n$ is a bireduction of the equivalence relations. $\square$

Next we show that $\ell^p <_B (\ell_p)^\omega$. This is done by noting that $E_0 \leq_B \ell^p$, and hence $(E_0)^\omega \leq_B (\ell^p)^\omega$ by the obvious combination of reductions, but $(E_0)^\omega \not\leq_B \ell^p$. A proof of this last fact can be found in [12]. But here we give a proof for a slightly more general fact. Hjorth [10] contains a proof of $c_0 \not\leq_B \ell^1$ by an argument similar to the following one.

Lemma 3.6. *If E is F_σ then $(E_0)^\omega \not\leq_B E$.*

Proof. Suppose $(E_0)^\omega \leq_B E$. Then by Lemma 1 of [12], there is a continuous reduction witnessing that $(E_0)^\omega$ is reducible to $E \times E$. Note that $E \times E$ is still F_σ. In particular each $E \times E$-equivalence class is F_σ. It follows that every $(E_0)^\omega$-equivalence class is also F_σ. But this is a contradiction since there are $\mathbf{\Pi}_3^0$-complete classes of $(E_0)^\omega$. $\square$

All the ℓ^p equivalence relations are F_σ. Hence $(E_0)^\omega \not\leq_B \ell^p$ and $\ell^p <_B (\ell_p)^\omega$. We do not know if it is the case that $\ell^q \not\leq_B (\ell^p)^\omega$ when $p < q$. The proof of the Dougherty-Hjorth theorem might give some information.

4. The O notation

Before we move on to the next classical Banach space ℓ^∞ let us consider a related equivalence relation arising from a familiar context in mathematics and computer science. This will eventually lead to a meaningful application and will help motivate the study of the equivalence relation ℓ^∞, even if it is not a Polish group. The material mainly comes from the last chapter of [7], which has not appeared elsewhere.

The O notation is widely used in computer science (algorithm analysis) and in mathematical analysis (estimation of orders). When a function comes up in these fields, it is usually compared with some well understood functions first. Examples of such canonical functions include n^a, $n^a(\log n)^b$, $n^a e^{bn}$, etc. Sometimes the analyzed function can be shown to have exactly the same order as one of these canonical functions; but more often no such equivalence can be obtained. Of course, obtaining the exact order of a function might not be very important in these fields, depending on the motive of the study and the applications in mind. However, from a theoretical point of view, it is certainly very desirable if the list of canonical functions can be completed. Intuitively, it would mean that we understand the functions so well that, as far as its order is concerned, we have captured all possibilities in this list. The list itself may be long, even infinite, and each form above is actually representing uncountably many different functions, so the current list is already uncountable. We are willing to understand the word "list" in its weakest sense as long as an empirical such list shows up and is proved to be complete. Probably the only restriction is that all of these have to be done in some definable fashion, since one can immediately think of an abstract set of representatives by applying the Axiom of Choice. One way to formulate the ideal here in rigorous terms is the following: find a Polish space X such that each point in the space is coding some function and for any function there is a point in X provably in the same order. Note that the above listed functions fit in this framework. For example, the functions n^a can be coded by $\mathbb{R}$ (from which the constant a is chosen), and similarly $n^a(\log n)^b$ can be coded by a copy of $\mathbb{R}^2$. Therefore, $\mathbb{R}^3$ is enough to code all of the functions of the form n^a and $n^a(\log n)^b$. This formulation is still vague in two senses. First, it is not clear what is in general an acceptable coding for the functions. Second, it is not clear in what sense a function is provably equivalent to another function. The existence of such a list is dubious no matter how the question is understood. The point is to give a *proof* in some sense that there is no such list. In this section we try to formulate this question in a workable form and an answer to the general question will be given in the next section.

Recall the definition of O. For two functions f and g, we say that f is $O(g)$, denoted by $f \in O(g)$, if there is some constant $c > 0$ such that

$f(x) \leq cg(x)$ for sufficiently large x.

In view of the possible application in computer science we make the following more strict definition. Consider the Polish space $(\mathbb{R}^+)^\omega$, where $\mathbb{R}^+$ is the space of all positive real numbers. For $f, g \in (\mathbb{R}^+)^\omega$, $f \in O(g)$ if there is $c > 0$ such that $f(n) \leq cg(n)$ for all $n \in \omega$. We define the equivalence relation Θ on $(\mathbb{R}^+)^\omega$ by

$$f\Theta g \iff f \in O(g) \wedge g \in O(f), \ \forall f, g \in (\mathbb{R}^+)^\omega.$$

Conventionally $f\Theta g$ is written as $f \in \Theta(g)$. Here we intend to emphasize that it is an equivalence relation.

Let us justify that considerations in mathematical analysis would end up with the same "discrete" version of the equivalence relation. It would be reasonable to assume that functions of interest in analysis are continuous. Suppose f and g are positive continuous functions with domain $\mathbb{R}^+$ (thus functions without definition at 0, such as $1/x$ are included in this consideration). Note that the continuity of the function $f(x)/g(x)$ implies that $f \in O(g)$ iff there is $c > 0$ such that $f(x) \leq cg(x)$ for all $x \geq 1$. Now let $\{q_n\}$ enumerate all rational numbers ≥ 1. Then again by continuity of $f(x)/g(x)$ we have that $f \in O(g)$ iff there is $c > 0$ such that for all $n \in \omega$, $f(q_n) \leq cg(q_n)$. This is essentially a reduction to the notion defined in the preceding paragraph.

Recall that an equivalence relation is *smooth* if $E \leq_B \mathrm{id}(X)$ for some Polish space X, where $\mathrm{id}(X)$ is the identity relation on X. We would like to understand the above question as whether Θ is smooth. We will not argue that this is the only reasonable formulation of the above question. Nevertheless, it is very close in spirit and its answer will give us a somewhat deep understanding of the situation. We will show in the next section that Θ is not smooth. In fact, our results give more information about how high above Θ is located in the Borel reducibility hierarchy.

Let us remark that the above definition of Θ also makes sense for the space $(\mathbb{N}^+)^\omega$ instead of $(\mathbb{R}^+)^\omega$, where $\mathbb{N}^+$ is the set of positive natural numbers. The space $(\mathbb{N}^+)^\omega$ is essentially the same as the Baire space ω^ω. We denote this alternate equivalence relation $(\mathbb{N}^+)^\omega/\Theta$. If a distinction needs to be made, the original equivalence relation will be denoted $(\mathbb{R}^+)^\omega/\Theta$.

5. Θ and ℓ^∞

Recall that

$$\ell^\infty = \{f \in \mathbb{R}^\omega \mid \exists M > 0 \, \forall n \, |f(n)| < M\}.$$

Again the notation can be used to address an equivalence relation on $\mathbb{R}^\omega$ and an additive group ℓ^∞.

In these definitions if X is a subspace of $\mathbb{R}^\omega$ we denote by X/ℓ^∞ the equivalence relation $\ell^\infty \upharpoonright X$. Then we have the following simple fact.

Lemma 5.1. *The following equivalence relations are pairwise Borel bireducible:*

    (i) $(\mathbb{N}^+)^\omega/\Theta$,
    (ii) $(\mathbb{R}^+)^\omega/\Theta$,
    (iii) $\mathbb{R}^\omega/\ell^\infty$,
    (iv) $\mathbb{Z}^\omega/\ell^\infty$,
    (v) $\omega^\omega/\ell^\infty$.

Proof. We show that (i)$\Rightarrow$(ii)$\Rightarrow$(iii)$\Rightarrow$(iv)$\Rightarrow$(v)$\Rightarrow$(i) by a series of simple observations. (i)$\Rightarrow$(ii) is obvious. (ii)$\Rightarrow$(iii) is witnessed by the following reduction $\theta : (\mathbb{R}^+)^\omega \to \mathbb{R}^\omega$ given by

$$\theta(f)(n) = \log f(n), \ \forall f \in (\mathbb{R}^+)^\omega.$$

(iii)$\Rightarrow$(iv): Let $\theta : \mathbb{R}^\omega \to \mathbb{Z}^\omega$ be defined by

$$\theta(f)(n) = \lfloor f(n) \rfloor, \ \forall f \in \mathbb{R}^\omega,$$

where $\lfloor x \rfloor$ is the largest integer $\leq x$. It is easy to verify that θ works.

(iv)$\Rightarrow$(v): Define $\theta : \mathbb{Z}^\omega \to \omega^\omega$ by

$$\theta(f)(2n) = \begin{cases} f(n) & \text{if } f(n) \geq 0 \\ 0 & \text{otherwise} \end{cases}$$

$$\theta(f)(2n+1) = \begin{cases} 0 & \text{if } f(n) \geq 0 \\ -f(n) & \text{otherwise} \end{cases}$$

for $f \in \mathbb{Z}^\omega$. Then when $f(n)g(n) > 0$, we have $|\theta(f)(n) - \theta(g)(n)| = |f(n) - g(n)|$; when $f(n)g(n) < 0$, we have

$$|\theta(f)(n) - \theta(g)(n)| \leq \max(|f(n)|, |g(n)|) \leq |f(n)| + |g(n)| = |f(n) - g(n)|.$$

These imply that $f\ell^\infty g \Rightarrow \theta(f)\ell^\infty\theta(g)$. For the reverse direction, suppose $M > 0$ is such that $|\theta(f)(n) - \theta(g)(n)| \leq M$ for all n. Then by the definition of θ we have that $|f(n)|, |g(n)| < M$ for all n. Therefore $|f(n) - g(n)| < 2M$ for all n.

(v)$\Rightarrow$(i): Let $\theta : \omega^\omega \to (\mathbb{N}^+)^\omega$ be given by

$$\theta(f)(n) = 2^{f(n)}, \ \forall f \in \omega^\omega.$$

Then θ works. $\qquad\square$

The lemma allows us to shift our focus from Θ to ℓ^∞. We extend the Dougherty-Hjorth theorem by taking ℓ^∞ into account.

Lemma 5.2. *For* $1 \le p < \infty$, $\ell^p <_B \ell^\infty$.

Proof. It suffices to show that for $1 \le p < \infty$, $\ell^p \le_B \ell^\infty$. Fix $1 \le p < \infty$. Let $\mathbb{Q}^{<\omega}$ be the set of all finite sequences of rational numbers and let $\{s_m\}_{m \in \omega}$ enumerate the elements of $\mathbb{Q}^{<\omega}$. For each $s \in \mathbb{Q}^{<\omega}$ let $l(s)$ denote the length of s. We define $\theta : \mathbb{R}^\omega \to \mathbb{R}^\omega$ by

$$\theta(f)(m) = \left(\sum_{n < l(s_m)} |f(n) - s_m(n)|^p \right)^{\frac{1}{p}}$$

for any $f \in \mathbb{R}^\omega$. We verify that θ works.

First, suppose $\theta(f)\ell^\infty\theta(g)$, that is, $\theta(f) - \theta(g) \in \ell^\infty$. Let $\{m_k\}_{k \in \omega}$ be a sequence such that for each k, $l(s_{m_k}) = k$ and

$$|g(n) - s_{m_k}(n)| \le \frac{1}{2^{n+1}}, \ \forall n < k.$$

Then for each k,

$$\theta(g)(m_k) = \left(\sum_{n < k} |g(n) - s_{m_k}(n)|^p \right)^{\frac{1}{p}}$$

$$\le \left(\sum_{n < k} 2^{-(n+1)p} \right)^{\frac{1}{p}} \le \left(\sum_{n=0}^{\infty} \frac{1}{2^{n+1}} \right)^{\frac{1}{p}} = 1;$$

and by Minkowski's inequality,

$$\left(\sum_{n < k} |f(n) - g(n)|^p \right)^{\frac{1}{p}} \le \left(\sum_{n < k} |f(n) - s_{m_k}(n)|^p \right)^{\frac{1}{p}}$$

$$+ \left(\sum_{n < k} |f(n) - s_{m_k}(n)|^p \right)^{\frac{1}{p}}$$

$$= \theta(f)(m_k) + \theta(g)(m_k)$$

$$= (\theta(f)(m_k) - \theta(g)(m_k)) + 2\theta(g)(m_k)$$

$$\le |\theta(f)(m_k) - \theta(g)(m_k)| + 2.$$

It follows that the (increasing) sequence $(\sum_{n < k} |f(n) - g(n)|^p)^{\frac{1}{p}}$ is bounded, hence $\|f - g\|_p < \infty$, or $f - g \in \ell^p$.

For the reverse direction, suppose $f - g \in \ell^p$. Then by Minkowski's inequality again, for each m,

$$
|\theta(f)(m) - \theta(g)(m)|
$$

$$
= \left| \left(\sum_{n < l(s_m)} |f(n) - s_m(n)|^p \right)^{\frac{1}{p}} - \left(\sum_{n < l(s_m)} |g(n) - s_m(n)|^p \right)^{\frac{1}{p}} \right|
$$

$$
\leq \left(\sum_{n < l(s_m)} |f(n) - g(n)|^p \right)^{\frac{1}{p}}
$$

$$
\leq \|f - g\|_p.
$$

It follows that $\theta(f) - \theta(g)$ is bounded, hence $\theta(f) - \theta(g) \in \ell^\infty$. $\qquad\square$

This proof is similar to Oliver's proof ([16]) that c_0 is Borel bireducible to E_Z for the density ideal Z (see next section for more). The proof for $\ell^1 \leq_B \ell^\infty$ was previously known to Casevitz ([2]) by a similar argument.

We have shown enough to answer the question about the O notation.

Corollary 5.3. Θ *is not smooth.*

Proof. Because $\mathrm{id}(2^\omega) \leq_B \ell^1 <_B \ell^\infty \sim_B \Theta$. $\qquad\square$

In the following section we summarize other known results for ℓ^∞.

6. More about ℓ^∞

We first note two simple facts about ℓ^∞.

Lemma 6.1. ℓ^∞ *is splitting.*

Proof. This is similar to the proof for ℓ^p, except that we need to choose countably many non-ℓ^∞-equivalent sequences instead of the constant sequences. $\qquad\square$

Also similar to the situation of ℓ^p is the fact that ℓ^∞ is an upper bound, but not the least upper bound for all ℓ^p, $1 \leq p < \infty$.

Lemma 6.2. $\ell^\omega \leq_B \ell^\infty$.

Proof. As before this follows from the fact that ℓ^∞ is splitting. $\qquad\square$

We will show that $\ell^\omega <_B \ell^\infty$. For this we need to recall a number of previous theorems and folklore facts on Polishability of groups. In particular it is well known that ℓ^∞ is not Polishable. For the convenience of the reader we give a proof here. Recall that a Borel subgroup of a Polish group is Polishable if it admits a Polish topology which induces the same Borel structure as the one inherited from the Polish group.

Lemma 6.3. *The Borel subgroups of $\mathbb{R}^\omega$ ℓ^∞ and $\ell^\infty \cap \mathbb{Z}^\omega$ are not Polishable.*

Proof. We show the result for ℓ^∞. The proof is the same for $\ell^\infty \cap \mathbb{Z}^\omega$.

Assume ℓ^∞ is Polishable. Work under the Polish topology of ℓ^∞ that gives the same Borel sets as before. For each $N \in \omega$, let

$$B_N = \{f \in \ell^\infty \mid \forall n(|f(n)| \leq N)\}.$$

Then $\ell^\infty = \bigcup B_N$. Note that each B_N is Borel, and hence has the Baire property. Thus there is some N such that B_N is nonmeager. By a theorem of Pettis (Theorem 1.2.5 of [1]) $B_N - B_N$ contains an open neighborhood of the identity. Since $B_N - B_N \subseteq B_{2N}$, B_{2N} contains an open neighborhood of the identity. Then from the separability of the Polish topology in assumption it follows that there are countably many elements $f_0, f_1, \ldots, f_k, \ldots$ such that

$$\ell^\infty = \bigcup_k f_k + B_{2N}.$$

We then construct a $g \in B_{5N}$ such that $g \notin f_k + B_{2N}$ for any k, arriving at a contradiction. The construction of g is by diagonalization:

$$g(k) = \begin{cases} 5N & \text{if } |f_k(k)| \leq 2N \\ 0 & \text{if } |f_k(k)| > 2N \end{cases}$$

For any k, since $|g(k) - f_k(k)| > 2N$, $g \notin f_k + B_{2N}$. $\qquad\square$

Recall that the equivalence relation E_1 on the space $\mathbb{R}^\omega$ defined by:

$$f E_1 g \Leftrightarrow \exists N \forall n > N(f(n) = g(n)).$$

In this definition $\mathbb{R}$ can be replaced by 2^ω, resulting in essentially the same equivalence relation. This important equivalence relation was studied by Kechris and Louveau in [14]. Their most significant findings are summarized in the following theorem.

Theorem 6.4 (Kechris-Louveau [14]).
    (i) *Let E be a Borel equivalence relation. If $E <_B E_1$, then $E \leq_B E_0$.*
    (ii) *Let G be a Polish group and X be a Borel G-space. Then $E_1 \nleq_B E_G^X$.*

In fact a conjecture of Kechris says that if E is an arbitrary Borel equivalence relation, then either $E_1 \leq_B E$ or else there is an orbit equivalence relation E_G^X such that $E \leq_B E_G^X$. This has been partially verified by Solecki (see [18] and [19]).

Theorem 6.5 (Solecki [19]).

(i) *Let I be an analytic ideal on ω. Then either I is Polishable or else $E_1 \leq_B E_I$.*
(ii) *Let G be an abelian Polish group and H a Π_3^0 subgroup of G. Let G/H be the orbit equivalence relation of the shift action of H on G. Then either H is Polishable or else $E_1 \leq_B G/H$.*

Coming back to ℓ^∞ we have the following immediate corollary.

Corollary 6.6. $E_1 <_B \ell^\infty$ *and* $\ell^\omega <_B \ell^\infty$.

Proof. Note that ℓ^∞ is an F_σ subgroup of $\mathbb{R}^\omega$. Clause (ii) of the above theorem and Lemma 6.3 imply that $E_1 \leq_B \ell^\infty$. To see that $E_1 <_B \ell^\infty$, note that $E_0 <_B \ell^1 <_B \ell^\infty$; if $E_1 \sim_B \ell^\infty$ there would be a contradiction to the theorem of Kechris-Louveau. Now $E_1 \not\leq_B \ell^\omega$ since the latter is an orbit equivalence relation, thus $\ell^\omega <_B \ell^\infty$. $\qquad\square$

It is also possible to construct a reduction directly for $E_1 \leq_B \ell^\infty$. In [2] the following reduction was noticed:

$$\theta(f)(n) = nf(n), \ \forall f \in \mathbb{R}^\omega.$$

Note also that Lemma 6.3 is an immediate corollary of Solecki's theorem.

In fact there are also $2^{\aleph_0}$ many Borel equivalence relations with complexity between E_1 and ℓ^∞. Mazur defined a system of $2^{\aleph_0}$ many Borel equivalence relations arising from ideals resembling ℓ^∞ ([15]). Oliver has shown that all of them are Borel reducible to ℓ^∞. The argument in Lemma 6.3 can be modified to show that all the ideals in the Mazur system are not Polishable, hence by Solecki's theorem, all these equivalence relations are above E_1.

The following result of Rosendal strengthens all known reductions to ℓ^∞.

Theorem 6.7 (Rosendal [17]). *Let E be a K_σ equivalence relation on a Polish space X. Then $E \leq_B \ell^\infty$.*

In particular, all ℓ^p equivalence relations, ℓ^ω, E_1, and E_∞ are all K_σ. As for products of ℓ^∞, we have exactly the same results as for ℓ^p, by the same proofs. For any $0 < n < \omega$, $(\ell^\infty)^n \sim_B \ell^\infty$. However, $\ell^\infty <_B (\ell^\infty)^\omega$.

7. c_0 and Kanovei's question

The study of the equivalence relation c_0 started simultaneously as that of ℓ^p. However, historically it took longer for similar questions about c_0 to be answered. This is partially due to the fact that c_0 is Π_3^0, whereas in contrast all ℓ^p are F_σ. A simple but significant result is the connection of c_0 to the density ideal, which allows well understood techniques on ideals to be applied.

Recall that the *density ideal* Z on $\mathbb{N}$ is defined by

$$Z = \left\{ A \subseteq \mathbb{N} \ : \ \lim_{n \to \infty} \frac{|A \cap n|}{n} = 0 \right\},$$

where $|A \cap n|$ denotes the cardinality of the set $A \cap \{m \ : \ m < n\}$ for a positive integer n.

Lemma 7.1 (Oliver [16]). $c_0 \sim_B E_Z$.

Comparing c_0 to ℓ^p directly, Hjorth also showed the basic result below.

Theorem 7.2 (Hjorth [10]). $c_0 \not\leq_B \ell^1$, and $\ell^1 \not\leq_B c_0$.

It is easy to adapt Hjorth's proof to show that $c_0 \not\leq_B \ell^p$ for any $1 \leq p \leq \infty$. Here we give an alternative proof.

Lemma 7.3. $(c_0)^\omega \sim_B c_0$.

Proof. It suffices to show that $(c_0)^\omega \leq_B c_0$. First note that $c_0 \sim_B c_0 \upharpoonright [0,1]^\omega$, so we may pretend that c_0 has underlying space $[0,1]^\omega$. Assume also $(c_0)^\omega$ has underlying space $[0,1]^{\omega \times \omega}$. Fix a recursive bijection $\langle \cdot, \cdot \rangle$ between $\omega \times \omega$ and ω. We define a reduction θ from $(c_0)^\omega$ to c_0 as follows. Given $x \in [0,1]^{\omega \times \omega}$, let

$$\theta(x)(\langle m, n \rangle) = 2^{-m} x(m, n).$$

Now if $x (c_0)^\omega y$, then for any $\epsilon > 0$ and $m \in \omega$, there is n_m such that for all $n \geq n_m$, $|x(m,n) - y(m,n)| < \epsilon$. Let m_0 be such that $2^{-m_0} < \epsilon$. Then for any $m > m_0$ and arbitrary $n \in \omega$, $|x(m,n) - y(m,n)| \leq 2^{-m_0} < \epsilon$. This shows that if (m,n) is a pair with $|x(m,n) - y(m,n)| \geq \epsilon$, then $m \leq m_0$ and $n \leq n_m$; thus there are only finitely many such pairs. It follows that $\theta(x) - \theta(y) \in c_0$.

Conversely, if $x - y \notin (c_0)^\omega$, then there is m_0 such that the sequence $(x(m_0,n) - y(m_0,n))$ indexed by n does not converge to 0. Thus there is a subsequence of $\theta(x) - \theta(y)$ which does not converge to 0, and hence $\theta(x) - \theta(y) \notin c_0$. $\qquad\square$

Since $E_0 \leq_B c_0$ trivially, it follows that $(E_0)^\omega \leq_B c_0$. (This last fact was proved by a similar reduction in [12].) Therefore by our Lemma 3.6 c_0 is not Borel reducible to any F_σ equivalence relation. In particular, $c_0 \not\leq_B \ell^\infty$.

It follows trivially from Hjorth's theorem that $\ell^\infty \not\leq_B c_0$.

Farah [6] has studied ideals and equivalence relations more general but similarly defined as c_0. He called them c_0-equalities and used them to obtain significant results about the basis problem for Borel equivalence relations. In an earlier study [5], he also used non-classical Banach spaces as inspirations for constructions of large classes of Borel equivalence relations with a complicated Borel reducibility structure (compare [8]).

In the remainder of this section we introduce and study some new equivalence relations related to c_0. The study of these equivalence relations was motivated by a question asked by Kanovei in his lecture delivered to the 9th Asian Logic Conference in Novosibirsk, Russia, in August 2005. We will state the question and provide some information on the equivalence relations involved.

First, let us define an equivalence relation $c_0^* = c_0(\omega^2)$ as follows. Given $x, y \in \mathbb{R}^{\omega \times \omega}$,

$$x c_0^* y \iff \forall \epsilon > 0 \exists m_0, n_0 \forall m \geq m_0, n \geq n_0 \, (\, |x(m,n) - y(m,n)| < \epsilon \,).$$

Apparently $c_0 \leq_B c_0^*$. The equivalence relation c_0^* is similar to $(c_0)^\omega$ but it is not immediately clear if $c_0^* \leq_B c_0$. In fact, we note that it is very different.

Lemma 7.4. $E_1 \leq_B c_0^*$. *Thus* $c_0 <_B c_0^*$.

Proof. Consider E_1 on $2^{\omega \times \omega}$. Fix again a recursive bijection $\langle \cdot, \cdot \rangle$ between $\omega \times \omega$ and ω. We define a reduction θ from E_1 to c_0^* as follows. Given $x \in 2^{\omega \times \omega}$, let

$$\theta(x)(m, \langle n, k \rangle) = x(m, n).$$

Note that the element on the right takes value in $\{0, 1\}$, understood as a subset of $\mathbb{R}$. Now if $x E_1 y$, then there is some m_0 such that for all $n \in \omega$, $x(m,n) = y(m,n)$. It follows that for all $m \geq m_0$ and $n \geq 0$, $\theta(x)(m,n) = \theta(y)(m,n)$, which guarantees that $\theta(x) c_0^* \theta(y)$. On the other hand, if x and y are not E_1-equivalent, then for any m_0 there is $m \geq m_0$ and $n \geq 0$ such that $x(m,n) \neq y(m,n)$. Let m and n be fixed. Now for any n_0 choose k sufficiently large (for all usual coding functions can choose $k = n_0$) so that $\langle n, k \rangle \geq n_0$, then $\theta(x)(m, \langle n, k \rangle) = x(m,n) \neq y(m,n) = \theta(y)(m, \langle n, k \rangle)$. And note that in fact their difference is 1. So we have that $\theta(x)$ and $\theta(y)$ are not c_0^*-equivalent. $\square$

Next we modify the definition of c_0^* to obtain another equivalence relation. It is named u_0^* for its similarity to c_0^* and its connection to uniform convergence. Given $x, y \in \mathbb{R}^{\omega \times \omega}$, let

$$x u_0^* y \iff \forall \epsilon > 0 \exists m_0 \forall m \geq m_0 \forall n \, (|x(m, n) - y(m, n)| < \epsilon).$$

Again $c_0 \leq_B u_0^*$. However, we still have $E_1 \leq_B u_0^*$, this time even by a simpler reduction, namely the identity embedding of $2^{\omega \times \omega}$ into $\mathbb{R}^{\omega \times \omega}$. Thus again $c_0 <_B u_0^*$. A similar reduction as above also witnesses that $u_0^* \leq_B c_0^*$. We do not know if $u_0^* \sim_B c_0^*$.

We are now ready to state the question asked by Kanovei.

Question 7.5 (Kanovei). *Consider $C(\mathbb{R}^+)$, the space of continuous functions on $\mathbb{R}^+$, and define an equivalence relation E_K for $f, g \in C(\mathbb{R}^+)$ by*

$$f E_K g \iff \lim_{x \to \infty} (f(x) - g(x)) = 0.$$

Is $E_K \sim_B c_0$?

This question is still open. Here we show that $c_0 \leq_B E_K \leq_B u_0^*$, hence providing an upper bound and a lower bound for its complexity. In fact we will define a discrete equivalence relation u_0 essentially on $\mathbb{R}^{\omega \times \omega}$ and show that $E_K \sim_B u_0$ and $c_0 \leq_B u_0 \leq_B u_0^*$.

The consideration of equivalence relation E_K is in some sense natural and parallel to our consideration of Θ on $C(\mathbb{R}^+)$. However, some technical difficulties arise in the current context. First note that E_K is in fact an orbit equivalence relation induced by an additive action of the group

$$C_0(\mathbb{R}^+) = \left\{ f \in C(\mathbb{R}^+) \; : \; \lim_{x \to \infty} f(x) = 0 \right\}.$$

Without changing the essence of the problem, we will pretend that our $\mathbb{R}^+$ refers to $\mathbb{R}^{\geq 0}$, the set of all nonnegative real numbers. With this convention $C_0(\mathbb{R}^+)$ becomes a separable Banach space with the complete metric

$$\|f\|_0 = \sup_{x \geq 0} |f(x)|.$$

It is not hard to show that $C_0(\mathbb{R}^+)$ is a universal separable Banach space, that is, it contains all other separable Banach spaces as closed subspaces. In particular, its additive group is a Polish group. If the Kechris-Louveau theorem applies then we would be able to conclude immediately that $E_1 \not\leq_B E_K$. However, the nuisance is that the underlying space $C(\mathbb{R}^+)$ for E_K is not a Polish space, and therefore the Kechris-Louveau theorem does not apply. In fact, the next thing we will do is to interpret the problem as an equivalence relation on a Polish space, as we did for Θ.

Let $f \in C(\mathbb{R}^+)$ be given. Note first that there is a piecewise linear continuous function $L_f \in C(\mathbb{R}^+)$ such that $L_f - f \in C_0(\mathbb{R}^+)$ and there are finitely many linear pieces of L_f in any bounded set. To obtain this L_f we consider each interval $[m, m+1]$ for $m \in \mathbb{N}$. Since f is uniformly continuous on $[m, m+1]$ there is $\delta_m > 0$ such that $|f(x) - f(y)| < 2^{-m}$ whenever $x, y \in [m, m+1]$ and $|x - y| < \delta_m$. Thus let $m = a_0 < a_1 < \cdots < a_i = m+1$ be a partition of $[m, m+1]$ with $a_{j+1} - a_j < \delta$ for $j = 0, \ldots, i - 1$, and define L_f on $[m, m+1]$ to be the piecewise linear continuous function with extremal points $(a_0, f(a_0)), (a_1, f(a_1)), \ldots, (a_i, f(a_i))$. It follows that $|L_f(x) - f(x)| < 2^{-m-1}$ for all $x \in [m, m+1]$. Thus eventually $L_f - f \in C_0(\mathbb{R}^+)$.

Next let D denote the set of nonnegative dyadic rationals, that is, rationals of the form $k2^{-l}$ for $k, l \in \mathbb{N}$. We need to modify L_f so that all linear pieces of L_f have elements in D as endpoints. From the above construction of L_f it is clear that this can be achieved by choosing the partition points $a_0 < a_1 < \cdots < a_i$ to be dyadic rationals.

We fix a canonical enumeration of $D \cap [0, 1)$ as $q_0 = 0, q_1 = 1/2, q_2 = 1/4, q_3 = 3/4, \ldots$. This enumeration can be obtained by first enumerating the set $\{k2^{-l} : k < 2^l\}$ according to the lexicographic order of (l, k) and then eliminating all the repetitions. We can now represent L_f by an element x_f of $\mathbb{R}^{\omega \times \omega}$ defined by

$$x_f(m, n) = L_f(m + q_n).$$

Since D is dense and L_f is continuous, we have that the map $L_f \mapsto x_f$ is one-one. Intuitively this means that x_f contains all the information about L_f. Note that, by our construction of L_f, for each m there is an $l_f(m)$ such that the L_f values on the finite set $\{m + k2^{-l} : k \le 2^l, l \le l_f(m)\}$ completely determine other values of L_f on $[m, m+1]$. We finally let the pair (x_f, l_f) represent the function f.

Now we consider the equivalence relation on pairs (x_f, l_f) corresponding to E_K on the functions f. For this let $f, g \in C(\mathbb{R}^+)$ and let L_f, L_g as well as $(x_f, l_f), (x_g, l_g)$ be given. Note that $f - g \in C_0(\mathbb{R}^+)$ iff $L_f - L_g \in C_0(\mathbb{R}^+)$. To further unravel the latter equivalence, let $a_0 = 0 < a_1 < \cdots < a_i < \ldots$ and $b_0 = 0 < b_1 < \cdots < b_i < \ldots$ be enumerations of all the extremal points used in the definitions of L_f and L_g respectively. Let $c_0 = 0 < c_1 < \ldots$ be an enumeration of the union $\{a_i : i < \omega\} \cup \{b_i : i < \omega\}$. Then

$$L_f - L_g \in C_0(\mathbb{R}^+) \iff \lim_{i \to \infty} (L_f(c_i) - L_g(c_i)) = 0.$$

For (x_f, l_f) and (x_g, l_g), the enumeration of the extremal points corresponds to the lexicographic order of the set

$$S_{f,g} = \left\{ (m, n) : n < 2^{\max\{l_f(m), l_g(m)\}} \right\}.$$

Thus we obtain that

$$fE_K g \iff \lim_{(m,n)\in S_{f,g}} (x_f(m,n) - x_g(m,n)) = 0,$$

where the limit on the right hand side is taken as a sequence according to this order of $S_{f,g}$.

We can thus define a relation u_0 on $\mathbb{R}^{\omega\times\omega} \times \omega^\omega$ (which is isomorphic to $\mathbb{R}^{\omega\times\omega}$) by

$$(x,l)u_0(y,h) \iff \lim_{(m,n)\in S_{l,h}} (x(m,n) - y(m,n)) = 0,$$

where the limit on the right hand side is taken as a sequence according to the lexicographic order of the set

$$S_{l,h} = \left\{ (m,n) : n < 2^{\max\{l(m),h(m)\}} \right\}.$$

Now u_0 as a relation on $\mathbb{R}^{\omega\times\omega} \times \omega^\omega$ is not an equivalence relation, so we need to restrict it to an appropriate subset. In view of the properties of (x_f, l_f), we define the underlying space for u_0 to be the set P of pairs $(x,l) \in \mathbb{R}^{\omega\times\omega} \times \omega^\omega$ such that

for all $m \in \omega$ and $n \geq 2^{l(m)}$, if $q_n = (2k+1)2^{-l}$, $q_p = k2^{-l+1}$ and $q_s = (k+1)2^{-l+1}$, then $2x(m,n) = x(m,p) + x(m,s)$ if $q_s < m+1$ and $2x(m,n) = x(m,p) + x(m+1,0)$ if $q_s = m+1$.

We claim that any $(x,l) \in P$ gives rise to a unique piecewise linear continuous function $f^{x,l} \in C(\mathbb{R}^+)$. In fact, simply define, for all $m \in \omega$ and $n < 2^{l(m)}$, $f^{x,l}(m + q_n) = x(m,n)$. Then the values of $f^{x,l}$ on all other dyadic rationals are completely determined by the definition of P, whose closure is the graph of a piecewise linear continuous function. It follows from this observation that u_0 on P is an equivalence relation, and moreover, for $(x,l), (y,h) \in P$,

$$(x,l)u_0(y,h) \iff f^{x,l} E_K f^{y,h}.$$

Note from the definition of P that it is a closed subspace of $\mathbb{R}^{\omega\times\omega} \times \omega^\omega$, and hence it is Polish. Also note that for any $(x,l) \in P$,

$$x_{f^{x,l}} = x \quad \text{and} \quad l_{f^{x,l}} \leq l.$$

This is not essential but curious.

To summarize, we have in effect shown the following fact.

Lemma 7.6. $E_K \sim_B u_0$.

It is clear that $c_0 \sim_B u_0 \upharpoonright P_0$, where P_0 is the set of $(x, l) \in P$ where l is the constant 0 function. Elements in P_0 correspond to elements of $\mathbb{R}^\omega$ in an obvious way. For an upper bound of u_0, we have the following lemma.

Lemma 7.7. $u_0 \sim_B u_0^* \upharpoonright \pi(P)$, where $\pi : \mathbb{R}^{\omega \times \omega} \times \omega^\omega \to \mathbb{R}^{\omega \times \omega}$ is the projection.

Proof. In fact the witness is just the projection map π. Given $(x, l), (y, h) \in P$, we have noted that $(x, l)u_0(y, h)$ iff $f^{x,l} E_K f^{y,h}$. Suppose $f^{x,l} E_K f^{y,h}$. Then for any $\epsilon > 0$ there is $m \in \omega$ such that for all $t \in \mathbb{R}$ and $t \geq m$, $|f^{x,l}(t) - f^{y,h}(t)| < \epsilon$. In particular, for all $t \in D$ with $t \geq m$, $|f^{x,l}(t) - f^{y,h}(t)| < \epsilon$. However, for any n, $f^{x,l}(m + q_n) = x(m, n)$. Thus we have that for all $n \in \omega$, $|x(m, n) - y(m, n)| < \epsilon$. This shows that $xu_0^* y$. The converse is similar. $\square$

It is clear that u_0 is a $\mathbf{\Pi}_3^0$ equivalence relation on P. It is unlikely that u_0 is literally induced by a Polish group action on P, but we do not know whether it is Borel bireducible to an orbit equivalence relation on a Polish space. On the other hand, we do not even know that there is no continuous reduction from u_0 to c_0. Our study thus has come to an abrupt end with the following question equivalent to Kanovei's.

Question 7.8. Is $u_0 \leq_B c_0$?

References

1. H. Becker and A. S. Kechris, The Descriptive Set Theory of Polish Group Actions, London Mathematical Society Lecture Notes Series 232, Cambridge University Press, 1996.
2. P. Casevitz, Dichotomies pour les espaces de suites reelles, Fundamenta Mathematicae 165 (2000), 249–284.
3. R. Dougherty and G. Hjorth, Reducibility and non-reducibility between ℓ^p equivalence relations, Transactions of the American Mathematical Society 351 (1999), no. 5, 1835–1844.
4. R. Dougherty, S. Jackson and A. S. Kechris, The structure of hyperfinite Borel equivalence relations, Transactions of the American Mathematical Society 34 (1994), no. 1, 193–225.
5. I. Farah, Basis problem for turbulent actions. I. Tsirelson submeasures. Proceedings of XIth Latin American Symposium in Mathematical Logic (Merida, 1998). Annals of Pure and Applied Logic 108 (2001), no. 1–3, 189–203.
6. I. Farah, Basis problem for turbulent actions. II. c_0-equalities. Proceedings of the London Mathematical Society (3) 83 (2001), no. 1, 1–30.
7. S. Gao, The isomorphism relation bewteen countable models and definable equivalence relations, PhD dissertation, UCLA, 1998.

8. S. Gao, Some applications of the Adams-Kechris technique, Proceedings of the American Mathematical Society 130 (2002), no. 3, 863–874.

9. S. Gao and V. Pestov, On a universality property of some abelian Polish groups, Fundamenta Mathematicae 179 (2003), no. 1, 1–15.

10. G. Hjorth, Actions by the classical Banach spaces, Journal of Symbolic Logic 65 (2000), no. 1, 392–420.

11. G. Hjorth, Classification and Orbit Equivalence Relations, Mathematical Surveys and Monographs 75, American Mathematical Society, 2000.

12. V. Kanovei, Ideals and Equivalence Relations, manuscript, 2005. Available at arXiv:math.LO/0603506.

13. A. S. Kechris, Classical Descriptive Set Theory, Graduate Texts in Mathematics, Springer-Verlag, 1995.

14. A. S. Kechris and A. Louveau, The classification of hypersmooth Borel equivalence relations, Journal of the American Mathematical Society 10 (1997), no. 1, 215–242.

15. K. Mazur, A modification of Louveau and Veličkovič construction for F_σ-ideals, Proceedings of the American Mathematical Society 128 (2000), no. 5, 1475–1479.

16. M. Oliver, Borel cardinalities below c_0, Proceedings of the American Mathematical Society 134 (2006), no. 8, 2419–2425.

17. C. Rosendal, Cofinal families of Borel equivalence relations and quasiorders, Journal of Symbolic Logic 70 (2005), no. 4, 1325–1340.

18. S. Solecki, Analytic ideals, Bulletin of Symbolic Logic 2 (1996), no. 3, 339–348.

19. S. Solecki, Analytic ideals and their applications, Annals of Pure and Applied Logic 99 (1999), no. 1–3, 51–72.

NEGATIVE DATA IN LEARNING LANGUAGES

SANJAY JAIN

School of Computing, National University of Singapore, Singapore 117543.
Email: sanjay@comp.nus.edu.sg

EFIM KINBER

Department of Computer Science,
Sacred Heart University, Fairfield, CT 06432-1000, U.S.A.
Email: kinbere@sacredheart.edu

The paper is a survey of recent results on algorithmic learning (inductive inference) of languages from full collection of positive examples and some negative
data. Different types of negative data are considered. We primarily concentrate on learning using (1) carefully chosen finite negative data (2) negative
counterexamples provided when conjectures contain data not in the target language (3) negative counterexamples obtained from a teacher (formally, oracle),
when a learner queries the oracle if an hypothesis is contained in the target
language. We also explore how least counterexamples and counterexamples of
bounded size fair against arbitrary counterexamples. The effects of random
negative data are also briefly considered.

1. Introduction

Based on motivations from theories of language acquisition by children,
Gold [10] developed an algorithmic model of learning (in the limit) from
examples. This model may be described as follows. A learner receives as
input, one by one, $x_0, x_1, \ldots$, where, $\{x_0, x_1, \ldots\}$ is exactly the target language, except possibly for a special pause symbol (which is useful for dealing
with empty language). Note that there is no particular order among the elements $x_0, x_1, \ldots$, and repetitions are allowed. As the learner is receiving
this data, it conjectures a sequence of grammars, $g_0, g_1, \ldots$ which are intended as descriptors of the target language. The learner can be regarded as
successful if eventually the sequence of grammars stabilizes to a grammar
g which generates/enumerates/accepts the target language. This model of
learning is called **TxtEx** in the literature (**Txt** stands for "text", which
is a complete positive data presentation, and "**Ex**" stands for explana-
tory learning). Note that it is more interesting to consider learnability of
a class of languages by a single learner (since, if we are only interested in
learning one fixed language, then some learner — which just outputs the

grammar for the fixed language — can easily learn it). The influence of Gold's paradigm to understanding human language learning is discussed in Pinker [24], Wexler and Culicover [28], Wexler [27] and Osherson, Stob and Weinstein [21].

Note that in the above model, the learner only receives elements of the language as input. It is not given any explicit information about elements not in the target language. This was based on the studies by linguists which hypothesised that children rarely, if ever, get negative information (see for example, Brown and Hanlon [5], Hirsh-Pasek, Treiman and Schneiderman [11] and Demetras, Post and Snow [8]).

Along with the above model of learning from positive data, Gold also studied learning from both positive and negative data. In this model, a learner is given all elements of the language, one by one, marked as positive, as well as all non-elements of the language, one by one, marked as negative. This criteria of learning is called **InfEx**. However based on studies about child learning, it is unrealistic to expect that children get all the negative data. On the other hand, as some studies point out, see Brown and Hanlon [5], Hirsh-Pasek, Treiman and Schneiderman [11] and Demetras, Post and Snow [8], children do get something more than just positive data.

The aim of the current paper is to survey some models of learning, where some amounts of negative data is provided to the learner. We will first consider two models of providing some *core* negative data to the learner. These models and results are based on work by Shinohara [26], Fulk [9], Motoki [18] and Baliga, Case and Jain [1]. We will then consider the case where negative data is provided to the learner via counterexamples to its conjectures. This is based on the philosophy that parents often correct their children by providing them counterexamples. This part is based on work done by the authors [13, 12]. We also introduce and briefly consider a model in which learners are provided with random negative examples.

Before we study different models of negative data, it is useful to also consider some variants of the basic model of learnability from text as described above. Case and Lynes [6] (see also Osherson and Weinstein [22]) studied the case where the final hypothesis of the learner may not be accurate, but have upto n errors (finite number of errors). This criteria of learning is called **TxtEx**^{n} (**TxtEx***). This was motivated by the fact that humans rarely, if ever, learn a language perfectly. Case and Lynes [6] (also see Osherson and Weinstein [22]) considered the case when learner need not syntactically converge to a grammar, but eventually output only correct grammars (i.e., semantically converge rather than syntactically converge). In this model for all but finitely many n, the grammar g_n is a grammar for the target language. This criteria of learning is called **TxtBc**-learning. **Bc** here stands for behaviorally correct. **TxtBc**^{n} and **TxtBc*** can be naturally defined.

Fulk [9] considered the case when, in addition to positive data, the learner is provided with a grammar for the complement of the language. Note that one can generate complete negative data using the grammar for the complement of the language. Fulk went on to show that this allows the learner to learn more than what can be learned using informants, that is using both complete positive and complete negative data. Though interesting, this model is quite unrealistic in the sense that children are definitely not given a grammar for the complement of the language. Most of the literature (see for example, Brown and Hanlon [5], Hirsh-Pasek, Treiman and Schneiderman [11] and Demetras, Post and Snow [8]) also argues that children do not get complete negative data. What is more realistic is that a learner is provided with some negative data, probably carefully selected or based on what the child has learnt (that is in a way based on child's current conjecture). Jain and Sharma [15] considered a modification where the learner instead of being given a grammar for the complete $\overline{L}$, is given only a grammar for a subset of $\overline{L}$, where this subset satisfies some density constraints. Despite being somewhat weaker than Fulk's model, it still seems unrealistic to expect that children are provided with grammars for any parts of the complement of the target language.

Based on this, Shinohara [26] considered the case where the learner is given $\geq n$ (n fixed beforehand) arbitrary negative examples along with the complete positive data about the language. Clearly, this is possible only when complement of the language does contain at least n elements. Shinohara showed that this method of presenting negative data is not useful, in the sense that it does not give any learnability advantages over just positive data. Extending this work, Baliga, Case and Jain [1] considered the case that the learner is given upto n carefully chosen elements of the complement of the language. These negative examples may be considered as *core* negative data. Intuitively, this was aimed to model the situation when a teacher carefully selects the negative examples to be provided to the student. Indeed, as expected this model turned out to be quite powerful. For example, it can be shown that the class of all recursively enumerable sets, $\mathcal{E}$, can be learned by some learner in **TxtEx** sense, when it additionally receives upto two carefully selected negative examples. Even one carefully selected negative example is enough if one allows upto one error in the final grammar, or allows behaviourally correct learning. In contrast one carefully selected negative example is not enough to learn the class $\mathcal{E}$ according to **TxtEx** crtieria, though it still can be shown to be quite useful.

The reason for this apparent gains by having only one or two negative examples in the above model is based on the fact that one can "code" information into these negative data, allowing the learner to essentially extract a grammar for the target language from the negative data. To avoid such coding, Baliga, Case and Jain (motivated by a model considered by Motoki [18]) considered the following modification. For each possible target language, besides the core negative data, the learner may be given some further negative data. This model of learning is called open negative data, reminding one of the basic open sets for the topology with respect to which enumeration operators are continuous. As the learner may not be able to distinguish core negative data from the other negative data, the effects of "coding" are somewhat eliminated. This model turned out to be quite useful in studying the effects of negative data. In particular, above criteria lie strictly between **TxtEx** and **InfEx** models of learning. Let **NegO^{n}I** (**NegO*I**) denote the criteria of learning formed when the core negative information is of size at most n (the core negative information is of finite size), and **I** is the basic model of learning (such as **Ex^a**, or **Bc^a**). It can be shown that **NegO*I** turns out to be of the same power as **InfI**. Furthermore, each additional element allowed in the core, gives learnability advantages (that is **NegO^{n+1}I** allows learning strictly more classes compared to **NegO^{n}I**). On the other hand, the finite negative core information is not enough to overcome extra errors (that is, one can learn something in **TxtEx^{n+1}** model of learning, but cannot in **NegO*TxtEx^n** model of learning). Additionally, it was shown that small packets of negative information also lead to increased *speed* of learning. This result agrees with a psycholinguistic hypothesis of McNeill correlating the availability of parental expansions with the speed of child language development. McNeill [17] posits that there is *faster* learning of language for children in homes in which more corrections (usually in the form of *possibly exemplary* expansions) are given. These corrections are, in part, a form of negative information.

Note that in both models considered above, one selects carefully negative examples based on the language being learned. However, in reality often negative examples are formed more as "counterexamples" based on errors done by child, rather than being preselected. To model such a situation, authors [13] considered a criteria of learning where the learner is given a negative counterexample to each of its conjectures, if it exists. This model of learning is called **NCEx**. This model turned out to be robust with respect to different variations (giving least counterexamples, or the counterexamples being delayed). Besides the usual hierarchy results showing the advantages of having counterexamples, the paper [13] contrasts this criteria with **TxtEx** and **InfEx**, showing that in some cases structurally it behaves more like "InfEx" rather than like "TxtEx". For example, results such as (a) if $\mathcal{L} \in$ **NCEx** then so is $\mathcal{L} \cup \mathcal{S}$, for any finite class $\mathcal{S}$ of recursive languages, (b) **NCEx**$^* \subseteq$ **NCBc** follow more along the lines of results in learning from informants. On the other hand, it is shown that in some cases full negative data, informant, is needed for learning, and just counterexamples are not enough. A surprising result, in the case of behaviorally correct learning is that the whole class $\mathcal{E}$ can be learned in **NCBc**[1] model — making it more powerful than even learning from informants! (by contrast **NCBc** $\subset$ **InfBc** and **NCEx**$^a \subseteq$ **InfEx**^{a}).

An interesting complexity aspect is that, for **Ex** model, though **NCEx** is a strict subset of **InfEx**, it can sometimes give huge complexity advantages. That is, in some cases one can learn a class in **NCEx** model using only n mind changes, whereas learning with informants requires exponentially many mind changes. In a variation of **NCEx** model, where least negative counterexamples are given, one can even show that there are classes which are learnable using 1 mind change, though learning with informants requires unbounded number of mind changes! Though, as mentioned above, several variations of negative counterexample models do not give different learning power, there is often complexity advantages which may result from a particular variation.

Learning from counterexamples also addresses a general concern about overgeneralization in learning. When one only receives positive data, then overgeneralized hypothesis cannot be corrected based on input data alone. However, if negative counterexamples are provided to the learner, then one can address this issue.

One can view getting counterexamples, as asking a "subset query" about the conjecture to a teacher. However in the usual model of learning from subset queries, a learner is allowed to query about other languages (besides just the conjectured language) being subsets of target language. This led us [12] to consider learning with subset (and other kind of) queries. It can be shown that if a **TxtEx** learner is allowed finitely many (but unbounded) subset queries, then the learning ability is same as that in the **NCEx** model. If the learner is allowed infinitely many subset queries, then a learner (using texts) can learn all the recursively enumerable languages. Thus it is more interesting to study the case when the number of queries is bounded. Authors showed several results comparing the criteria of learning with negative counterexamples and subset queries, and giving hierarchies based on number of queries allowed. They also showed hierarchies based on variations of the query model where no answers are accompanied by least, arbitrary, or no counterexamples.

An interesting research work to consider would be to see how random negative examples work — this may be more closer to how humans learn languages. It can be shown that often random negative examples do help.

2. Preliminaries

2.1. *Notation*

Any unexplained recursion theoretic notation is from Rogers [25]. N denotes the set of natural numbers, $\{0, 1, 2, 3, \ldots\}$. $*$ denotes a non-member of N and is assumed to satisfy $(\forall n)[n < * < \infty]$. $\emptyset$ denotes the empty set. $\subseteq$, $\subset$, $\supseteq$ and $\supset$ respectively denote subset, proper subset, superset, and proper superset. $\mathrm{card}(S)$ denotes the cardinality of S. $S_1 =^n S_2$ denotes $\mathrm{card}((S_1 - S_2) \cup (S_2 - S_1)) \leq n$; $S_1 =^* S_2$ means that $\mathrm{card}((S_1 - S_2) \cup (S_2 - S_1))$ is finite. $\downarrow$ denotes defined and $\uparrow$ denotes undefined. $\max(\cdot), \min(\cdot)$ denote the maximum and minimum of a set, respectively, where $\max(\emptyset) = 0$ and $\min(\emptyset) = \uparrow$. $\langle i, j \rangle$ stands for an arbitrary, computable, one-to-one encoding of all pairs of natural numbers onto N (see for example [25]).

The quantifiers '$\overset{\infty}{\forall}$', and '$\overset{\infty}{\exists}$' essentially from Blum [4], mean 'for all but finitely many' and 'there exist infinitely many', respectively. The quantifier '$\exists!$' means 'there exists a unique'.

φ denotes a fixed *acceptable* programming system for the partial computable functions: $N \to N$ (see the books [25, 16]). φ_i denotes the partial computable function computed by program i in the φ-system.

W_i denotes $\mathrm{domain}(\varphi_i)$. W_i is, then, the r.e. set/language ($\subseteq N$) accepted (or equivalently, generated) by the φ-program i. $\mathcal{E}$ will denote the set of all r.e. languages. L, with or without decorations, ranges over $\mathcal{E}$. $\overline{L}$ denotes the complement of L. $\mathcal{L}$, with or without decorations, ranges over subsets of $\mathcal{E}$. $\mathcal{L} = \{L_i \mid i \in N\}$ is called an indexed family iff there exist a recursive function f such that $f(i, x) = 1$ iff $x \in L_i$.

2.2. *Some Notions from Language Learning*

We now consider some basic notions in language learning. Following definition gives the concepts of data that is presented to a learner. Part (a) considers the notion of positive data, and part (b) considers the case when both positive and negative data are given.

Definition 2.1. (Gold [10])

(a) A *text* T is a mapping from N into $(N \cup \{\#\})$. The *content* of a text T, denoted content(T), is the set of natural numbers in the range of T.

(b) An infinite information sequence I is a mapping from N to $(N \times \{0,1\}) \cup \{\#\}$, such that if (x, b) appears in the sequence, then $(x, 1 - b)$ does not appear in the sequence. The *content* of an information sequence I denoted content(I), is the set of pairs in the range of I. PosInfo$(I) = \{x \mid (x, 1) \in \text{content}(I)\}$, and NegInfo$(I) = \{x \mid (x, 0) \in \text{content}(I)\}$.

(c) T is a text for L iff content$(T) = L$. I is an information sequence for L iff PosInfo$(I) = L$ and NegInfo$(I) = \overline{L}$.

(d) $T[n]$ denotes the initial segment of T of length n. Similarly, $I[n]$ denotes the initial segment of I of length n.

We let T (I), with or without superscripts, range over texts (information sequences).

Intuitively, $\#$'s in the texts/information sequences denote pauses in the presentation of data. For example, the only text for the empty language is just an infinite sequence of $\#$'s. Note that by our convention on information sequences, PosInfo$(I) \cap$ NegInfo$(I) = \emptyset$.

A finite sequence σ is an initial segment of a text or an infinite information sequence. One can similarly define content(σ) (and PosInfo(σ), NegInfo(σ) in case of σ being initial segment of an information sequence).

SEQ denotes the set of all finite initial segments of texts. SEG denotes the set of all finite initial segments of information sequences. Note that SEQ and SEG can be coded onto N.

Definition 2.2. A *language learning machine* is an algorithmic device which computes a mapping from SEQ (or SEG) into N.

Later we will consider variation of learning machines. For convenience of exposition we avoid defining these variants until we need them.

We let **M**, with or without decorations, range over learning machines. We say that $\mathbf{M}(T){\downarrow} = i \Leftrightarrow (\overset{\infty}{\forall} n)[\mathbf{M}(T[n]) = i]$. Convergence on information sequences is similarly defined.

We now define some common criteria for learning. Our first criterion is based on learner, given a text for the language, converging to a grammar for the language.

Definition 2.3. (Gold [10], Case and Lynes [6], Osherson and Weinstein [22]) Let $a \in N \cup \{*\}$.

(a) $\mathbf{M}$ $\mathbf{TxtEx}^a$-*identifies* L (written: $L \in \mathbf{TxtEx}^a(\mathbf{M})$) $\Leftrightarrow$ ($\forall$ texts T for L)($\exists i \mid W_i =^a L$)$[\mathbf{M}(T)\!\downarrow = i]$.

(b) $\mathbf{TxtEx}^a = \{\mathcal{L} \mid (\exists \mathbf{M})[\mathcal{L} \subseteq \mathbf{TxtEx}^a(\mathbf{M})]\}$.

The criterion we call $\mathbf{TxtEx}^0$ is due to Gold [10]. The $a > 0$ case is from Case and Lynes [6] (Osherson and Weinstein [22] independently introduced the $a = *$ case). We refer the reader to Pinker [24], Wexler and Culicover [28], Wexler [27], Osherson, Stob, and Weinstein [19, 20, 21], and Jain *et al* [14] for further discussion on the paradigm.

The next definition is based on learner semantically rather than syntactically converging to the grammar(s) for the language.

Definition 2.4. (Case and Lynes [6]) Let $a \in N \cup \{*\}$.

(a) $\mathbf{M}$ $\mathbf{TxtBc}^a$-*identifies* L (written: $L \in \mathbf{TxtBc}^a(\mathbf{M})$) $\Leftrightarrow$ ($\forall$ texts T for L)($\overset{\infty}{\forall} n$)$[W_{\mathbf{M}(T[n])} =^a L]$.

(b) $\mathbf{TxtBc}^a = \{\mathcal{L} \mid (\exists \mathbf{M})[\mathcal{L} \subseteq \mathbf{TxtBc}^a(\mathbf{M})]\}$.

The $a \in \{0, *\}$ cases were independently introduced by Osherson and Weinstein [22, 23]. The corresponding notion in the case of learning functions was introduced by Bārzdiņš [2] and Case and Smith [7].

We now consider the corresponding learning criteria when information sequences are provided to the learner.

Definition 2.5. (Gold [10] and Case and Lynes [6]) Let $a \in N \cup \{*\}$.

(a) $\mathbf{M}$ $\mathbf{InfEx}^a$-*identifies* L (written: $L \in \mathbf{InfEx}^a(\mathbf{M})$) $\Leftrightarrow$ for all information sequences I for L, $\mathbf{M}(I)\!\downarrow$ and $W_{\mathbf{M}(I)} =^a L$.

$\mathbf{InfEx}^a = \{\mathcal{L} \mid (\exists \mathbf{M})[\mathcal{L} \subseteq \mathbf{InfEx}^a(\mathbf{M})]\}$.

(b) $\mathbf{M}$ $\mathbf{InfBc}^a$-*identifies* L (written: $L \in \mathbf{InfBc}^a(\mathbf{M})$) $\Leftrightarrow$ for all information sequences I for L, ($\overset{\infty}{\forall} n$)$[W_{\mathbf{M}(I[n])} =^a L]$.

$\mathbf{InfBc}^a = \{\mathcal{L} \mid (\exists \mathbf{M})[\mathcal{L} \subseteq \mathbf{InfBc}^a(\mathbf{M})]\}$.

We often write $\mathbf{TxtEx}$ (respectively, $\mathbf{TxtBc}$, $\mathbf{InfEx}$, $\mathbf{InfBc}$) for $\mathbf{TxtEx}^0$ (respectively, $\mathbf{TxtBc}^0$, $\mathbf{InfEx}^0$, $\mathbf{InfBc}^0$).

The following theorem gives some basic comparison between the criteria of inference discussed above. Note that by definition, for all $a \in N \cup \{*\}$, $\mathbf{TxtEx}^a \subseteq \mathbf{InfEx}^a \cap \mathbf{TxtBc}^a$, and $(\mathbf{TxtBc}^a \cup \mathbf{InfEx}^a) \subseteq \mathbf{InfBc}^a$.

Theorem 2.1. (Gold [10], Blum and Blum [3], Case and Lynes [6] and Case and Smith [7]) *For all $n \in N$, the following hold.*

(a) $\mathbf{TxtEx}^{n+1} - \mathbf{InfEx}^n \neq \emptyset$.

(b) $\mathbf{TxtEx}^* - \bigcup_{m \in N} \mathbf{InfEx}^m \neq \emptyset$.

(c) $\mathbf{TxtBc} - \mathbf{InfEx}^* \neq \emptyset$.

(d) $\mathbf{TxtBc}^{n+1} - \mathbf{InfBc}^n \neq \emptyset$.

(e) $\mathbf{TxtBc}^* - \bigcup_{m \in N} \mathbf{InfBc}^m \neq \emptyset$.

(f) $\mathbf{TxtEx}^{2n} \subset \mathbf{TxtBc}^n$.

(g) $\mathbf{TxtEx}^{2n+1} - \mathbf{TxtBc}^n \neq \emptyset$.

(h) $\mathbf{InfEx}^* \subseteq \mathbf{InfBc}^0$.

(i) $\mathbf{InfEx} - \mathbf{TxtBc}^* \neq \emptyset$.

(j) $\mathcal{E} \in \mathbf{InfBc}^*$.

3. Identification with Finite Negative Information

We first consider the model where an apparently small finite set of negative information is given in addition to text. In part (a) of both Definitions 3.1 and 3.2 just below, S is the *core* of negative information. The learner gets (besides the positive data) exactly this core negative data (marked as such) and no other negative data.

Definition 3.1. (Baliga, Case and Jain [1]) Suppose $a, b \in N \cup \{*\}$.

(a)
$\mathbf{M}$ $\mathbf{NegF}^b\mathbf{TxtEx}^a$-*identifies* $L \in \mathcal{E}$ (written: $L \in \mathbf{NegF}^b\mathbf{TxtEx}^a(\mathbf{M})$)
$\Leftrightarrow (\exists S \subseteq \overline{L} \mid \mathrm{card}(S) \leq b)(\forall I \mid \mathrm{PosInfo}(I) = L \ \& \ \mathrm{NegInfo}(I) = S)[\mathbf{M}(I)\!\downarrow$
and $W_{\mathbf{M}(I)} =^a L]$.

(b) $\mathbf{NegF}^b\mathbf{TxtEx}^a = \{\mathcal{L} \subseteq \mathcal{E} \mid (\exists \mathbf{M})[\mathcal{L} \subseteq \mathbf{NegF}^b\mathbf{TxtEx}^a(\mathbf{M})]\}$.

Definition 3.2. (Baliga, Case and Jain [1]) Suppose $a, b \in N \cup \{*\}$.

(a) $\mathbf{M}$ $\mathbf{NegF}^b\mathbf{TxtBc}^a$-*identifies* $L \in \mathcal{E}$ (written:
$L \in \mathbf{NegF}^b\mathbf{TxtBc}^a(\mathbf{M})) \Leftrightarrow (\exists S \subseteq \overline{L} \mid \mathrm{card}(S) \leq b)(\forall I \mid \mathrm{PosInfo}(I) =$
$L \ \& \ \mathrm{NegInfo}(I) = S)(\overset{\infty}{\forall} n)[W_{\mathbf{M}(I[n])} =^a L]$.

(b) $\mathbf{NegF}^b\mathbf{TxtBc}^a = \{\mathcal{L} \subseteq \mathcal{E} \mid (\exists \mathbf{M})[\mathcal{L} \subseteq \mathbf{NegF}^b\mathbf{TxtBc}^a(\mathbf{M})]\}$.

By definition, for all a, $\mathbf{NegF}^0\mathbf{TxtEx}^a = \mathbf{TxtEx}^a$ and $\mathbf{NegF}^0\mathbf{TxtBc}^a = \mathbf{TxtBc}^a$.

The next theorem illustrates the gain in learning power obtained by using sets of negative information with cardinality at most one/two.

Theorem 3.1. (Baliga, Case and Jain [1])
$\mathcal{E} \in \mathbf{NegF}^2\mathbf{TxtEx} \cap \mathbf{NegF}^1\mathbf{TxtEx}^1 \cap \mathbf{NegF}^1\mathbf{TxtBc}$.

In contrast to the above result, we have:

Theorem 3.2. (Baliga, Case and Jain [1])
$$\mathcal{E} \notin \mathbf{NegF^1TxtEx}.$$

However $\mathbf{NegF^1TxtEx}$ is still quite powerful as shown by the following theorem.

Theorem 3.3. (Baliga, Case and Jain [1])
(a) $\{L \in \mathcal{E} \mid \overline{L} \text{ is infinite}\} \in \mathbf{NegF^1TxtEx}$.
(b) $\mathbf{NegF^1TxtEx} - \mathbf{TxtBc}^* \neq \emptyset$.
(c) $\mathbf{NegF^1TxtEx} - \mathbf{InfBc}^n \neq \emptyset$.
(d) $\mathbf{TxtEx}^1 \subset \mathbf{NegF^1TxtEx}$.
(e) $\mathbf{InfEx} \subseteq \mathbf{NegF^1TxtEx}$.

For $i \geq 2$, it is open at present whether $\mathbf{TxtEx}^i \subset \mathbf{NegF^1TxtEx}$.

4. Some other Negative Information Models

Shinohara [26] considered giving to the learner atleast (but arbitrary) n negative data items.

Definition 4.1. (Shinohara [26]) Let $n \in N$.
(a) Suppose $\overline{L}$ has at least n elements. $\mathbf{M}$ $\mathbf{PP}^n$-*identifies* L (written $L \in \mathbf{PP}^n(\mathbf{M})$), iff for all information sequences I such that $\mathrm{PosInfo}(I) = L$ and $\mathrm{card}(\mathrm{NegInfo}(I)) \geq n$, $\mathbf{M}(I)$ converges to a grammar for L.
(b) $\mathbf{PP}^n = \{\mathcal{L} \mid (\exists \mathbf{M})[\mathcal{L} \subseteq \mathbf{PP}^n(\mathbf{M})]\}$.

Theorem 4.1. (Shinohara [26]) *Let $n \in N$. Suppose for any $L \in \mathcal{L}$, $\overline{L}$ contains at least n elements. Then $\mathcal{L} \in \mathbf{TxtEx}$ iff $\mathcal{L} \in \mathbf{PP}^n$.*

Fulk considered giving the grammar for the complement of L to the learner. For this notion consider $\mathbf{M}$ as being given two inputs: (a) a grammar, and (b) a text. Convergence of $\mathbf{M}(i, T)$ can be defined as usual.

Definition 4.2. (Fulk [9])
(a) $\mathbf{M}$ $\mathbf{CTxtEx}$-identifies L (written: $L \in \mathbf{CTxtEx}(\mathbf{M})$) iff for all i such that $W_i = \overline{L}$, for all texts T for L, $\mathbf{M}(i, T)$ converges to a grammar for L.
(b) $\mathbf{CTxtEx} = \{\mathcal{L} \mid (\exists \mathbf{M})[\mathcal{L} \subseteq \mathbf{CTxtEx}(\mathbf{M})]\}$.

Fulk showed that having a grammar for the complement gives tremendous advantages.

Theorem 4.2. (Fulk [9]) *Let $n \in N$. $\mathbf{CTxtEx} - \mathbf{InfBc}^n \neq \emptyset$.*

Fulk also considered the case when instead of being given a grammar for complement of L, the learner is given a sequence of grammars all but finitely many of which are grammars for L. It is not known at present whether this gives any advantages over informants.

Jain and Sharma [15] considered giving a grammar for a subset of the complement of the language being learned, where this subset has certain density.

Motoki [18] considered a form of open negative information as follows.

Definition 4.3. (Motoki [18]) $\mathbf{M}$ *identifies L using advisor A_L* iff for all information sequences I such that $\mathrm{PosInfo}(I) = L$ and $\mathrm{NegInfo}(L) \supseteq A_L$, $\mathbf{M}(I)$ converges to a grammar for L.

We use a general definition, though Motoki was mainly interested in indexed families.

Motoki showed that there exists a class $\mathcal{L} \notin \mathbf{TxtEx}$, such that a learner $\mathbf{M}$ can identify each $L \in \mathcal{L}$ using some advisor A_L, where $\mathrm{card}(A_L) \leq 1$. Motoki also gave a characterization of indexed families which can be learned using some advisor. We will be discussing a general form of open negative information in the next section.

5. Identification with Open Negative Information

We now consider another model of presenting negative information to learning machines. Here the negative information is supplied in a manner reminding one of the basic open sets for the topology with respect to which enumeration operators are continuous. This is the first topology described in Exercise 11–35, page 217 of Rogers [25]. These models were motivated in part by those considered by Motoki [18] (see Definition 4.3 above) and those in Section 3 above. Basically, this model allows the possibility of more negative information being supplied in addition to the finite cores of negative information.

Definition 5.1. (Baliga, Case and Jain [1]) Suppose $a, b \in N \cup \{*\}$.

(a) $\mathbf{M}$ $\mathbf{NegO}^b\mathbf{TxtEx}^a$-*identifies* $L \in \mathcal{E}$ (written: $L \in \mathbf{NegO}^b\mathbf{TxtEx}^a(\mathbf{M})$) $\Leftrightarrow (\exists S \subseteq \overline{L} \mid \mathrm{card}(S) \leq b)(\forall I \mid \mathrm{PosInfo}(I) = L \ \& \ S \subseteq \mathrm{NegInfo}(I) \subseteq \overline{L})[\mathbf{M}(T){\downarrow}$ and $W_{\mathbf{M}(I)} =^a L]$.

(b) $\mathbf{NegO}^b\mathbf{TxtEx}^a = \{\mathcal{L} \subseteq \mathcal{E} \mid (\exists \mathbf{M})[\mathcal{L} \subseteq \mathbf{NegO}^b\mathbf{TxtEx}^a(\mathbf{M})]\}$.

Thus, in contrast with Definition 3.1, in above model the learner must satisfy the stronger constraint that it needs to learn when the negative information present in the data given to it is any S' such that $S \subseteq S' \subseteq \overline{L}$ (here S' may be infinite).

Definition 5.2. (Baliga, Case and Jain [1]) Suppose $a, b \in N \cup \{*\}$.

(a) $\mathbf{M}$ $\mathbf{NegO}^b\mathbf{TxtBc}^a$-*identifies* $L \in \mathcal{E}$ (written:

$L \in \mathbf{NegO}^b\mathbf{TxtBc}^a(\mathbf{M})) \Leftrightarrow (\exists S \subseteq \overline{L} \mid \mathrm{card}(S) \leq b)(\forall I \mid \mathrm{PosInfo}(I) =$
L & $S \subseteq \mathrm{NegInfo}(I) \subseteq \overline{L})(\overset{\infty}{\forall} n)[W_{\mathbf{M}(I[n])} =^a L].$

(b) $\mathbf{NegO}^b\mathbf{TxtBc}^a = \{\mathcal{L} \subseteq \mathcal{E} \mid (\exists \mathbf{M})[\mathcal{L} \subseteq \mathbf{NegO}^b\mathbf{TxtBc}^a(\mathbf{M})]\}.$

Clearly, for all a, $\mathbf{NegO}^0\mathbf{TxtEx}^a = \mathbf{TxtEx}^a$ and $\mathbf{NegO}^0\mathbf{TxtBc}^a = \mathbf{TxtBc}^a$.

Theorem 5.1 below shows that the $\mathbf{NegO}^*$ criteria are equivalent to supplying *all* the negative (as well as the positive) information to a learning machine.

Theorem 5.1. (Baliga, Case and Jain [1]) *For all* $a \in N \cup \{*\}$, $\mathbf{NegO}^*\mathbf{TxtEx}^a = \mathbf{InfEx}^a$ *and* $\mathbf{NegO}^*\mathbf{TxtBc}^a = \mathbf{InfBc}^a$.

Thus, in particular we have $\mathcal{E} \in \mathbf{NegO}^*\mathbf{TxtBc}^*$, and $\mathbf{NegO}^*\mathbf{TxtEx} \subseteq \mathbf{NegF}^1\mathbf{TxtEx}$.

Note that if we consider languages such that informant for a language can be effectively obtained from its text, then above theorem shows that $\mathbf{NegO}$ type negative data does not help.

As a corollary to Theorem 5.1, using Theorems 2.1 and 3.3, we have

Corollary 5.1. (Baliga, Case and Jain [1])
(a) *For all* $n \in N$, $\mathbf{TxtEx}^{n+1} - \mathbf{NegO}^*\mathbf{TxtEx}^n \neq \emptyset$;
(b) *For all* $n \in N$, $\mathbf{TxtBc}^{n+1} - \mathbf{NegO}^*\mathbf{TxtBc}^n \neq \emptyset$;
(c) $\mathbf{TxtBc} - \mathbf{NegO}^*\mathbf{TxtEx}^* \neq \emptyset$;
(d) $\mathbf{NegF}^1\mathbf{TxtEx} - \mathbf{NegO}^*\mathbf{TxtBc}^n \neq \emptyset$;
(e) $\mathbf{NegF}^1\mathbf{TxtEx} - \mathbf{NegO}^*\mathbf{TxtEx}^* \neq \emptyset$.

The above Corollary shows that there are classes of languages which can be learned with $n+1$ mistakes, but not with n, no matter how much open negative information is provided in the n mistake case. In other words, the gap left by the possible extra anomaly can be greater in information content than the information provided by open negative information.

The following theorem generalizes Theorem 2.1(f).

Theorem 5.2. (Baliga, Case and Jain [1]) *For all* $a \in N \cup \{*\}$ *and* $j \in N$, $[\mathbf{NegO}^a\mathbf{TxtEx}^{2j} \subset \mathbf{NegO}^a\mathbf{TxtBc}^j]$.

The following result contrasts with Theorem 2.1(g).

Theorem 5.3. (Baliga, Case and Jain [1]) $\mathbf{TxtEx}^* \subset \mathbf{NegO}^1\mathbf{TxtBc}$.

The next theorem contrasts nicely with Theorem 5.1 above. It provides classes of languages which can be learned with $n + 1$ pieces of core open negative information, but not with n, no matter how many anomalies are permitted in the n piece case. In other words, the extra possible negative information can be greater in information content than the information that may be omitted by the anomalies.

Theorem 5.4. (Baliga, Case and Jain [1])
    (a) $\mathbf{NegO}^1\mathbf{TxtEx} - \mathbf{NegO}^0\mathbf{TxtBc}^* \neq \emptyset$.
    (b) For all $n \in N$, $\mathbf{NegO}^{n+1}\mathbf{TxtEx} - \mathbf{NegO}^n\mathbf{TxtEx}^* \neq \emptyset$.
    (c) For all $n \in N$, $\mathbf{NegO}^{n+1}\mathbf{TxtEx} - \bigcup_{j \in N} \mathbf{NegO}^n\mathbf{TxtBc}^j \neq \emptyset$.

The previous theorem has the following straightforward corollary.

Corollary 5.2. (Baliga, Case and Jain [1]) *For all* $a \in N \cup \{*\}$ *and* j, $n \in N$,
    (a) $\mathbf{NegO}^n\mathbf{TxtEx}^a \subset \mathbf{NegO}^{n+1}\mathbf{TxtEx}^a$ *and*
    (b) $\mathbf{NegO}^n\mathbf{TxtBc}^j \subset \mathbf{NegO}^{n+1}\mathbf{TxtBc}^j$.

5.1. *Complexity Advantages of Open Negative Information*

McNeill [17] posits that there is faster learning of language for children in homes in which more corrections (usually in the form of, possibly exemplary, expansions) are given. These corrections are, in part, a form of negative information. Theorem 5.5 below shows that an improvement in *speed* (measured by mind-changes) can result from the presence of open negative information even when the classes themselves can be learned without the negative information.

For this section it is convenient to modify the definition of the learning machine to the following.

Definition 5.3. A *language learning machine* is an algorithmic device which computes a mapping from SEQ (or SEG) into $N \cup \{?\}$.

Intuitively the outputted ?s represent the machine not yet committing to an output. This avoids biasing the number of mind changes before a learning machine converges.

In the next definition, the subscript b represents a bound on the number of mind changes allowed before convergence.

Definition 5.4. (Case and Smith [7], Case and Lynes [6]) Suppose $a, b \in N \cup \{*\}$. We say that $\mathbf{M}$ $\mathbf{TxtEx}_b^a$-*identifies* $L \Leftrightarrow [[L \in \mathbf{TxtEx}^a(\mathbf{M})] \wedge (\forall \text{ texts } T \text{ for } L)[\mathrm{card}(\{x \mid [? \neq \mathbf{M}(T[x])] \wedge [\mathbf{M}(T[x]) \neq \mathbf{M}(T[x+1])]\}) \leq b]]$.

One can similarly define $\mathbf{NegO}^c\mathbf{TxtEx}_b^a$.

Next theorem shows the speed advantage of having open negative information.

Theorem 5.5. (Baliga, Case and Jain [1]) *There exists a class of languages* $\mathcal{L}$ *such that,*

    (a) $\mathcal{L} \in \mathbf{TxtEx}$,

    (b) $\mathcal{L} \in \mathbf{NegO}^1\mathbf{TxtEx}_0$, *and*

    (c) $\mathcal{L} \notin \bigcup_{n \in N} \mathbf{TxtEx}_n^*$.

We now list some of the open problems regarding this model.

    (a) For $i \geq 1$, $\mathcal{E} \in \mathbf{NegO}^i\mathbf{TxtBc}^*$? Here note that $\mathcal{E} \in \mathbf{NegO}^*\mathbf{TxtBc}^*$.

    (b) By Theorem 2.1(g), $\mathbf{TxtEx}^{2j+1} - \mathbf{TxtBc}^j \neq \emptyset$. Similarly, can it be shown that, for $i \geq 1$, $\mathbf{NegO}^i\mathbf{TxtEx}^{2j+1} - \mathbf{NegO}^i\mathbf{TxtBc}^j \neq \emptyset$?

    (c) For $i \geq 1$, is $\mathbf{NegO}^i\mathbf{TxtEx}^* \subset \mathbf{NegO}^{i+1}\mathbf{TxtBc}$? So far we know that $\mathbf{NegO}^*\mathbf{TxtEx}^* \subset \mathbf{NegO}^*\mathbf{TxtBc}$.

6. Learning with Negative Counterexamples

We now consider providing negative data to the learner via counterexamples to the conjectures of the learner. We will be considering three variants of the model. Intuitively, for learning with negative counterexamples, we may consider the learner being provided a text, one element at a time, along with a negative counterexample to the latest conjecture, if any. The list of negative counterexamples may be modeled as a second text provided to the learner. Thus the learning machines get as input two texts, one for positive data, and other for negative counterexamples. We say that $\mathbf{M}(T, T')$ converges to a grammar i, iff for all but finitely many n, $\mathbf{M}(T[n], T'[n]) = i$.

In the basic model of learning from positive data and negative counterexamples, if a conjecture contains elements not in the target language, then a negative counterexample is provided to the learner. $\mathbf{NC}$ in the definition below stands for negative counterexample.

Definition 6.1. (Jain and Kinber [13]) Suppose $a \in N \cup \{*\}$.

    (a) $\mathbf{M}$ $\mathbf{NCEx}^a$-*identifies a language* L (written: $L \in \mathbf{NCEx}^a(\mathbf{M})$) iff for all texts T for L, and for all T' satisfying the condition:

$$T'(n) \in S_n, \text{ if } S_n \neq \emptyset \text{ and } T'(n) = \#, \text{ if } S_n = \emptyset,$$

where $S_n = \overline{L} \cap W_{\mathbf{M}(T[n],T'[n])}$

$\mathbf{M}(T,T')$ converges to a grammar i such that $W_i =^a L$.

(b) $\mathbf{NCEx}^a = \{\mathcal{L} \mid (\exists \mathbf{M})[\mathcal{L} \subseteq \mathbf{NCEx}^a(\mathbf{M})]\}$.

We also consider two variants of above definition as follows:

— the learner gets least negative counterexample instead of any counterexample. This criteria is denoted $\mathbf{LNCEx}^a$.

— the negative counterexample is provided only if there exists one such counterexample $\leq$ the maximum positive element seen in the input so far (otherwise the learner gets $\#$). This criteria is denoted by $\mathbf{BNCEx}^a$. (Essentially S_n in the definition of $T'(n)$ in part (a) is replaced by $S_n = \overline{L} \cap W_{\mathbf{M}(T[n],T'[n])} \cap \{x \mid x < \max(\text{content}(T[n]))\}$). The $\mathbf{BNC}$ model essentially addresses some complexity constraints.

Similarly, we can define $\mathbf{NCBc}^a$, $\mathbf{LNCBc}^a$ and $\mathbf{BNCBc}^a$ criteria of inference.

It is easy to see that $\mathbf{TxtEx}^a \subseteq \mathbf{BNCEx}^a \subseteq \mathbf{NCEx}^a \subseteq \mathbf{LNCEx}^a$. All of these containments, except the last one, are proper.

Part (a) of the following theorem shows that every indexed family can be learned using positive data and negative counterexamples. This improves a classical result that every indexed family is learnable from informants. Since there exist indexed families not in $\mathbf{TxtEx}$, this illustrates a difference between $\mathbf{NCEx}$ learning and learning without negative counterexamples.

Part (b) of the following theorem illustrates another difference between $\mathbf{NCEx}$ learning and $\mathbf{TxtEx}$ learning. Such a result does not hold for $\mathbf{TxtEx}$ (for example, $\{F \mid F \text{ is finite }\} \cup \{L\} \notin \mathbf{TxtEx}$, for any infinite language L).

Theorem 6.1. (Jain and Kinber [13])

(a) *Suppose $\mathcal{L}$ is an indexed family. Then $\mathcal{L} \in \mathbf{NCEx}$.*

(b) *Suppose $\mathcal{L} \in \mathbf{NCEx}$ and L is a recursive language. Then $\mathcal{L} \cup \{L\} \in$* $\mathbf{NCEx}$.

Part (b) of the above theorem does not generalize to taking r.e. language (instead of recursive language) L, as witnessed by $\mathcal{L} = \{\{A \cup \{x\}\} \mid x \notin A\}$, and $L = A$, where A is any non-recursive r.e. set. Here note that $\mathcal{L} \in \mathbf{TxtEx}$, but $\mathcal{L} \cup \{L\}$ is not in $\mathbf{NCEx}$.

The following theorem shows that using least negative counterexamples, rather than arbitrary negative counterexamples, does not enhance power of a learner.

Theorem 6.2. (Jain and Kinber [13]) *Let $a \in N \cup \{*\}$. Then,* $\mathbf{NCEx}^a = \mathbf{LNCEx}^a \subseteq \mathbf{InfEx}^a$.

For **Bc**-style learning, a limited version of above holds. Though, the equality $\mathbf{NCBc} = \mathbf{LNCBc}$ can be generalized to learning with anomalies (see Corollary 6.2 below), $\mathbf{LNCBc} \subseteq \mathbf{InfBc}$, cannot be generalized to learning with anomalies.

Proposition 6.1. (Jain and Kinber [13]) $\mathbf{NCBc} = \mathbf{LNCBc} \subseteq \mathbf{InfBc}$.

Part (a) of the following theorem shows that all classes of languages learnable in the basic **Ex**-style model with arbitrary finite number of errors in almost all conjectures can be learned without errors in the basic **Bc**-style model. This contrasts with learning from texts where $\mathbf{TxtEx}^{2j+1} - \mathbf{TxtBc}^j \neq \emptyset$ (Theorem 2.1(g)).

Part (b) of the following theorem is somewhat surprising. It shows that sometimes negative counterexamples are not enough: to learn a language, the learner must have access to *all* negative examples.

Theorem 6.3. (Jain and Kinber [13])
    (a) $\mathbf{NCEx}^* \subseteq \mathbf{NCBc}$.
    (b) $\mathbf{InfEx} - \mathbf{NCBc} \neq \emptyset$.

We now show advantages of having negative counterexamples. Part (a) of the following theorem shows that the model **BNCEx** is quite powerful: there are classes of languages learnable in this model that cannot be learned in the classical **Bc**-style model even when an arbitrary finite number of errors is allowed in almost all conjectures. Part (b) of the following theorem shows that there are classes of languages learnable in the basic model that cannot be learned in any of the models that use negative counterexamples of limited size.

Theorem 6.4. (Jain and Kinber [13])
    (a) $\mathbf{BNCEx} - \mathbf{TxtBc}^* \neq \emptyset$.
    (b) $\mathbf{NCEx} - \mathbf{BNCBc}^* \neq \emptyset$.

Note that the diagonalizations in Theorem 6.4 can be shown using indexed families of languages. Thus, in contrast to Theorem 6.1, there exists an indexed family not in $\mathbf{BNCBc}^*$.

In contrast to Theorem 6.4 (b), the following shows that if attention is restricted to only infinite languages, then **NCEx** and **BNCEx** behave similarly.

Theorem 6.5. (Jain and Kinber [13]) *Suppose $a \in N \cup \{*\}$. Suppose $\mathcal{L}$ consists of only infinite languages. Then*
    (a) $\mathcal{L} \in \mathbf{NCEx}^a$ *iff* $\mathcal{L} \in \mathbf{BNCEx}^a$.
    (b) $\mathcal{L} \in \mathbf{NCBc}^a$ *iff* $\mathcal{L} \in \mathbf{BNCBc}^a$.

We now consider the error hierarchy for learning with negative counterexamples. That is, learning with at most $n + 1$ errors in almost all conjectures in the basic model is stronger than learning with at most n errors. The hierarchy easily follows from the following theorem.

Theorem 6.6. (Jain and Kinber [13]) *Suppose $n \in N$.*
    (a) $\mathbf{TxtEx}^{n+1} - \mathbf{NCEx}^n \neq \emptyset$.
    (b) $\mathbf{TxtEx}^* - \bigcup_{n \in N} \mathbf{NCEx}^n \neq \emptyset$.
    (c) $\mathbf{TxtBc} - \mathbf{NCEx}^* \neq \emptyset$.
    (d) $\mathbf{TxtBc}^1 - \mathbf{NCBc} \neq \emptyset$.

As, $\mathbf{TxtEx}^{n+1} \subseteq \mathbf{BNCEx}^{n+1} \subseteq \mathbf{NCEx}^{n+1} \subseteq \mathbf{LNCEx}^{n+1}$, the following corollary follows from Theorem 6.6.

Corollary 6.1. (Jain and Kinber [13]) *Suppose $n \in N$. Then, for $\mathbf{I} \in \{\mathbf{NCEx}, \mathbf{LNCEx}, \mathbf{BNCEx}\}$, we have $\mathbf{I}^n \subset \mathbf{I}^{n+1}$.*

Now we consider another surprising result. There exists a $\mathbf{Bc}^1$-style learner with negative counterexamples, with the "ultimate power" - it can learn the class of all recursively enumerable languages!

Theorem 6.7. (Jain and Kinber [13]) $\mathcal{E} \in \mathbf{NCBc}^1$.

Since $\mathcal{E} \in \mathbf{InfBc}^*$, we have

Corollary 6.2. (Jain and Kinber [13]) (a) $\mathbf{NCBc}^1 = \mathbf{InfBc}^*$.
    (b) *For all $a \in N \cup \{*\}$, $\mathbf{NCBc}^a = \mathbf{LNCBc}^a$.*

The following corollary shows a contrast with respect to the case when there are no errors in conjectures (Proposition 6.1 and Theorem 6.3(c)). What a difference just one error can make!

Corollary 6.3. (Jain and Kinber [13]) *For all $n \in N$, $n > 0$, $\mathbf{InfBc}^n \subset \mathbf{NCBc}^n = \mathbf{NCBc}^1$.*

Based on the ideas similar to the ones used for proving Theorem 6.7, one can show

Theorem 6.8. (Jain and Kinber [13]) (a) *Let $\mathcal{L} = \{L \in \mathcal{E} \mid L$ is infinite$\}$. Then $\mathcal{L} \in \mathbf{BNCBc}^1$.*

(b) *For all $n \in N$,* $\mathbf{TxtBc}^n \subseteq \mathbf{BNCBc}^1$.
(c) $\mathbf{TxtEx}^* \subseteq \mathbf{BNCBc}^1$.

As there exists a class of infinite languages which does not belong to $\mathbf{InfBc}^n$ (see Case and Smith [7]), we have

Corollary 6.4. (Jain and Kinber [13])
For all $n \in N$, $\mathbf{BNCBc}^1 - \mathbf{InfBc}^n \neq \emptyset$.

Thus, $\mathbf{BNCBc}^m$ and $\mathbf{InfBc}^n$ are incomparable for $m > 0$, $m, n \in N$. The above result does not generalize to $\mathbf{InfBc}^*$, as $\mathbf{InfBc}^*$ contains the class $\mathcal{E}$.

We now mention some of the open questions regarding behaviourally correct learning when the size of the negative counterexamples is bounded.
(a) Is $\mathbf{BNCBc}^n$ hierarchy strict?
(b) Is $\mathbf{TxtBc}^* \subseteq \mathbf{BNCBc}^1$?

6.1. *Complexity Issues*

We now consider the complexity advantages of having negative counterexamples. This section is based on the paper [13].

The class $\mathcal{L}_1 = \{L \mid \mathrm{card}(N - L) = 1\}$ is in $\mathbf{TxtEx}$, but requires unbounded number of mind changes to learn. On the other hand, $\mathcal{L}_1$ can be easily learned using one mind change if negative counterexamples are available. Thus, not only does $\mathbf{NCEx}$ model give learnability advantages over $\mathbf{TxtEx}$, it also gives complexity advantages over $\mathbf{TxtEx}$ for some classes in $\mathbf{TxtEx}$. Note that if one does not allow mind changes, then $\mathbf{NCEx}$ and $\mathbf{TxtEx}$ are both the same — thus the above result is the best mind change complexity advantage possible.

The class $\mathcal{L}_2 = \{L \mid (\exists i)[L = \{x \mid x \leq i\}]\} \cup \{N\}$, is learnable in $\mathbf{NCEx}$ model, but the number of mind changes is unbounded. However, $\mathcal{L}_2$ can be learned by using at most one mind change in the model $\mathbf{LNCEx}$. Thus, even though $\mathbf{LNCEx}$ does not give learnability advantages over $\mathbf{NCEx}$, it does give complexity advantages.

Let $a \dot{-} b = a - b$, if $a \geq b$; $a \dot{-} b = 0$ otherwise; Consider the class:
$\mathcal{L}_3 = \{L \mid (\exists! e)[\langle 0, e \rangle \in L \ \wedge \ L - \{\langle 0, e \rangle\} \subseteq \{\langle x, y \rangle \mid x > 1\} \ \wedge \ \mathrm{card}(L - \{\langle 0, e \rangle\}) = e \dot{-} \min(W_e)]\}$

$\mathcal{L}_3$ is in $\mathbf{LNCEx}$ with at most one mind change. However $\mathcal{L}_3$ cannot be learned in $\mathbf{InfEx}$ using bounded number of mind changes. Note that $\mathbf{LNCEx} \subset \mathbf{InfEx}$. So getting negative counterexamples gives complexity advantages over informants, despite informant being more advantageous for learning as a whole.

The situation is more complex in considering the complexity advantages of **NCEx**-model compared to **InfEx** model. There exist classes which can be **NCEx**-identifies using $n - 1$ mind changes, but cannot be **InfEx**-identified using $(2^n - 1) - 2$ mind changes. This is optimal as it can be shown that any class which can be **NCEx**-identified using $n - 1$ mind changes can also be identified using $(2^n - 1) - 1$ mind changes in **InfEx**-model. We omit the details.

7. Learning With Subset Queries

We now consider learning with subset queries, which turn out to be another mechanism for providing negative examples. In this model learner is allowed to ask queries of the form "is $Q \subseteq L$?", where L is the language being learned.

If the answer to query is "no", we additionally can have the following possibilities:

(a) Learner is given an arbitrary counterexample (a member of $Q - L$);

(b) Learner is given the least counterexample;

(c) Learner is just given the answer 'no', without any counterexample.

We would often also consider bounds on the number of queries. We first formalize the definition of a learner which uses queries.

Definition 7.1. (Jain and Kinber [12]) A learner using queries can ask a query of form "$W_j \subseteq L$?" on any input σ. Answer to the query is "yes" or "no" (along with a possible counterexample). Then, based on input σ and answers received for queries made on prefixes of σ, **M** outputs a conjecture (from N).

Note that the queries are for recursively enumerable languages, which are posed to the teacher using a grammar (index) for the language. Many of the diagonalization results stand even if one uses arbitrary type of query language. However simulation results often crucially depend on the queries being made only via grammars for the queried languages.

Here, if one allows infinite number of subset queries, then one can learn the whole class $\mathcal{E}$ of recursively enumerable languages in **Ex**-model of learning. Furthermore, as we will see below (Proposition 7.2) if one allows finite, but unbounded, number of queries, then for **Ex**-model of learning the notion coincides with learning from negative counterexamples.

We now formalize learning via subset queries.

Definition 7.2. (Jain and Kinber [12]) Suppose $a \in N \cup \{*\}$.

(a) **M SubQ^{a}Ex**-*identifies* a language L (written: $L \in$ **SubQ^{a}Ex(M)**) iff for any text T for L, it behaves as follows:

(i) The number of queries **M** asks on prefixes of T is bounded by a (if $a = *$, then the number of such queries is finite). Furthermore, all the queries are of the form "$W_j \subseteq L$?"

(ii) Suppose the answers to the queries are made as follows. For a query "$W_j \subseteq L$?", the answer is "yes" if $W_j \subseteq L$, and the answer is "no" if $W_j - L \neq \emptyset$. For "no" answers, **M** is also provided with a counterexample, $x \in W_j - L$. Then, for some k such that $W_k = L$, for all but finitely many n, **M**($T[n]$) outputs the grammar k.

(b) **SubQ^{a}Ex** $= \{\mathcal{L} \mid (\exists M)[\mathcal{L} \subseteq$ **SubQ^{a}Ex(M)**$]\}$.

LSubQ^{a}Ex-identification and **ResSubQ^{a}Ex**-identification can be defined similarly, where for **LSubQ^{a}Ex**-identification the learner gets the least counterexample for "no" answers, and for **ResSubQ^{a}Ex**-identification, the learner does not get any counterexample along with the "no" answers.

For $a, b \in N \cup \{*\}$, for $\mathbf{I} \in \{\mathbf{Ex}^b, \mathbf{Bc}^b\}$, one can similarly define **SubQ^{a}I**, **LSubQ^{a}I**, and **ResSubQ^{a}I**.

Next two propositions show a close correspondence between learning via negative counterexamples and learning via subset queries. In particular, learning via finite number of subset queries coincides with learning via negative counterexamples for **Ex**-model of learning.

Proposition 7.1. (Jain and Kinber [12]) *For any* $a \in N \cup \{*\}, \mathbf{I} \in \{\mathbf{Ex}^a, \mathbf{Bc}^a\}$,

(a) **SubQ*I** $\subseteq$ **NCI**.
(b) **LSubQ*I** $\subseteq$ **LNCI**.
(c) **ResSubQ*I** $\subseteq$ **ResNCI**.

Proposition 7.2. (Jain and Kinber [12]) *Suppose* $a \in N \cup \{*\}$.
NCEx^a $=$ **SubQ*Ex^a** $=$ **LNCEx^a** $=$ **LSubQ*Ex^a** $=$ **ResNCEx^a** $=$ **ResSubQ*Ex^a**.

Next theorem establishes a hierarchy of learning capabilities with respect to the number of subset queries.

Theorem 7.1. (Jain and Kinber [12]) *Suppose* $n \in N$. *Then,*
ResSubQ^{n+1}Ex $-$ **LSubQ^{n}Bc*** $\neq \emptyset$.

We now consider relationship between various types of subset queries. When only a single query or an unbounded but finite number of queries are used, different types of counterexamples do not make a difference.

Theorem 7.2. (Jain and Kinber [12]) *Suppose $a \in N \cup \{*\}$, $b \in \{0, 1, *\}$, and* $\mathbf{I} \in \{\mathbf{Ex}^a, \mathbf{Bc}^a\}$. *Then,* $\mathbf{ResSubQ}^b\mathbf{I} = \mathbf{SubQ}^b\mathbf{I} = \mathbf{LSubQ}^b\mathbf{I}$.

Thus, one needs to consider at least two queries when showing differences between various types of subset queries. The following theorem establishes the relationship between different types of subset queries.

Theorem 7.3. (Jain and Kinber [12]) *For all $n \in N$,*
    (a) $\mathbf{LSubQ}^2\mathbf{Ex} - \mathbf{SubQ}^n\mathbf{Bc}^* \neq \emptyset$.
    (b) $\mathbf{SubQ}^2\mathbf{Ex} - \mathbf{ResSubQ}^n\mathbf{Bc}^* \neq \emptyset$.

We next consider the anomaly hierarchy for the subset query learning criteria.

Theorem 7.4. (Jain and Kinber [12]) (a) *For all $n \in N$,* $\mathbf{TxtEx}^{n+1} - \mathbf{LSubQ}^*\mathbf{Ex}^n \neq \emptyset$.
    (b) *For all $n \in N$,* $\mathbf{TxtBc}^{n+1} - \mathbf{LSubQ}^*\mathbf{Bc}^n \neq \emptyset$.
    (c) $\mathbf{LSubQ}^*\mathbf{Ex}^* \subseteq \mathbf{ResSubQ}^*\mathbf{Bc}$.

As a corollary we get:

Corollary 7.1. (Jain and Kinber [12]) *Let $a \in N \cup \{*\}$, and $n \in N$.*
    (a) $\mathbf{SubQ}^a\mathbf{Ex}^n \subset \mathbf{SubQ}^a\mathbf{Ex}^{n+1}$.
    (b) $\mathbf{LSubQ}^a\mathbf{Ex}^n \subset \mathbf{LSubQ}^a\mathbf{Ex}^{n+1}$.
    (c) $\mathbf{ResSubQ}^a\mathbf{Ex}^n \subset \mathbf{ResSubQ}^a\mathbf{Ex}^{n+1}$.

Similar corollary exists for **Bc**-criteria of learning with **Ex** being replaced by **Bc** in the above.

8. Random Negative Examples

In this section we briefly consider the impact of having random negative examples. It would be interesting to explore in general how random negative examples effect learning compared to other kind of negative examples as discussed in this paper.

When considering giving random negative examples, one may consider any measure theoretic method of selecting a random negative example. The only property used in the following is that if A is infinite and B is a finite subset of A, then measure of $A - B$ (with respect to A) is 1. Let

$\mathbf{Rand}_p^1\mathbf{TxtEx}$ denote the class of languages that can be identified using positive data and one random negative example with probability p.

Theorem 8.1. *Consider the following class of languages:* $\mathcal{L} = \{L \mid (\exists i)[W_i = L \ \& \ \mathrm{card}(\overline{L} - \{\langle i, x \rangle \mid x \in N\}) < \infty \ \& \ \mathrm{card}(\overline{L} \cap \{\langle i, x \rangle \mid x \in N\}) = \infty]\}$. *Then,* $\mathcal{L} \in \mathbf{Rand}_1^1\mathbf{TxtEx} - \mathbf{TxtEx}$.

Note here that by Theorem 4.1, if one considers having arbitrary counterexamples, then for any class of languages which consists only of coinfinite languages, k arbitrary negative examples do not help in learning. So above theorem also shows that random negative examples are more useful for learning compared to arbitrary negative examples.

Acknowledgements

Sanjay Jain was supported in part by NUS grant number R252-000-127-112.

References

1. G. Baliga, J. Case, and S. Jain. Language learning with some negative information. *Journal of Computer and System Sciences*, 51(5):273–285, 1995.
2. J. Bārzdiņš. Two theorems on the limiting synthesis of functions. In *Theory of Algorithms and Programs, vol. 1*, pages 82–88. Latvian State University, 1974. In Russian.
3. L. Blum and M. Blum. Toward a mathematical theory of inductive inference. *Information and Control*, 28:125–155, 1975.
4. M. Blum. A machine-independent theory of the complexity of recursive functions. *Journal of the ACM*, 14:322–336, 1967.
5. R. Brown and C. Hanlon. Derivational complexity and the order of acquisition in child speech. In J. R. Hayes, editor, *Cognition and the Development of Language*. Wiley, 1970.
6. J. Case and C. Lynes. Machine inductive inference and language identification. In M. Nielsen and E. M. Schmidt, editors, *Proceedings of the 9th International Colloquium on Automata, Languages and Programming*, volume 140 of *Lecture Notes in Computer Science*, pages 107–115. Springer-Verlag, 1982.
7. J. Case and C. Smith. Comparison of identification criteria for machine inductive inference. *Theoretical Computer Science*, 25:193–220, 1983.
8. M. Demetras, K. Post, and C. Snow. Feedback to first language learners: The role of repetitions and clarification questions. *Journal of Child Language*, 13:275–292, 1986.
9. M. Fulk. *A Study of Inductive Inference Machines*. PhD thesis, SUNY/Buffalo, 1985.
10. E. M. Gold. Language identification in the limit. *Information and Control*, 10:447–474, 1967.

11. K. Hirsh-Pasek, R. Treiman, and M. Schneiderman. Brown and Hanlon revisited: Mothers' sensitivity to ungrammatical forms. *Journal of Child Language*, 11:81–88, 1984.

12. S. Jain and E. Kinber. Learning languages from positive data and a finite number of queries. In Kamal Lodaya and Meena Mahajan, editors, *Foundations of Software Technology and Theoretical Computer Science*, volume 3328 of *Lecture Notes in Computer Science*, pages 360–372. Springer-Verlag, 2004.

13. S. Jain and E. Kinber. Learning languages from positive data and negative counterexamples. *Journal of Computer and System Sciences*, 2005. To appear.

14. S. Jain, D. Osherson, J. Royer, and A. Sharma. *Systems that Learn: An Introduction to Learning Theory*. MIT Press, Cambridge, Mass., second edition, 1999.

15. S. Jain and A. Sharma. Learning in the presence of partial explanations. *Information and Computation*, 95:162–191, 1991.

16. M. Machtey and P. Young. *An Introduction to the General Theory of Algorithms*. North Holland, New York, 1978.

17. D. McNeill. Developmental psycholinguistics. In F. Smith and G. Miller, editors, *The Genesis of Language*, pages 15–84. MIT Press, 1966.

18. T. Motoki. Inductive inference from all positive and some negative data. *Information Processing Letters*, 39(4):177–182, 1991.

19. D. Osherson, M. Stob, and S. Weinstein. Ideal learning machines. *Cognitive Science*, 6:277–290, 1982.

20. D. Osherson, M. Stob, and S. Weinstein. Learning theory and natural language. *Cognition*, 17:1–28, 1984.

21. D. Osherson, M. Stob, and S. Weinstein. *Systems that Learn: An Introduction to Learning Theory for Cognitive and Computer Scientists*. MIT Press, 1986.

22. D. Osherson and S. Weinstein. Criteria of language learning. *Information and Control*, 52:123–138, 1982.

23. D. Osherson and S. Weinstein. A note on formal learning theory. *Cognition*, 11:77–88, 1982.

24. S. Pinker. Formal models of language learning. *Cognition*, 7:217–283, 1979.

25. H. Rogers. *Theory of Recursive Functions and Effective Computability*. McGraw-Hill, 1967. Reprinted, MIT Press 1987.

26. T. Shinohara. *Studies on Inductive Inference from Positive Data*. PhD thesis, Kyushu University, Kyushu, Japan, 1986.

27. K. Wexler. On extensional learnability. *Cognition*, 11:89–95, 1982.

28. K. Wexler and P. Culicover. *Formal Principles of Language Acquisition*. MIT Press, 1980.

EFFECTIVE CARDINALS IN THE NONSTANDARD UNIVERSE

VLADIMIR KANOVEI*

Institute for the information transmission problems (IITP RAS)
Bol. Karetnyj Per. 19 GSP-4, Moscow 127994 Russia
E-mail: kanovei@mccme.ru and vkanovei@math.uni-wuppertal.de

MICHAEL REEKEN

Department of Mathematics, University of Wuppertal,
Gauss Strasse 20, Wuppertal 42097, Germany,
E-mail: reeken@math.uni-wuppertal.de

We study the structure of effective cardinals in the nonstandard set universe of Hrbaček set theory **HST**. Some results resemble those known in descriptive set theory in the domain of Borel reducibility of equivalence relations.

Introduction

Nonstandard analysis as a branch of mathematics[a] emerged in the beginning of 1960s when A. Robinson [26] demonstrated that nonstandard models (that is, proper elementary extensions) of the real continuum lead to a mathematically rigorous system including infinitesimals and infinitely large numbers. In the course of 1960s, the model theoretic tools used by Robinson were shown to be applicable to a variety of mathematical structures, and that such an applicability was based on a few general properties of nonstandard extensions, in particular, elementarity and saturation. For instance any $\aleph_1$-saturated elementary extension $^*\mathbb{N}$ of the integers $\mathbb{N}$ contains an infinitely large number. Several nonstandard axiomatical systems were proposed, beginning with the mid-1970s, based on those general principles. Unlike the model-theoretic approach, such theories as Nelson's internal set theory [23], two theories of [8, 9], bounded set theory [13], axiomatically described nonstandard extensions of the whole standard set universe of **ZFC** rather than extensions of any particular structure.

In the mid-1990s we formulated *Hrbaček set theory* **HST** [14], based on

---

*Contact author. Partially supported by RFBR 03-01-00757, 06-01-00608, and DFG 436 RUS 17/68/05.
[a]See [5, 29] on the early history of *infinitesimal analysis*.

earlier theories in [8, 9]. This theory combined achievements of different nonstandard set theories and avoided their faults. The set universe of **HST** is axiomatized as a von Neumann superstructure $\mathbb{H}$ over a fully saturated elementary extension $\mathbb{I}$ ($\mathbb{I}$ = internal sets) of the class $\mathbb{WF}$ of all well-founded sets, see more on this in Section 1. Our monograph [17] presents in detail the structure of the **HST** universe and metamathematical properties of **HST** and some other popular nonstandard set theories.

This paper is devoted to the structure of cardinalities in the nonstandard set universe of **HST**. Note that **HST** does not include the axioms of Power Set, Choice, and Regularity. In fact these axioms contradict **HST**. This is why methods of study of the structure of cardinalities known from **ZFC** are not always applicable in **HST**. Nevertheless there are two rather regular families of cardinalities in **HST**: $\mathbb{WF}$-cardinals and $\mathbb{I}$-cardinals. Either family behaves in **ZFC**-like manner simply because both $\mathbb{WF}$ and $\mathbb{I}$ satisfy **ZFC**. The intersection of the two families consists of finite cardinals. But little is known beyond this. Some independence results have been obtained. For instance, the hypothesis that all infinite sets in $\mathbb{I}$ are equinumerous in the whole universe $\mathbb{H}$, and the hypothesis that $\mathbb{I}$-cardinals are preserved in $\mathbb{H}$ (except for hyperfinite cardinalities $m < n$ such that $\frac{m}{n}$ is not infinitesimal, [19]) are consistent with **HST**, see [15] or [17], Chapter 7.

Yet an alternative approach seems to be much more promising in the context of **HST**. Instead of abstract "cantorial" cardinalities, we consider here those induced by *effective* embeddings, *i.e.* those definable in some way or given by a certain construction. In this we follow earlier works in nonstandard analysis. For instance studies on collapse of hyperfinite cardinalities by Borel and countably determined maps were carried out in 1980s, see [12, 19, 27]. Further studies revealed a complicated structure of "Borel" and "countably determined" cardinalities of hyperfinite sets [16].

However **HST** admits a much more general concept of effective cardinality than those based on Borel or countably determined maps. This concept involves the class $\mathbb{L}[\mathbb{I}]$ of all sets *constructible over* $\mathbb{I}$, and the class $\mathbf{\Delta}_2^{ss}$ of all sets $x \in \mathbb{L}[\mathbb{I}]$, $x \subseteq \mathbb{I}$ (see details below), which includes and greatly exceeds Borel and countably determined sets.

The first part of the paper is devoted to effective cardinalities of internal sets and, generally, sets that consist of internal elements. We prove that effective cardinalities of internal sets are just their $\mathbb{I}$-cardinals in the $\mathbb{I}$-infinite domain, and resemble multiplicative galaxies in the hyperfinite domain. Effective cardinalities of $\mathbf{\Sigma}_1^{ss}$ sets ($\mathbb{WF}$-size unions of internal sets) are still linearly ordered and admit characterization in terms of cuts (initial segments) in the class $^{*}\mathbf{Card}$ of all $\mathbb{I}$-cardinals. Some results for cardinalities in more complicated classes $\mathbf{\Pi}_1^{ss}$ and $\mathbf{\Delta}_2^{ss}$ will be presented, too.

The second part of the paper considers effective cardinalities in full generality. Fortunately there is a reduction down to $\mathbb{I}$: any set in $\mathbb{L}[\mathbb{I}]$ admits an effective bijection onto the quotient structure of the form X/E, where E is a $\mathbf{\Delta}_2^{ss}$ relation on a $\mathbf{\Delta}_2^{ss}$ set X (by necesity $X \subseteq \mathbb{I}$).

And this brings us to an analogy with modern descriptive set theory, where cardinality problems for Borel quotient structures in Polish spaces became the focal point since early 1990s — especially in the form of *Borel reducibility* of quotients and the corresponding equivalence relations, see e.g. [6, 7, 18]. We pursue essentially the same idea, with $\mathbf{\Delta}_2^{ss}$ reduction maps in the same role as Borel reductions in descriptive set theory.

Inspired by this analogy, we prove several results related to dichotomy of "large"–"small" sets, a nonstandard form of the Ramsey theorem, a theorem saying that quotients with rather small (for instance countable) classes are "smooth" in a sense similar to the smoothness for quotients in descriptive set theory, and finally consider effective reducibility within the family of *monadic* equivalence relations. Those readers with an experience in descriptive set theory may be interested to recognize similarities and differences with the set-up they are accustomed to.

1. Structure of the nonstandard universe

The language of *Hrbaček set theory* **HST** contains two basic predicates, the membership $\in$ and the standardness st, hence it is called *the $\mathsf{st}\text{-}\in\text{-}language$*. The axioms of **HST** describe a set universe $\mathbb{H}$ where the following classes are defined,

$$\mathbb{S} = \{x : \mathsf{st}\,x\} \qquad - \text{ standard sets;}$$

$$\mathbb{I} = \{y : \exists^{\mathsf{st}} x\,(y \in x)\} \quad - \text{ internal set;}$$

$$\mathbb{WF} \qquad\qquad\qquad - \text{ well-founded}^{\,b} \text{ sets;}$$

so that $\mathbb{S} \subseteq \mathbb{I}$, $\mathbb{I}$ is an elementary extension of $\mathbb{S}$ in the $\in$-language, $\mathbb{S}$ (and $\mathbb{I}$ as well) satisfies **ZFC** in the $\in$-language, the class $\mathbb{I}$ is transitive, and the universe $\mathbb{H}$ is a von Neumann superstructure over $\mathbb{I}$. The universe $\mathbb{H}$ satisfies all **ZFC** axioms except for Regularity (weakened to Regularity over $\mathbb{I}$), Choice (weakened to Standard Size Choice) and Power Set axioms. The axioms of Separation and Replacement are accepted in the $\mathsf{st}\text{-}\in$-language.

---

^{a} $\exists^{\mathsf{st}}$ and $\forall^{\mathsf{st}}$ are shorthands for "there is a standard", "for all standard".
^{b} A set x is well-founded iff its transitive closure has no infinite $\in$-decreasing chains.

Metamathematically, **HST** is equiconsistent with **ZFC**, and **HST** is a conservative extension of **ZFC** in the sense that any $\in$-formula Φ is a theorem of **ZFC** iff Φ^{st} (the relativization of Φ to $\mathbb{S}$) is a theorem of **HST**. See [17] on axioms, metamathematics, basic set theoretic structures, and the structure of hyperreals in the **HST** universe.

Convention 1.1. We argue in **HST** below unless otherwise stated. $\qquad\square$

Asterisks. An $\in$-isomorphism $x \mapsto {}^*x$ of $\mathbb{WF}$ onto $\mathbb{S}$ is defined in **HST** so that ${}^*x \cap \mathbb{S} = \{{}^*y : y \in x\}$ for all $x \in \mathbb{WF}$. The map $*$ is an elementary embedding of $\mathbb{WF}$ in $\mathbb{I}$ in the $\in$-language. The classes $\mathbb{S}$ and $\mathbb{WF}$ are $\in$-isomorphic and satisfy **ZFC**. Each of them can be informally identified with the conventional set theoretic universe. The class $\mathbb{WF}$ is somewhat more convenient in this role as it is transitive and contains all its subsets, hence some important set theoretic operations are absolute for $\mathbb{WF}$ in **HST**.

Integers and reals. The sets $\mathbb{N}, \mathbb{Q}, \mathbb{R}$ (integers, rationals, reals) belong to $\mathbb{WF}$ and are equal to resp. $(\mathbb{N})^{\mathbb{WF}}$ (*i.e.* $\mathbb{N}$ defined in $\mathbb{WF}$), $(\mathbb{Q})^{\mathbb{WF}}$, $(\mathbb{R})^{\mathbb{WF}}$. In addition ${}^*n = n$ for all $n \in \mathbb{N}$, therefore $\mathbb{N} \subseteq {}^*\mathbb{N}$, moreover $\mathbb{N}$ is an initial segment in ${}^*\mathbb{N}$. The set ${}^*\mathbb{N}$ coincides with the set $(\mathbb{N})^{\mathbb{I}}$ of all $\mathbb{I}$-*natural numbers*, similarly ${}^*\mathbb{Q}$ and ${}^*\mathbb{R}$ are equal to, resp., $(\mathbb{Q})^{\mathbb{I}}$ and $(\mathbb{R})^{\mathbb{I}}$. Elements of ${}^*\mathbb{N}, {}^*\mathbb{Q}, {}^*\mathbb{R}$ are often called resp. *hyperintegers*, *hyperrationals*, *hyperreals*.

A hyperreal $x \in {}^*\mathbb{R}$ is *infinitesimal*, $x \simeq 0$ in symbols, if $|x| < {}^*r$ in ${}^*\mathbb{R}$ for all $r \in \mathbb{R}$, $r > 0$, and *infinitely large*, if $x^{-1} \simeq 0$, *i.e.* $|x| > {}^*r$ for all $r \in \mathbb{R}$. A hyperreal x is *limited*, if it is not infinitely large. In this case there exists a unique $r \in \mathbb{R}$ such that $x \simeq {}^*r$ (that is, $x - {}^*r \simeq 0$). Such a real r is denoted by ${}^\circ x$ (the *shadow*, or standard part, of $x \in {}^*\mathbb{R}$).

Ordinals and cardinals. The operation $*$ extends to proper classes $X \subseteq \mathbb{WF}$ by ${}^*X = \bigcup_{x \in \mathbb{WF},\, x \subseteq X} {}^*x$, and this does not yield contradiction provided $X \in \mathbb{WF}$. Then ${}^*\mathbb{WF} = \mathbb{I}$. In **HST**, the classes $\mathtt{Card}$ and $\mathtt{Ord}$ (all cardinals, resp., ordinals) satisfy $\mathtt{Card} \subseteq \mathtt{Ord} \subseteq \mathbb{WF}$ and $\mathtt{Ord} = (\mathtt{Ord})^{\mathbb{WF}}$ (that is, ordinals = $\mathbb{WF}$-ordinals), $\mathtt{Card} = (\mathtt{Card})^{\mathbb{WF}}$. Thus classes ${}^*\mathtt{Card} \subseteq {}^*\mathtt{Ord} \subseteq \mathbb{I}$ are defined (all $\mathbb{I}$-cardinals, resp., $\mathbb{I}$-ordinals). Note that ${}^*\mathbb{N} \subseteq {}^*\mathtt{Card}$.

Sets of standard size. Sets equinumerous with sets in $\mathbb{WF}$ are called *sets of standard size*. Note that $\mathtt{card}\, X \in \mathtt{Card}$ is defined then for any set X of standard size. In **HST**, sets of standard size is the same as well-orderable sets, 1.3.1 in [17]. The axiom of **Saturation** claims that every non-empty $\cap$-closed set $X \subseteq \mathbb{I} \smallsetminus \{\varnothing\}$ of standard size has a non-empty intersection $\bigcap X$. The axiom of **Standard Size Choice** claims the existence of a choice function f for any set X of standard size (*i.e.* $f(x) \in x$ for all $x \in X$, $x \neq \varnothing$). An easy consequence is the axiom of **Power Set** for sets X of standard size: $\mathscr{P}(X)$ is a set of standard size for any such X. Finite sets are sets of standard size. On the other hand any infinite set $X \in \mathbb{I}$, for instance any set of the form $\{0, 1, 2, \ldots, h\}$, where $h \in {}^*\mathbb{N} \smallsetminus \mathbb{N}$, is not a set of standard size.

2. Classes $\mathbf{\Delta}_2^{\mathrm{ss}}$ and $\mathbb{L}[\mathbb{I}]$: effective sets

Which sets should be viewed as effective in **HST**? Following the examples of recursive, Borel, constructible sets, we have to choose an initial class of sets and a set of operations applying to the initial sets. The sets obtained this way are considered as effective. In nonstandard set theoretic systems, internal sets are usually considered as the initial sets, because of their special role in the construction of nonstandard universes. (In particular $\mathbb{I}$ is the von Neumann basis of the **HST** universe of sets.) As for the operations, let us take unions and intersections of families of standard size.

We immediately obtain the classes $\mathbf{\Sigma}_1^{\mathrm{ss}}$, $\mathbf{\Pi}_1^{\mathrm{ss}}$ of all sets of the form resp. $\bigcup_{a \in A} X_a$, $\bigcap_{a \in A} X_a$, where $A \in \mathsf{WF}$ and all sets X_a belong to $\mathbb{I}$, or, that is the same, of the form resp. $\bigcup \mathscr{X}$, $\bigcap \mathscr{X}$, where $\mathscr{X} \subseteq \mathbb{I}$ is a set of standard size. (The index ^{ss} indicates that unions and intersections of sets of standard size are taken.) We further define the class $\mathbf{\Delta}_2^{\mathrm{ss}}$ of all sets that can be represented both in the form $\bigcup_{a \in A} \bigcap_{b \in B} X_{ab}$, where $A, B \in \mathsf{WF}$ and all X_{ab} belong to $\mathbb{I}$, and in the dual form (possibly with different sets A, B, X_{ab}). Note that taking, say, three operations of union and intersection no new sets appear according to the following result (1.4.2, 1.4.3 in [17]).

Proposition 2.1. *If $\mathscr{X} \subseteq \mathbf{\Delta}_2^{\mathrm{ss}}$ is a set of standard size then the sets $\bigcup \mathscr{X}$ and $\bigcap \mathscr{X}$ belong to $\mathbf{\Delta}_2^{\mathrm{ss}}$. In addition, any set $X \subseteq \mathbb{I}$ defined in $\mathbb{I}$ by a st-$\in$-formula with sets in $\mathbb{I}$ as parameters belongs to $\mathbf{\Delta}_2^{\mathrm{ss}}$.* $\qquad\square$

Thus $\mathbf{\Delta}_2^{\mathrm{ss}}$ is a rather large class of sets. [c] Yet it consists only of those sets X satisfying $X \subseteq \mathbb{I}$. The class $\mathbb{L}[\mathbb{I}]$ of all sets *constructible over* $\mathbb{I}$ extends $\mathbf{\Delta}_2^{\mathrm{ss}}$ on higher levels of the von Neumann hierarchy over $\mathbb{I}$.

Definition 2.2. $\mathbb{L}[\mathbb{I}]$ consists of all sets x which admit a transfinite construction determined by a well-founded tree T with sets in $\mathbb{I}$ attached to all endpoints of T. The tree T itself and the map which attaches internal sets

---

[c] There are meaningful subclasses within $\mathbf{\Delta}_2^{\mathrm{ss}}$, namely *countably determined* sets, *i.e.* those of the form $X = \bigcup_{b \in B} \bigcap_{n \in b} X_n$, where $B \subseteq \mathscr{P}(\mathbb{N})$ and all sets X_n are internal (there are different but equivalent formulations), and *Borel* sets that belong to the closure of $\mathbb{I}$ under countable operations of $\bigcup$ and $\bigcap$. These classes are considered within model theoretic nonstandard analysis under the assumption of $\aleph_1$-Saturation, [19].

to the endpoints of T belong to $\mathbf{\Delta}_2^{\mathbf{ss}}$. In every node t of T that is not an endpoint, the set of all sets, attached to immediate successors of t in T is defined. The final set x is obtained in the root of T. $\qquad\square$

Thus sets in $\mathbb{L}[\mathbb{I}]$ are obtained via effectively coded (in $\mathbf{\Delta}_2^{\mathbf{ss}}$) transfinite iterations of the operation of assembling of a set from its elements. This enables us to view sets in $\mathbb{L}[\mathbb{I}]$ as effectively definable. Conversely, any effective (informally) set belongs to $\mathbb{L}[\mathbb{I}]$. Indeed it follows from theorem 2.3 (ii) below that effective constructions have to be absolute for $\mathbb{L}[\mathbb{I}]$, hence the results of such constructions are necessarily sets in $\mathbb{L}[\mathbb{I}]$.

Identifying the informal notion of effectivity in **HST** with $\mathbb{L}[\mathbb{I}]$, we put

$$\left. \begin{array}{l} x \leq^{\mathbf{eff}} y\,,\ \ \text{iff}\ \ \text{there is an injection}\ \ f \in \mathbb{L}[\mathbb{I}]\ \text{of}\ x\ \text{into}\ y\,; \\[4pt] x \equiv^{\mathbf{eff}} y\,,\ \ \text{iff}\ \ \text{there is a bijection}\ \ f \in \mathbb{L}[\mathbb{I}]\ \text{of}\ x\ \text{onto}\ y \end{array} \right\} . \qquad (1)$$

and $x <^{\mathbf{eff}} y$ iff $x \leq^{\mathbf{eff}} y$ but $y \not\leq^{\mathbf{eff}} x$. The ordinary Cantor – Bernstein argument proves $x \leq^{\mathbf{eff}} y \wedge y \leq^{\mathbf{eff}} x \iff x \equiv^{\mathbf{eff}} y$ for any sets $x, y \in \mathbb{L}[\mathbb{I}]$. Define $|x|^{\mathbf{eff}}$, *the effective cardinality* of $x \in \mathbb{L}[\mathbb{I}]$, to be the $\equiv^{\mathbf{eff}}$-equivalence class $\{y \in \mathbb{L}[\mathbb{I}] : x \equiv^{\mathbf{eff}} y\}$. The inequalities $|x|^{\mathbf{eff}} \leq |y|^{\mathbf{eff}}$ and $|x|^{\mathbf{eff}} < |y|^{\mathbf{eff}}$ will be understood as synonimous to resp. $x \leq^{\mathbf{eff}} y$ and $x <^{\mathbf{eff}} y$.

Theorem 2.3. (i) *If* $x \subseteq \mathbb{I}$ *then* $x \in \mathbf{\Delta}_2^{\mathbf{ss}} \iff x \in \mathbb{L}[\mathbb{I}]$.

(ii) $\mathbb{L}[\mathbb{I}]$ *is a transitive class satisfying* **HST** [d] *and* $\mathsf{WF} \cup \mathbf{\Delta}_2^{\mathbf{ss}} \subseteq \mathbb{L}[\mathbb{I}]$.

(iii) *For any set* $A \in \mathbb{L}[\mathbb{I}]$ *there is a set* $X \in \mathbb{I}$ *and an equivalence relation* E *on* X, $\mathsf{E} \in \mathbf{\Delta}_2^{\mathbf{ss}}$, *such that* $A \equiv^{\mathbf{eff}} X/\mathsf{E}$.

Proof. On (i), (ii) see 5.5.4 in [17] where the class $\mathbf{\Delta}_2^{\mathbf{ss}}$ is denoted by $\mathbb{E}$.

(iii) According to 5.5.4(8) in [17], there exist a set $X \in \mathbb{I}$ and a map $h \in \mathbb{L}[\mathbb{I}]$, $h : X \xrightarrow{\text{onto}} A$. Define, for $x, y \in X$, $x \mathsf{E} y$ iff $h(x) = h(y)$, and consider the map $a \in A \mapsto f(a) = \{x \in X : h(x) = a\}$. $\qquad\square$

Theorem 2.3 allows to suitably replace $\mathbb{L}[\mathbb{I}]$ by $\mathbf{\Delta}_2^{\mathbf{ss}}$ in the context of $|\cdot|^{\mathbf{eff}}$. For instance we conclude from 2.3(i) that (1) is equivalent to the following in the domain of subsets of $\mathbb{I}$:

$$\left. \begin{array}{l} x \leq^{\mathbf{eff}} y\,,\ \ \text{iff}\ \ \text{there is a}\ \mathbf{\Delta}_2^{\mathbf{ss}}\ \text{injection}\ f : x \to y \\[4pt] x \equiv^{\mathbf{eff}} y\,,\ \ \text{iff}\ \ \text{there is a}\ \mathbf{\Delta}_2^{\mathbf{ss}}\ \text{bijection}\ f : x \xrightarrow{\text{onto}} y \end{array} \right\}\ \text{for}\ x, y \subseteq \mathbb{I}. \qquad (2)$$

We begin the study of the structure of effective cardinalities $|\cdot|^{\mathbf{eff}}$ with rather simple classes, internal sets and sets of standard size.

---

[d] In fact the least class with these properties. A suitable version of Gödel's definition of relative constructibility leads to exactly the same class $\mathbb{L}[\mathbb{I}]$ in **HST**. See 5.5.6 in [17].

3. Effective cardinalities of internal sets

Generally elements of $^*\mathbf{Card}$, that is, $\mathbb{I}$-cardinals, behave like $\mathbf{ZFC}$ cardinals since $\mathbb{I}$ is a $\mathbf{ZFC}$ universe (in the $\in$-language). Let $|x|^{\mathtt{int}} \in {}^*\mathbf{Card}$ denote the $\mathbb{I}$-cardinality of a set $x \in \mathbb{I}$. Obviously $|x|^{\mathtt{int}} = |y|^{\mathtt{int}}$ implies $|x|^{\mathtt{eff}} = |y|^{\mathtt{eff}}$ since $\mathbb{I} \subseteq \boldsymbol{\Delta}_2^{\mathtt{ss}}$. This implication is partially reversible according to Corollary 3.2 below. To figure out the effect of non-internal maps in the domain of internal sets, let us give some definitions. Define, for any x,

$$\left.\begin{aligned}
\|x\|_* &= \{|y|^{\mathtt{int}} : x \supseteq y \in \mathbb{I}\} - \text{the } \textit{interior spectrum} \text{ of } x \\
\|x\|^* &= \{|y|^{\mathtt{int}} : x \subseteq y \in \mathbb{I}\} - \text{the } \textit{exterior spectrum} \text{ of } x
\end{aligned}\right\}. \tag{3}$$

Then $\|x\|_*$ is a $\textit{cut}$ (initial segment) in $^*\mathbf{Card}$ while $\|x\|^*$ is a proper class and a final segment in $^*\mathbf{Card}$. Further, for any $\kappa \in {}^*\mathbb{N}$ define the cuts

$$\kappa\mathbb{N} = \{\lambda \in {}^*\mathbb{N} : \exists\, n \in \mathbb{N}\, (\lambda < n\kappa)\}, \quad \kappa/\mathbb{N} = \{\lambda \in {}^*\mathbb{N} : \forall\, n \in \mathbb{N}\, (\lambda < \kappa/n)\}$$

in $^*\mathbb{N}$, and the $\textit{multiplicative galaxy}$ $\operatorname{gal}\kappa = \kappa\mathbb{N} \smallsetminus \kappa/\mathbb{N}$ of κ. Then $\lambda \in \operatorname{gal}\kappa$ iff neither of the fractions $\frac{\kappa}{\lambda}$, $\frac{\lambda}{\kappa}$ is infinitesimal. To preserve the unity of notation put $\kappa\mathbb{N} = \kappa$, $\operatorname{gal}\kappa = \{\kappa\}$ for any $\kappa \in {}^*\mathbf{Card} \smallsetminus {}^*\mathbb{N}$.

Define, for $K, L \subseteq {}^*\mathbf{Card}$, $K \leq L$ iff $\forall\, \kappa \in K\, \exists\, \lambda \in L\, (\kappa \leq \lambda)$. Accordingly, $K < L$ iff $K \leq L$ but $L \not\leq K$. In particular, in two cases when one of the sets K, L is a singleton, we obtain

$$\kappa \leq L \ \text{ iff } \ \exists\, \lambda \in L\, (\kappa \leq \lambda), \quad \text{and} \quad K < \lambda \ \text{ iff } \ \forall\, \kappa \in K\, (\kappa < \lambda). \tag{4}$$

Note that galaxies are pairwise disjoint intervals in $^*\mathbf{Card}$ (singletons outside of $^*\mathbb{N}$), thus for any two galaxies Γ_1, Γ_2, $\Gamma_1 < \Gamma_2$ means that $\kappa_1 < \kappa_2$ for any (equivalently, for all) $\kappa_1 \in \Gamma_1$, $\kappa_2 \in \Gamma_2$.

See 1.4.9 and 9.6.12 in [17], or [19], on the next theorem. In the case of $\mathbb{I}$-infinite sets the factors $\mathbb{N}$ and h in 3.1 vanish by obvious reasons.

Theorem 3.1. (i) $\textit{Suppose that } X, Y \in \mathbb{I} \textit{ and } f : X \to Y \textit{ is a } \boldsymbol{\Delta}_2^{\mathtt{ss}} \textit{ map.}$ $\textit{Then } |X|^{\mathtt{int}}h \in \|\operatorname{ran} f\|^* \textit{ for any } h \in {}^*\mathbb{N} \smallsetminus \mathbb{N}. \textit{ In addition, } (a) \textit{ if } \operatorname{ran} f = Y$ $\textit{then } |Y|^{\mathtt{int}} \leq |X|^{\mathtt{int}}\mathbb{N}, \textit{ and } (b) \textit{ if } f \textit{ is an injection then } |X|^{\mathtt{int}} \leq |Y|^{\mathtt{int}}\mathbb{N}.$

(ii) $\textit{Suppose that } X \in \mathbb{I} \textit{ is infinite. Then } |X|^{\mathtt{eff}} = |X \times \mathbb{N}|^{\mathtt{eff}}, \textit{ in partic-}$ $\textit{ular, } |Y|^{\mathtt{eff}} \leq |X|^{\mathtt{eff}} \textit{ for any internal } Y \textit{ with } |Y|^{\mathtt{int}} \leq |X|^{\mathtt{int}}\mathbb{N}. \qquad \square$

Corollary 3.2. $\textit{If } x, y \in \mathbb{I} \textit{ then } |x|^{\mathtt{eff}} \leq |y|^{\mathtt{eff}} \textit{ is equivalent to } |x|^{\mathtt{int}} \leq$ $|y|^{\mathtt{int}}\mathbb{N} \textit{ provided } |y|^{\mathtt{int}} \in {}^*\mathbb{N} \smallsetminus \mathbb{N} \textit{ and to just } |x|^{\mathtt{int}} \leq |y|^{\mathtt{int}} \textit{ otherwise.} \qquad \square$

Thus $|x|^{\mathtt{eff}} = |y|^{\mathtt{eff}}$ is equivalent to $\operatorname{gal}|x|^{\mathtt{int}} = \operatorname{gal}|y|^{\mathtt{int}}$ in the domain $^*\mathbb{N} \smallsetminus \mathbb{N}$, and equivalent to just $|x|^{\mathtt{int}} = |y|^{\mathtt{int}}$ outside of the domain $^*\mathbb{N} \smallsetminus \mathbb{N}$. In the $\mathbb{I}$-infinite domain $^*\mathbf{Card} \smallsetminus {}^*\mathbb{N}$, the two characterizations coincide.

4. Effective cardinalities of sets of standard size

By definition sets of standard size, or *s. s. sets*, are those equinumerous (that is, admit a bijection onto) with sets in $\mathbb{WF}$. For any s.s. set X define $\operatorname{card} X = \operatorname{card} W \in \mathbf{Card}$, where W is a set in $\mathbb{WF}$ equinumerous with X.

Lemma 4.1. (i) *Any s. s. set $X \subseteq \mathbb{I}$ is $\boldsymbol{\Sigma}_1^{\mathrm{ss}}$ and $\boldsymbol{\Delta}_2^{\mathrm{ss}}$.*

(ii) *Any s. s. set W is equinumerous with an s. s. set $X \subseteq \mathbb{I}$.*

(iii) *If $X \subseteq \mathbb{I}$ is a s. s. set then ${}^*\mathbf{Card} \smallsetminus \mathbb{N} \subseteq \|X\|^*$ and $\|X\|^* \subseteq \mathbb{N}$.*

(iv) *If $X, Y \subseteq \mathbb{I}$ are s. s. sets then $\operatorname{card} X = \operatorname{card} Y$ iff $|X|^{\mathrm{eff}} = |Y|^{\mathrm{eff}}$, thus $|X|^{\mathrm{eff}}$ can be identified with $\operatorname{card} X$.*

Proof. (ii) We may assume that $W \in \mathbb{WF}$. Then the map $w \mapsto {}^*w$ is a bijection of W onto $X = \{{}^*w : w \in W\}$ and X is a set of standard size, too.

(iii) To prove ${}^*\mathbf{Card} \smallsetminus \mathbb{N} \subseteq \|X\|^*$ fix $h \in {}^*\mathbb{N} \smallsetminus \mathbb{N}$ and apply **Saturation** to the family of all sets $C_u = \{c \in \mathbb{I} : u \subseteq c \wedge |c|^{\mathrm{int}} = h\}$, where $u \subseteq X$ is finite.

(iv) Any bijection f between two sets $X, Y \subseteq \mathbb{I}$ of standard size is itself a set of standard size, then apply (i). $\qquad\square$

It follows that s. s. sets are adequately represented among $\boldsymbol{\Sigma}_1^{\mathrm{ss}}$ sets in the context of $\mathbf{card}$, and on the other hand $|\cdot|^{\mathrm{eff}}$ and $\mathbf{card}$ coincide on s. s. sets. The next theorem shows that effective cardinalities of $\boldsymbol{\Delta}_2^{\mathrm{ss}}$ sets begin with sets of standard size, where they coincide with well-founded cardinals, followed by the domain of $\boldsymbol{\Delta}_2^{\mathrm{ss}}$ sets not of standard size. It will be demonstrated below that the structure of effective cardinalities in the latter is connected with ${}^*\mathbf{Card}$ in a certain way.

Theorem 4.2. (i) *Infinite internal sets are not s. s. sets.*

(ii) *Any $\boldsymbol{\Delta}_2^{\mathrm{ss}}$ set X not of standard size contains an infinite internal subset, that is formally $\mathbb{N} \subsetneqq \|X\|_*$.*

(iii) *If X a s. s. set and Y is a $\boldsymbol{\Delta}_2^{\mathrm{ss}}$ but not s. s. set then $|X|^{\mathrm{eff}} < |Y|^{\mathrm{eff}}$.*

Proof. (i) A simple corollary of Lemma 4.1(iii).

(ii) By definition $\boldsymbol{\Delta}_2^{\mathrm{ss}}$ sets are s. s. unions of $\boldsymbol{\Pi}_1^{\mathrm{ss}}$ sets. Yet it is another rather simple corollary of **Saturation** that any infinite $\boldsymbol{\Pi}_1^{\mathrm{ss}}$ set contains an infinite internal subset, see 1.4.11 in [17].

(iii) By (ii) some number $h \in {}^*\mathbb{N} \smallsetminus \mathbb{N}$ belongs to $\|Y\|_*$. On the other hand $h \in \|X\|^*$ by Lemma 4.1(iii). This implies $|X|^{\mathrm{eff}} \leq |Y|^{\mathrm{eff}}$. The inequality $|Y|^{\mathrm{eff}} \not\leq |X|^{\mathrm{eff}}$ follows from (i). $\qquad\square$

5. Exteriors and interiors

It turns out that internal approximations $\|X\|_*$, $\|X\|^*$ are very instrumental in the study of effective cardinalities of $\boldsymbol{\Sigma}_1^{ss}$ and partly $\boldsymbol{\Pi}_1^{ss}$ sets X. Now a few words on cuts (initial segments) in ${}^*\mathbf{Card}$.

Definition 5.1. A cut $U \subseteq {}^*\mathbf{Card}$ is *standard size (s. s.) cofinal* resp. *coinitial*, iff there exist a cardinal $\vartheta \in \mathbf{Card}$, infinite or equal to $1 = \{0\}$, and an increasing, resp. decresing sequence $\{\nu_\xi\}_{\xi<\vartheta}$, of $\nu_\xi \in {}^*\mathbf{Card}$ such that $U = \bigcup_{\xi<\vartheta}\{\kappa \in {}^*\mathbf{Card} : \kappa < \nu_\xi\}$, resp., $U = \bigcap_{\xi<\vartheta}\{\kappa \in {}^*\mathbf{Card} : \kappa < \nu_\xi\}$. $\square$

Note that s.s. cofinal cuts are $\boldsymbol{\Sigma}_1^{ss}$ while s.s. coinitial cuts are $\boldsymbol{\Pi}_1^{ss}$. Internal cuts, *i.e.* those of the form $U = \{\kappa \in {}^*\mathbf{Card} : \kappa < \nu\}$, $\nu \in {}^*\mathbf{Card}$, belong to either of the two "standard size" categories, for take $\vartheta = 1$ and $\nu_0 = \nu$. See 1.4b in [17] on the next result:

Proposition 5.2. *Any $\boldsymbol{\Delta}_2^{ss}$ cut in ${}^*\mathbf{Card}$ is s. s. cofinal or s. s. coinitial. If a cut is both s. s. cofinal and s. s. coinitial then it is internal.* $\square$

Coming back to $\|X\|_*$ and $\|X\|^*$, note that for any X the intersection $\|X\|_* \cap \|X\|^*$ contains at most one element. If $\kappa \in \|X\|_* \cap \|X\|^*$ then there exist internal sets Y, Z with $Y \subseteq X \subseteq Z$ and $|Y|^{int} = |Z|^{int} = \kappa$. In this case, if $\kappa \in {}^*\mathbb{N}$ then X itself is internal with $|X|^{int} = \kappa$, while if κ is $\mathbb{I}$-infinite then only $|X|^{eff} = |\kappa|^{eff}$ holds provided X is $\boldsymbol{\Delta}_2^{ss}$.

Lemma 5.3. (i) *If X is a set in $\boldsymbol{\Sigma}_1^{ss}$, resp., $\boldsymbol{\Pi}_1^{ss}$ then $\|X\|_*$ is a standard size cofinal, resp., standard size coinitial cut in ${}^*\mathbf{Card}$.*

(ii) *In both cases, $\|X\|_* \cup \|X\|^* = {}^*\mathbf{Card}$.*

(iii) *In both cases, if either $\|X\|_*$ contains a largest element κ, or $\|X\|^*$ contains a least element κ, then $\kappa \in \|X\|_* \cap \|X\|^*$.*

Proof. (i) Consider a set $\mathscr{X} \subseteq \mathbb{I}$ of standard size. Let $X = \bigcup \mathscr{X}$. Then by **Saturation** any internal set $Y \subseteq X$ is covered by a set of the form $\bigcup \mathscr{X}'$ where $\mathscr{X}' \subseteq \mathscr{X}$ is finite. On the other hand, by 1.3.3 in [17] the set $\mathscr{P}_{fin}(\mathscr{X}) = \{\mathscr{X}' \subseteq \mathscr{X} : \mathscr{X}'$ is finite$\}$ is still a set of standard size.

Prove (ii) for $\boldsymbol{\Sigma}_1^{ss}$. Let $X = \bigcup \mathscr{X}$ be as above. Show that any $\mathbb{I}$-cardinal $\kappa \notin \|X\|_*$ belongs to $\|X\|^*$. Take any set $Z \in \mathbb{I}$ such that $X \subseteq Z$. If $\mathscr{X}' \subseteq \mathscr{X}$ is finite then by definition $\bigcup \mathscr{X}'$ is covered by an internal set of $\mathbb{I}$-cardinality κ, hence the set $P_{\mathscr{X}'} = \{C \in \mathbb{I} : \bigcup \mathscr{X}' \subseteq C \subseteq Z \wedge |C|^{int} \leq \kappa\} \in \mathbb{I}$ is non-empty. Apply **Saturation** to the family of all these sets $P_{\mathscr{X}'}$.

(iii) Apply **Saturation**. $\square$

Example 5.4. The following example [e] of a $\mathbf{\Delta}_2^{\mathrm{ss}}$ set X such that $\|X\|_* \cup \|X\|^* \subsetneqq {}^*\mathsf{Card}$ employs a nontrivial ultrafilter $U \in \mathsf{WF}$ over $\mathbb{N}$. Let $h \in {}^*\mathbb{N} \smallsetminus \mathbb{N}$ and $D = \{1, 2, \ldots, h\}$. The set $P = \mathscr{P}^{\mathbb{I}}(D) = \mathscr{P}(D) \cap \mathbb{I}$ of all internal sets $x \subseteq D$ belongs to $\mathbb{I}$ and satisfies $|P|^{\mathrm{int}} = 2^h$. Then

$$U' = \{x \in P : x \cap \mathbb{N} \in U\} = \bigcup_{b \in U} \bigcap_{n \in b} \{x \in P : n \in x\}$$

is an ultrafilter in P and a $\mathbf{\Delta}_2^{\mathrm{ss}}$ set. [f] *We claim that* $\|U'\|_* = 2^h/\mathbb{N}$.

Let $Z' \subseteq P$ be an internal set. By Saturation (see e.g. 9.2.15 in [17] or 1.6 in [19]), $Z = \{x \cap \mathbb{N} : x \in Z'\}$ is a closed subset of U. It follows that the Lebesgue measure of Z in $\mathscr{P}(\mathbb{N})$ (identified with $2^{\mathbb{N}}$) is 0. Then easily the Loeb measure of Z in $\mathscr{P}^{\mathbb{I}}(D)$ is 0, so that $|Z'|^{\mathrm{int}} \in 2^h/\mathbb{N}$. Thus $\|U'\|_* \subseteq 2^h/\mathbb{N}$. To prove the converse note that for any $u \in U$ the set $X = \{x \in P : x \cap \mathbb{N} = u\}$ is a $\mathbf{\Pi}_1^{\mathrm{ss}}$ subset of U' that surely satisfies $\|X\|_* = 2^h/\mathbb{N}$.

It follows from $\|U'\|_* = 2^h/\mathbb{N}$ that $\|U'\|^* = 2^h$ — by the symmetry of the sets U' and $P \smallsetminus U' = \{D \smallsetminus x : x \in P\}$ within P. $\qquad\square$

One can easily transform the set U' as in 5.4 to a $\mathbf{\Delta}_2^{\mathrm{ss}}$ set $X \subseteq {}^*\mathbb{N}$ such that $\|X\|_* = 2^h/\mathbb{N}$ and $\|X\|^* = 2^h\mathbb{N}$. The gap ${}^*\mathsf{Card} \smallsetminus (\|X\|^* \cup \|X\|_*)$ consists, in this case, of the whole galaxy $\mathsf{gal}\, 2^h = 2^h\mathbb{N} \smallsetminus 2^h/\mathbb{N}$ in ${}^*\mathbb{N}$. The next theorem shows that this is a maximal possible gap!

Theorem 5.5. *If X is $\mathbf{\Delta}_2^{\mathrm{ss}}$ and $\kappa \in {}^*\mathsf{Card}$, $\kappa \notin \|X\|^* \cup \|X\|_*$, then $\kappa \in {}^*\mathbb{N}$ and the difference ${}^*\mathsf{Card} \smallsetminus (\|X\|^* \cup \|X\|_*)$ is a subset of $\mathsf{gal}\,\kappa$.*

Thus if X is $\mathbf{\Delta}_2^{\mathrm{ss}}$ and ${}^*\mathbb{N} \subseteq \|X\|_*$ then $\|X\|_* \cup \|X\|^* = {}^*\mathsf{Card}$.

Proof. By definition $X = \bigcup_{a \in A} X_a$ where $A \in \mathsf{WF}$ and every X_a is a $\mathbf{\Pi}_1^{\mathrm{ss}}$ set. Take any $\mathbb{I}$-cardinal $\kappa \in \|X\|^* \smallsetminus \|X\|_*$. Obviously $\bigcup_{a \in A} \|X_a\|_* \subseteq \|X\|_*$, thus $\kappa \in \bigcap_{a \in A} \|X_a\|^*$ by Lemma 5.3. It suffices to prove that any $\lambda \in {}^*\mathsf{Card}$ belongs to $\|X\|^*$ in either of the two cases: 1) $\lambda = \kappa \notin {}^*\mathbb{N}$, 2) $\lambda \in {}^*\mathbb{N} \smallsetminus \kappa\mathbb{N}$. Note that $n\kappa \leq \lambda$ holds for all $n \in \mathbb{N}$ in both cases.

Using Standard Size Choice, choose, for any $a \in A$, a set $Y_a \in \mathbb{I}$ such that $X_a \subseteq Y_a$ and $|Y_a|^{\mathrm{int}} = \kappa$. Thus X is covered by the union $\bigcup_{a \in A} Y_a$. For any finite $A' \subseteq A$, the finite union $Y_{A'} = \bigcup_{a \in A'} Y_a$ is an internal set satisfying $|Y_{A'}|^{\mathrm{int}} \leq \lambda$ by the above. The same application of Saturation as in the proof of Lemma 5.3 yields an internal set Y still with $|Y|^{\mathrm{int}} \leq \lambda$, satisfying $\bigcup_{a \in A} Y_a \subseteq Y$, and hence $X \subseteq Y$ and $\lambda \in {}^*\mathsf{Card}$. $\qquad\square$

---

[e] Essentially given in [24], see also [19], p. 1172, but with a more complicated proof based on a rather nontrivial combinatorial theorem in [4].

[f] The set U' is even countably determined.

The following corollary belongs to the "small–large dichotomy" type. (B) witnesses that a given $\boldsymbol{\Delta}_2^{ss}$ set is rather large w.r.t. a given cut U (has rather large internal subsets), while (A1) and (A2) witness that X is rather small (can be covered by rather small internal sets). The proof is easy: if $U \subsetneqq \|X\|_*$ then (B) holds by definition, otherwise apply Theorem 5.5 and get (A1) or (A2) (or Lemma 5.3(ii) – in the case of $\boldsymbol{\Sigma}_1^{ss}$ and $\boldsymbol{\Pi}_1^{ss}$ sets).

Corollary 5.6. *If $X \subseteq \mathbb{I}$ is a $\boldsymbol{\Delta}_2^{ss}$ set and $U \subseteq {}^*\mathbf{Card}$ is a $\boldsymbol{\Delta}_2^{ss}$ cut then at least one of the following conditions holds, and moreover (A2) can be excluded for $\boldsymbol{\Sigma}_1^{ss}$ and $\boldsymbol{\Pi}_1^{ss}$ sets X:*

(A1) *for any $\kappa \notin U$ there is an internal set $Y \supseteq X$ such that $|Y|^{int} = \kappa$;*

(A2) *there exists $h \in {}^*\mathbb{N} \smallsetminus \mathbb{N}$ such that $h/\mathbb{N} \subseteq U \subseteq h\mathbb{N}$, and for any $\kappa \in {}^*\mathbf{Card} \smallsetminus h\mathbb{N}$ there exists an internal set $Y \supseteq X$ such that $|Y|^{int} = \kappa$;*

(B) *there exists an internal set $Y \subseteq X$ such that $|Y|^{int} \notin U$.* $\qquad\square$

6. Effective cardinalities of $\boldsymbol{\Sigma}_1^{ss}$ sets

One may expect that the bigger $\|X\|_*$ (or the smaller $\|X\|^*$) is the bigger $|X|^{eff}$ should be. According to the next theorem, such a connection holds for $\boldsymbol{\Sigma}_1^{ss}$ sets X except those satisfying $\|X\|_* \subseteq \mathbb{N}$.

Following the notation in Section 3, we define, for any $K \subseteq {}^*\mathbf{Card}$, a cut $K\mathbb{N} = \{\lambda : \exists\, \kappa \in K \; \exists\, n \in \mathbb{N} \,(\lambda \leq n\kappa)\}$ in ${}^*\mathbf{Card}$.

Theorem 6.1. *If X, Y are $\boldsymbol{\Sigma}_1^{ss}$ sets and $\mathbb{N} \subsetneqq \|Y\|_*$ then $|X|^{eff} \leq |Y|^{eff}$ is equivalent to $\|X\|_* \subseteq \|Y\|_*\mathbb{N}$, and also to $\|X\|_* \subseteq \|Y\|_*$ if ${}^*\mathbb{N} \subseteq \|Y\|_*$.*

The case $\|Y\|_* \subseteq \mathbb{N}$ will be considered below.

Proof. Suppose that $X = \bigcup \mathscr{X}$ and $Y = \bigcup \mathscr{Y}$, where $\mathscr{X}, \mathscr{Y} \subseteq \mathbb{I}$ are sets of standard size. There is a set $D \in \mathbb{I}$ such that $X \cup Y \subseteq D$. Assume w.l.o.g. that $\mathscr{X}, \mathscr{Y}$ are $\cap$-closed families. By **Saturation**, the sets of $\mathbb{I}$-cardinals $\{|X'|^{int} : X' \in \mathscr{X}\}$, $\{|Y'|^{int} : Y' \in \mathscr{Y}\}$ are cofinal in resp. $\|X\|_*$, $\|Y\|_*$.

Direction $\Longrightarrow$. Suppose otherwise. Then there is an internal set $X' \subseteq X$ such that $|Y'|^{int}m < |X'|^{int}/n$ for any internal $Y' \subseteq Y$ and $k, n \in \mathbb{N}$. As $\|Y\|_*$ is a s.s. cofinal cut in ${}^*\mathbf{Card}$ by Lemma 5.3(i), there exists, by **Saturation**, $\kappa \in {}^*\mathbf{Card}$ such that $|Y'|^{int}m < \kappa < |X'|^{int}/n$ for any internal $Y' \subseteq Y$ and $k, n \in \mathbb{N}$. Thus $\|Y\|_*\mathbb{N} < \kappa$, hence $\kappa \in \|Y\|^*$ by Lemma 5.3(ii). In other words, there is an internal set Z such that $Y \subseteq Z$ and $|Z|^{int} = \kappa$. On the other hand, $\kappa\mathbb{N} < |X'|^{int}$ while by $|X|^{eff} \leq |Y|^{eff}$ there exists a $\boldsymbol{\Delta}_2^{ss}$ injection $X' \to Z$, a contradiction to Theorem 3.1(i).

124

Direction $\Longleftarrow$, *in a stronger assumption that simply* $\|X\|_* \subseteq \|Y\|_*$.

Case 1: $\|Y\|_*$ contains a maximal element $\kappa = |Y_0|^{\mathtt{int}}$, where $Y_0 \in \mathscr{Y}$, hence $Y_0 \subseteq Y$. Then for any $X' \in \mathscr{X}$ the set

$$H_{X'} = \{h \in \mathbb{I} : h : D \to D \wedge h \restriction X' \text{ is an injection } \wedge h"X' \subseteq Y_0\}$$

is non-empty. In addition, $H_{X'' \cup X'} = H_{X''} \cap H_{X'}$. Saturation yields an element $h \in \bigcap_{X' \in \mathscr{X}} H_{X'}$. Clearly $h \restriction X$ is an injection of X into Y_0, and hence $|X|^{\mathtt{eff}} \leq |Y|^{\mathtt{eff}}$, as required.

Case 2: $\|Y\|_*$ does not contain a maximal element, and for every $\alpha \in \|Y\|_* \cap {}^*\mathbb{N}$ there exists $\gamma \in \|Y\|_* \cap {}^*\mathbb{N}$ such that $\alpha\mathbb{N} < \gamma$ – meaning that $\gamma > \alpha n$ for any $n \in \mathbb{N}$. By **Standard Size Choice** there is a map $f : \mathscr{X} \to \mathscr{Y}$ such that $|X'|^{\mathtt{int}} < |f(X')|^{\mathtt{int}}$ for all $X' \in \mathscr{X}$, and even $|X'|^{\mathtt{int}}\mathbb{N} < |f(X')|^{\mathtt{int}}$ provided $|f(X')|^{\mathtt{int}}$ (then also $|X'|^{\mathtt{int}}$) belongs to ${}^*\mathbb{N}$. Then

$$H_{X'} = \{h \in \mathbb{I} : h : D \to D \wedge h \restriction X' \text{ is an injection } \wedge h"X' \subseteq f(X')\}$$

is non-empty for any $X' \in \mathscr{X}$. Then argue as in Case 1.

Case 3: the negation of cases 1, 2. Then there is a number $c \in {}^*\mathbb{N} \smallsetminus \mathbb{N}$ such that $c \in \|Y\|_*$ but $2c \notin \|Y\|_*$. Then $|[0,c)|^{\mathtt{eff}} \leq |Y|^{\mathtt{eff}}$ while $|X|^{\mathtt{eff}} \leq |[0,2c)|^{\mathtt{eff}}$ (see case 1). However $|[0,2c)|^{\mathtt{eff}} = |[0,c)|^{\mathtt{eff}}$ by Corollary 3.2.

Direction $\Longleftarrow$, *general case*. If $\|X\|_* \subseteq \|Y\|_*\mathbb{N}$, but $\|X\|_* \subseteq \|Y\|_*$ does **not** hold then there exist numbers $c \in {}^*\mathbb{N} \smallsetminus \mathbb{N}$ and $n \in \mathbb{N}$ such that $\|X\|_* \subseteq [0,nc)$ and $[0,c) \subseteq \|Y\|_* \subseteq [0,2c)$. We have $|X|^{\mathtt{eff}} \leq |[0,nc)|^{\mathtt{eff}}$ by the above, and $|[0,c)|^{\mathtt{eff}} \leq |Y|^{\mathtt{eff}}$. It remains to apply Corollary 3.2. $\square$

It remains to consider the case $\|Y\|_* \subseteq \mathbb{N}$ avoided in the theorem. It leads to sets of standard size!

Lemma 6.2. *For a set* $X \subseteq \mathbb{I}$ *to be of standard size each of the conditions* $\|X\|_* \subseteq \mathbb{N}$, ${}^*\mathbb{N} \smallsetminus \mathbb{N} \subseteq \|X\|^*$ *is necessary and, if* X *is* $\mathbf{\Delta}_2^{\mathtt{ss}}$, *also sufficient.*

Proof. By Theorem 4.2(i) $\|X\|_* \subseteq \mathbb{N}$. On the other hand ${}^*\mathbb{N} \smallsetminus \mathbb{N} \subseteq \|X\|^*$ by Lemma 4.1(iii). The sufficiency follows from Theorem 4.2(ii). $\square$

Thus Theorem 6.1 *fails* in the case $\|X\|_* = \mathbb{N}$: take any pair of infinite sets $X, Y \subseteq \mathbb{I}$ of standard size with $\mathbf{card}\, X \neq \mathbf{card}\, Y$ and apply 6.2 to show that $\|X\|^* = \|Y\|^* = \mathbb{N}$, and Lemma 4.1 to show that $|X|^{\mathtt{eff}} \neq |Y|^{\mathtt{eff}}$. Nevertheless we easily obtain the following corollary.

Corollary 6.3. *If* X, Y *are* $\mathbf{\Sigma}_1^{\mathtt{ss}}$ *sets then their effective cardinalities are comparable in the sense that at least one of the following inequalities holds:* $|X|^{\mathtt{eff}} \leq |Y|^{\mathtt{eff}}$ *or* $|Y|^{\mathtt{eff}} \leq |X|^{\mathtt{eff}}$. $\square$

7. Effective cardinalities of Π_1^{ss} sets

The proof of $\Longrightarrow$ in Theorem 6.1 does not work for Π_1^{ss} sets since $\|Y\|_*$ is now s. s. coinitial and the **Saturation** argument does not work. On the other hand there is a suitable counterexample.

Example 7.1. Fix $h \in {}^*\mathbb{N} \smallsetminus \mathbb{N}$ and let S be the set of all internal maps $s : \{0, 1, 2, \ldots, h - 1, h\} \to \{0, 1\} = 2$. Define $a_s, b_s \in 2^\mathbb{N}$ (hence $\in \mathsf{WF}$) so that $a_s(k) = s(k)$ and $b_s(k) = s(h - k)$ for all $k \in \mathbb{N}$. For $a, b \in 2^\mathbb{N}$ put $S_{ab} = \{s : a_s = a \wedge b_s = b\}$ and $S_a = \{s : a_s = a\}$. Then S is internal, $|S|^{\mathrm{int}} = 2^{h+1}$, while each S_a is a Π_1^{ss} set with $\|S_a\|_* = 2^h/\mathbb{N}$. Obviously $(2^h/\mathbb{N})\mathbb{N} = (2^h/\mathbb{N})$. To see that S and S_a lead to a counterexample to $\Longrightarrow$ of Theorem 6.1, it suffices to prove that $|S|^{\mathrm{eff}} = |S_a|^{\mathrm{eff}}$ for some a.

Since either of S, S_a is a union of $2^\mathbb{N}$-many sets of the form S_{ab}, it remains to show that $|S_{ab}|^{\mathrm{eff}} = |S_{a'b'}|^{\mathrm{eff}}$ for all a, b, a', b'. By **Saturation** there is $\sigma \in S$ such that $a(n) = a'(n) \oplus \sigma(n)$ and $b(n) = b'(n) \oplus \sigma(h - n)$ for all $n \in \mathbb{N}$, where $\oplus$ is addition modulo 2. Finally the internal map $s \mapsto s \oplus \sigma$ (in the termwise sense) easily maps S_{ab} onto $S_{a'b'}$ in 1-1 way. $\qquad\square$

In fact 7.1 is the only possible counterexamle for Π_1^{ss} sets in the following sense: if X, Y are Π_1^{ss} sets, $\mathbb{N} \subsetneqq \|Y\|_*$, and $|X|^{\mathrm{eff}} \leq |Y|^{\mathrm{eff}}$ then either $\|X\|_* \subseteq \|Y\|_*\mathbb{N}$ or there is a number $\kappa \in \|Y\|_*$, $\kappa \in {}^*\mathbb{N} \smallsetminus \mathbb{N}$, such that $\|X\|_* = \kappa/\mathbb{N}$ while $\|Y\|_* \subseteq \kappa\mathbb{N}$. We skip the proof.

Our further goal is to present what looks like a near-counterexample, (ii) of Theorem 7.2, to $\Longleftarrow$ of Theorem 6.1 in the field of Π_1^{ss} sets.

If X is a Π_1^{ss} set then $\|X\|^*$ is standard size coinitial by Lemma 5.3. If $\|X\|^*$ contains a least element κ then κ is simultaneously the largest element in $\|X\|_*$ still by Lemma 5.3, and then easily $|X|^{\mathrm{eff}} = |\kappa|^{\mathrm{eff}}$. It follows that if in this case Y is another Π_1^{ss} set with $\|Y\|^* = \|X\|^*$ then $|X|^{\mathrm{eff}} = |Y|^{\mathrm{eff}}$. But if $\|X\|^*$ does not contain a least element then there is an infinite coinitial sequence with standard size many terms. This case is considered by the next theorem. It follows from (ii) that there are sets of the largest effective cardinality among all Π_1^{ss} sets X with the same $\|X\|^*$, while (iii) presents a rather nontrivial partial counterexample to Theorem 6.1 for Π_1^{ss} sets. We deal with $\mathbb{I}$-infinite cardinals here, but similar results can be obtained in the hyperfinite domain — we leave it to the reader.

Theorem 7.2. (i) *If* $\mathscr{X}, \mathscr{Y} \subseteq \mathbb{I}$ *are sets of standard size,* $X = \bigcap \mathscr{X}$, $Y = \bigcap \mathscr{Y}$, $\vartheta = \mathsf{card}\, \mathscr{X} \in \mathsf{WF}$ *is an infinite regular cardinal,* $\|X\|^* = \|Y\|^*$, *and the coinitiality of* $\|X\|^*$ *is exactly* ϑ, *then* $|Y|^{\mathrm{eff}} \leq |X|^{\mathrm{eff}}$.

(ii) *There exist* $\mathbf{\Pi}_1^{\mathrm{ss}}$ *sets* X, Y *as in* (ii) *such that* $|X|^{\mathrm{eff}} \leq |Y|^{\mathrm{eff}}$ *fails via* $\mathbf{\Delta}_2^{\mathrm{ss}}$ *injections* g *of the form* $g = \bigcup_{w \in W} \bigcap_{\xi < \vartheta} g_{w\xi}$, *where all* $g_{w\xi}$ *are internal and* W *is a set of standard size.*

Proof. (i) Assume w. l. o. g. that there exist sets $X_0 \in \mathscr{X}$, $Y_0 \in \mathscr{Y}$ such that $X \subseteq X_0$ and $Y \subseteq Y_0$ for all $X \in \mathscr{X}$, $Y \in \mathscr{Y}$, and the families $\mathscr{X}, \mathscr{Y}$ are $\cap$-closed. We claim that there exists a function $\psi : \mathscr{X} \to \mathscr{Y}$ satisfying

$$\forall A \in \mathscr{P}_{\mathrm{fin}}(\mathscr{X}) \, \exists f \in F \, \forall X \in A \, (\psi(X) \subseteq \operatorname{dom} f \wedge f"\psi(X) \subseteq X), \qquad (5)$$

where $F \in \mathbb{I}$ is the set of all 1–1 functions $f \in \mathbb{I}$ with $\operatorname{dom} f \subseteq Y_0$ and $\operatorname{ran} f \subseteq X_0$. To define ψ fix an enumeration $\mathscr{X} = \{X_\alpha : \alpha < \vartheta\}$. Suppose that $\alpha' < \vartheta$, and the values $\psi(X_\alpha) \in \mathscr{Y}$, $\alpha < \alpha'$, have been defined. In our assumptions, there is a set $Y \in \mathscr{Y}$ such that $|Y|^{\mathrm{int}} < |\bigcap_{\alpha \in A} X_\alpha|^{\mathrm{int}}$ for every finite $A \subseteq [0, \alpha']$. To complete the inductive step put $\psi(X_{\alpha'}) = Y$.

To prove (5) consider a finite set $A = \{\alpha_1 < \cdots < \alpha_n\} \subseteq \vartheta$. By the construction $|\psi(X_{\alpha_k})|^{\mathrm{int}} < |\bigcap_{1 \leq i \leq k} X_{\alpha_k}|^{\mathrm{int}}$ for all $k = 1, \ldots, n$. Arguing in $\mathbb{I}$, we easily find a map $f \in F$ such that $f"(\psi(X_{\alpha_k})) \subseteq \bigcap_{1 \leq i \leq k} X_{\alpha_k}$ for every $k = 1, \ldots, n$, hence (5) holds.

Yet by **Saturation** (5) is equivalent to the following:

$$\exists f \in F \, \forall X \in \mathscr{X} \, (\psi(X) \subseteq \operatorname{dom} f \wedge f"(\psi(X)) \subseteq X). \qquad (6)$$

Thus $f"Y \subseteq X$, for such an f, and hence $|Y|^{\mathrm{eff}} \leq |X|^{\mathrm{eff}}$ holds even by means of an internal map f.

(ii) Fix an infinite cardinal ϑ in WF. It easily follows from **Saturation** that there exists a strictly decreasing sequence $\vec{\nu} = \{\nu_\xi\}_{\xi < \vartheta}$ of $\mathbb{I}$-cardinals $\nu_\xi \in {}^*\mathsf{Card} \setminus {}^*\mathbb{N}$. A $\vec{\nu}$-*large set* will be any $X \in \mathbb{I}$ such that $\exists \xi < \vartheta \, (|X|^{\mathrm{int}} \geq \nu_\xi)$.

Let $\tau = \vartheta^+$ (the next cardinal in WF). The counterexample is based on a sequence $\{Y_\gamma\}_{\gamma < \tau}$ of internal sets Y_γ such that

(a) for any pair of disjoint finite sets $u, v \subseteq \tau$, $u \neq \varnothing$, the set $Y_{uv} = \bigcap_{\alpha \in u} Y_\alpha \setminus \bigcup_{\beta \in v} Y_\beta$ is $\vec{\nu}$-large;

(b) $|Y_{uv}|^{\mathrm{int}} = |Y_{u\varnothing}|^{\mathrm{int}}$ for any disjoint finite $u, v \subseteq \tau$;

(c) if $\xi < \vartheta$, $A \subseteq \tau$, and $|Y_{\{\alpha,\beta\},\varnothing}|^{\mathrm{int}} \geq \nu_\xi$ (that is, $|Y_\alpha \cap Y_\beta|^{\mathrm{int}} \geq \nu_\xi$) for all $\alpha, \beta \in A$ then $\operatorname{card} A \leq \vartheta$.

We define Y_γ by induction. To begin with put $Y_0 = [0, \nu_0)$ (an initial segment in ${}^*\mathsf{Ord}$). Now suppose that $\gamma < \tau$ and a set $Y_\delta \in \mathbb{I}$ has been defined for every $\delta < \gamma$ so that (a) and (b) hold below γ. Re-enumerate $\{Y_\delta : \delta < \gamma\} = \{Z_\alpha : \alpha < \lambda\}$, where $\lambda = \min\{\gamma, \vartheta\}$, without repetitions.

For any pair of disjoint finite sets $u, v \subseteq \lambda$, $u \neq \varnothing$, define the internal set $Z_{uv} = \bigcap_{\alpha \in u} Z_\alpha \setminus \bigcup_{\beta \in v} Z_\beta$. In our assumptions, the $\mathbb{I}$-cardinals $\kappa_{uv} =$

$|Z_{uv}|^{\text{int}}$ satisfy $\kappa_{uv} = \kappa_u$, where $\kappa_u = \kappa_{u\varnothing}$, and $\exists \xi < \vartheta \, (\kappa_u \geq \nu_\xi)$. For any finite $u \subseteq \lambda$ let $\xi(u)$ be the least ordinal $\xi < \vartheta$ such that $\nu_\xi < \kappa_u$ and $\xi > \sup u$. We assert that there is an internal set Z satisfying

(d) $|Z \cap Z_{uv}|^{\text{int}} = \nu_{\xi(u)}$ and $|Z_{uv} \smallsetminus Z|^{\text{int}} = \kappa_u$ for any pair of disjoint finite sets $u, v \subseteq \lambda$, $u \neq \varnothing$, and

(e) $|Z \smallsetminus \bigcup_{\beta \in v} Z_\beta|^{\text{int}} = |Z|^{\text{int}} \geq \nu_0$ for each finite set $v \subseteq \lambda$.

Indeed as ϑ is a set of standard size it suffices to prove that for any finite $d \subseteq \vartheta$ there is a set $Z \in \mathbb{I}$ satisfying (d), (e) for all $u, v \subseteq d$.

Note that the sets of the form Z_{uv}, where $u \cup v = d$ and $u \cap v = \varnothing$, are mutually disjoint, and by definition satisfy $\nu_{\xi(u)} \leq \kappa_u = |Z_{uv}|^{\text{int}}$. This allows us to define an internal Z satisfying (d) for all pairs u, v with $u \cup v = d$, $u \cap v = \varnothing$, $u \neq \varnothing$, and, adding a sufficient portion out of $\bigcup_{\beta \in d} Z_\beta$, also $|Z \smallsetminus \bigcup_{\beta \in d} Z_\beta|^{\text{int}} = |Z|^{\text{int}} \geq \nu_0$. It remains to show (d) for all disjoint sets $u, v \subseteq d$ not necessarily with $u \cup v = d$.

We show this by backward induction on the cardinality of $u \cup v$. Suppose that $u \cup v \subsetneq d$. Take any $\alpha \in d \smallsetminus (u \cup v)$. Let $u' = u \cup \{\alpha\}$ and $v' = v \cup \{\alpha\}$. Then by the inductive hypothesis $|Z \cap Z_{u'v}|^{\text{int}} = \nu_{\xi(u')}$ and $|Z \cap Z_{uv'}|^{\text{int}} = \nu_{\xi(u)}$. Since $Z_{uv} = Z_{u'v} \cup Z_{uv'}$ and easily $\xi(u) \leq \xi(u')$ whenever $u \subseteq u'$, we conclude that $|Z \cap Z_{uv}|^{\text{int}} = \nu_{\xi(u)}$ as required. Similarly, $|Z_{u'v} \smallsetminus Z|^{\text{int}} = \kappa_{u'}$ and $|Z_{uv'} \smallsetminus Z|^{\text{int}} = \kappa_u$, therefore $|Z_{uv} \smallsetminus Z|^{\text{int}} = \kappa_{u'} + \kappa_u = \kappa_u$ as required.

Take as Y_γ any set $Z \in \mathbb{I}$ satisfying (d), (e). We have to demonstrate that (a), (b) remain true for the sequence $\{Y_\delta\}_{\delta \leq \gamma}$, or, that is equivalent, for the sequence $\{Z_\alpha\}_{\alpha \leq \lambda}$, where $Z_\lambda = Y_\gamma = Z$.

Take any pair of disjoint sets $u, v \subseteq \lambda \cup \{\lambda\}$. If $\lambda \notin u \cup v$ then the set $Z_{uv} = \bigcap_{\alpha \in u} Z_\alpha \smallsetminus \bigcup_{\beta \in v} Z_\beta$ is the same as above so there is nothing to prove. Suppose that $\lambda \in u$; put $u' = u \smallsetminus \{\lambda\}$. Then $Z_{uv} = Z \cap Z_{u'v}$, and hence Z_{uv} is $\vec{\nu}$-large by (d) (applied for the pair u', v). Separately if $u = \{\lambda\}$ then $u' = \varnothing$, hence $Z_{u'v}$ is not defined, but obviously $Z_{uv} = Z \smallsetminus \bigcup_{\beta \in v} Z_\beta$, therefore Z_{uv} is $\vec{\nu}$-large by (e). Suppose that $\lambda \in v$; put $v' = v \smallsetminus \{\lambda\}$. Then $Z_{uv} = Z_{uv'} \smallsetminus Z$, and hence Z_{uv} is $\vec{\nu}$-large still by (d). This proves (a); the derivation of (b) from (d), (e) is similar.

This ends the recursive construction of the sets Y_γ.

Show that such a sequence $\{Y_\gamma\}_{\gamma < \tau}$ also satisfies (c). We prove not only that $\operatorname{card} A \leq \vartheta$ for any set A as in (c), but even more the order type of A in τ is $\leq \vartheta$. Suppose that $\gamma \in A$. Let us come back to the renumerated system $\{Y_\delta : \delta < \gamma\} = \{Z_\alpha : \alpha < \lambda\}$, where $\lambda = \min\{\gamma, \vartheta\}$, and to the construction of $Y_\gamma = Z_\lambda = Z$ satisfying (d). It follows from (d) that, for any $\alpha < \lambda$, $|Y_\gamma \cap Z_\alpha|^{\text{int}} = \nu_{\xi(\{\alpha\})} < \nu_\alpha$. In other words, for any $\xi < \vartheta$ the

128

inequality $|Y_\gamma \cap Z_\alpha|^{\text{int}} \geq \nu_\xi$ can be true only for $\alpha < \xi$. Thus there exist $(<\vartheta)$-many sets Y_δ, $\delta < \gamma$, satisfying $|Y_\gamma \cap Y_\delta|^{\text{int}} \geq \nu_\xi$, as required.

Coming back to the proof of (ii) of Theorem 7.2, we fix a sequence $\{Y_\gamma\}_{\gamma<\tau}$ satisfying (a), (b), (c), and put $\mathscr{Y} = \{Y_\gamma : \gamma < \tau\}$. Then $Y = \bigcap_{\gamma<\tau} Y_\gamma$ is a $\mathbf{\Pi}_1^{\text{ss}}$ set. Note that every $\mathbb{I}$-cardinal ν_ξ, $\xi < \vartheta$, belongs to $\|Y\|^*$ by (c), and on the other hand it follows by **Saturation** that every internal superset H of Y contains a subset of the form $\bigcap_{\alpha \in u} Y_\alpha = Y_{u\varnothing}$, where $u \subseteq \tau$ is finite, and hence $|H|^{\text{int}} \geq \nu_\xi$ for some $\xi < \vartheta$ by (a). It follows that the sequence $\{\nu_\xi\}_{\xi<\vartheta}$ is coinitial in $\|Y\|^*$. It follows from Lemma 5.3 that $\|Y\|_*$ coincides with the set $\Omega = \{\kappa \in {}^*\mathtt{Card} : \forall \xi < \vartheta \, (\kappa < \nu_\xi)\}$.

The other side of the counterexample will be the $\mathbf{\Pi}_1^{\text{ss}}$ set $X = \bigcap_{\xi<\vartheta} X_\xi$, where $X_\xi = \{\kappa \in {}^*\mathtt{Ord} : \kappa < \nu_\xi\} \in \mathbb{I}$. Easily $|X_\xi|^{\text{int}} = \nu_\xi$, therefore the sequence $\{\nu_\xi\}_{\xi<\vartheta}$ is coinitial in $\|Y\|^*$, too. We conclude that $\|X\|^* = \|Y\|^*$, hence $\|X\|_* = \Omega = \|Y\|_*$ by Lemma 5.3.

To accomplish (ii), suppose towards the contrary that there is an injection $g \in \mathbf{\Delta}_2^{\text{ss}}$, $g : X \to Y$ of the form $g = \bigcup_{w \in W} \bigcap_{\xi<\vartheta} g_{w\xi}$, where all $g_{w\xi}$ are internal and W a set of standard size. Then each $g_w = \bigcap_{\xi<\vartheta} g_{w\xi}$ is still an injection into Y, whose domain $D_w = \operatorname{dom} g_w \subseteq X$ is still a $\mathbf{\Pi}_1^{\text{ss}}$ set, moreover, an intersection of $(\leq\vartheta)$-many internal sets. (The combination of quantifiers $\exists \, \forall^{\text{st}} \xi < \vartheta$ converts to $\forall^{\text{st}} p \in \mathscr{P}_{\text{fin}}(\vartheta) \, \exists$ by **Saturation**.)

We claim that $\|D_w\|_* = \Omega$ for at least one $w \in W$.

(Indeed otherwise choose any $\kappa_\alpha \in \Omega \smallsetminus \|D_w\|_*$ for every $w \in W$; here **Standard Size Choice** is applied. Recall that $\|X\|^*$ is a standard size coinitial final segment in ${}^*\mathtt{Card}$, therefore the complement Ω of is not standard size cofinal by **Saturation**. It follows that there is an $\mathbb{I}$-cardinal $\kappa \in \Omega$ bigger than each κ_w. Then $\kappa \in \|D_w\|^*$ for any $w \in W$, thus any D_w is covered by an internal set of $\mathbb{I}$-cardinality κ. Still by **Saturation**, the union $\bigcup_{w \in W} D_w$ can be covered by an internal set C, $|C|^{\text{int}} = \kappa$. Then $X \subseteq C$, contradiction.)

This result allows us to replace X by D_a, or, in different words, reduce the task to the case when g, a given injection $X \to Y$, is equal to $\bigcap_{\xi<\vartheta} g_\xi$, each g_ξ being an internal set. An easy application of **Saturation** shows that there is a finite set $u \subseteq \vartheta$ such that $h = \bigcap_{\xi \in u} g_\xi$ is an injective function. On the other hand h is an internal function extending g, thus $X = \operatorname{dom} g \subseteq h$ and $h"X \subseteq Y$. Let $D = \operatorname{dom} h$ (an internal superset of X).

Recall that $Y = \bigcap_{\gamma<\tau} Y_\gamma$ where all Y_γ are internal. Thus, for any γ, $h"X \subseteq Y_\gamma$, and hence, as $X = \bigcap_{\xi<\vartheta} X_\xi$ and the family of all sets X_ξ is $\cap$-closed, **Saturation** yields an ordinal $\xi(\gamma) < \vartheta$ such that $X_{\xi(\gamma)} \subseteq D$ and $h"X_{\xi(\gamma)} \subseteq Y_\gamma$. As $\tau = \vartheta^+$, there is at least one $\xi < \vartheta$ such that $G = \{\gamma < \tau : h"X_\xi \subseteq Y_\gamma\}$ is unbounded in τ. Thus all sets Y_γ, $\gamma \in G$ include as a subset one and the same internal set $R = h \, "X_\xi$. Note that $|R|^{\text{int}} = |X_\xi|^{\text{int}}$ because h is an injection. But X_ξ is a $\vec{\nu}$-large set, a contradiction with (c). $\qquad \square$

Question 7.3. Is Corollary 6.3 still true for sets in $\mathbf{\Delta}_2^{\text{ss}}$ or in $\mathbf{\Pi}_1^{\text{ss}}$? $\qquad \square$

Theorem 10.2 below shows that a wider category of $\mathbf{\Delta}_2^{\text{ss}}$ *quotients* has plenty of incomparable sets. Note that the existence of countably determined sets incomparable in the sense of countably determined injections, is also an open problem. A counterexample defined in [2] in the **AST** frameworks makes use of the hypothesis that there exist only $\aleph_1$-many internal sets, and hence is irreproducible in **HST**.

On the other hand all Borel sets (in the sense of Footnote [c]) are Borel-comparable. This result was first obtained by **AST**-followers, see *e.g.* [12], and then reproved in [27]. See more on this in [17], 9.6 and 9.7.

8. Effective sets in the form of quotients

Sets of the form X/E, where X is $\mathbf{\Delta}_2^{\text{ss}}$ while E is a $\mathbf{\Delta}_2^{\text{ss}}$ equivalence relation on X will be called $\mathbf{\Delta}_2^{\text{ss}}$ *quotients*. These $\mathbf{\Delta}_2^{\text{ss}}$ quotients include the class $\mathbf{\Delta}_2^{\text{ss}}$ itself, for take E to be just the equality on a given $\mathbf{\Delta}_2^{\text{ss}}$ set X, so that the map sending any $x \in X$ to $\{x\}$ is a bijection of X onto X/E.

On the other hand, it follows from Theorem 2.3(iii) that every set in $\mathbb{L}[\mathbb{I}]$, that is, every effective set in the sense explained in Section 2, admits an effective bijection onto a $\mathbf{\Delta}_2^{\text{ss}}$ quotient. Thus $\mathbf{\Delta}_2^{\text{ss}}$ quotients exhaust, in the context of effective cardinalities, all effective $(= \mathbb{L}[\mathbb{I}])$ sets in general.

One may ask whether $\mathbf{\Delta}_2^{\text{ss}}$ quotients produce more effective cardinalities than just $\mathbf{\Delta}_2^{\text{ss}}$ sets. Call *smooth* any $\mathbf{\Delta}_2^{\text{ss}}$ quotient that admits a $\mathbf{\Delta}_2^{\text{ss}}$ bijection onto a $\mathbf{\Delta}_2^{\text{ss}}$ set. We show in Section 9 that every $\mathbf{\Delta}_2^{\text{ss}}$ quotient X/E, such that all E-classes $[x]_\mathsf{E} = \{y \in X : x \mathrel{\mathsf{E}} y\}$, $x \in X$, are sets of standard size, is smooth. A family of non-smooth $\mathbf{\Delta}_2^{\text{ss}}$ quotients, those defined by means of *monadic partitions* of $^*\mathbb{N}$, will be studied in Sections 10, 11. We prove there that there exist incomparable effective cardinalities of monadic $\mathbf{\Delta}_2^{\text{ss}}$ quotients, still an open problem for $\mathbf{\Delta}_2^{\text{ss}}$ sets themselves. We also prove a "small–large" type theorem for $\mathbf{\Delta}_2^{\text{ss}}$ quotients in Section 12, similar to 5.6 but not so sharp, with an interesting Ramsey-like corollary.

Note that $\mathbf{\Delta}_2^{\text{ss}}$ quotients consist of subsets of $\mathbb{I}$ which are not necessarily internal sets themselves. Accordingly injections of $\mathbf{\Delta}_2^{\text{ss}}$ quotients are maps whose **dom** and **ran** not necessarily consist of internal sets. Still there is a way to pull the consideration down to the basic level.

Definition 8.1. Let E, F be equivalence relations on sets X, Y. A set $R \subseteq$

$X \times Y$ is a (E, F)-*invariant pre-injection of* X *into* Y iff 1) $\operatorname{dom} R = X$ [g] and 2) the equivalence $x \mathrel{\mathsf{E}} x' \iff y \mathrel{\mathsf{F}} y'$ holds for all $\langle x, y \rangle \in R$ and $\langle x', y' \rangle \in R$.

Such a set R is a *reduction of* X/E *to* Y/F (or just of E to F) if in addition 3) R is a (graph of a) function $X \to Y$.

Write $\mathsf{E} \leq_{\mathrm{eff}} \mathsf{F}$ iff there is a (E, F)-invariant pre-injection $P \subseteq X \times Y$, $P \in \boldsymbol{\Delta}_2^{\mathrm{ss}}$, of X into Y. Write $\mathsf{E} \leq_{\mathrm{eff}}^{+} \mathsf{F}$, in words: E *is effectively reducible to* F, iff there is a reduction $\rho \in \boldsymbol{\Delta}_2^{\mathrm{ss}}$, $\rho : X \to Y$ of E to F.

An equivalence relation E on a set X and the quotient X/E are $\boldsymbol{\Delta}_2^{\mathrm{ss}}$-*smooth* iff there is a $\boldsymbol{\Delta}_2^{\mathrm{ss}}$ set Y such that $\mathsf{E} \leq_{\mathrm{eff}} \mathsf{D}_Y$, where D_Y is the equality on Y considered as an equivalence relation. [h] $\qquad\square$

This definition resembles some central concepts in modern descriptive set theory, like Borel reducibility and "Borel cardinals" (see, for instance, [6, 7, 18]), where Borel maps are used in approximately the same role as $\boldsymbol{\Delta}_2^{\mathrm{ss}}$ maps in this paper.

Proposition 8.2. (i) *Suppose that* E, F *are* $\boldsymbol{\Delta}_2^{\mathrm{ss}}$ *equivalence relations on* $\boldsymbol{\Delta}_2^{\mathrm{ss}}$ *sets* X, Y. *Then* $|X/\mathsf{E}|^{\mathrm{eff}} \leq |Y/\mathsf{F}|^{\mathrm{eff}}$ *iff* $\mathsf{E} \leq_{\mathrm{eff}} \mathsf{F}$.

(ii) *An* $\boldsymbol{\Delta}_2^{\mathrm{ss}}$ *equivalence relation* E *on a* $\boldsymbol{\Delta}_2^{\mathrm{ss}}$ *set* X *is* $\boldsymbol{\Delta}_2^{\mathrm{ss}}$-*smooth iff there exists a* $\boldsymbol{\Delta}_2^{\mathrm{ss}}$ *set* Y *such that* $|X/\mathsf{E}|^{\mathrm{eff}} = |Y|^{\mathrm{eff}}$.

Proof. (i) Suppose that $f \in \mathbb{L}[\mathbb{0}]$ is an injection $X/\mathsf{E} \to Y/\mathsf{F}$. Then $P = \{\langle x, y \rangle \in X \times Y : f([x]_\mathsf{E}) = [y]_\mathsf{F}\}$ is a set in $\mathbb{L}[\mathbb{0}]$, hence a $\boldsymbol{\Delta}_2^{\mathrm{ss}}$ set by 2.3(i), and obviously an invariant pre-injection. The converse is equally simple: if P is an invariant pre-injection then to define an injection $f : X/\mathsf{E} \to Y/\mathsf{F}$ put $f([x]_\mathsf{E}) = [y]_\mathsf{F}$ for any $\langle x, y \rangle \in P$.

(ii) Suppose that $\mathsf{E} \leq_{\mathrm{eff}} \mathsf{D}_Z$, where Z is a $\boldsymbol{\Delta}_2^{\mathrm{ss}}$ set. Let this be witnessed by an invariant pre-injection $R \subseteq X \times Z$ of class $\boldsymbol{\Delta}_2^{\mathrm{ss}}$. Clearly $R = \rho$ is then a reduction (a map $X \to Z$ such that $x \mathrel{\mathsf{E}} x' \iff \rho(x) = \rho(x')$). The set $Y = \operatorname{ran} \rho \subseteq Z$ is as required. $\qquad\square$

9. Equivalence relations with standard size classes

In modern descriptive set theory, an equivalence relation E is *countable* iff all equivalence classes $[x]_\mathsf{E} = \{y : x \mathrel{\mathsf{E}} y\}$, $x \in \operatorname{dom} \mathsf{E}$, are at most countable. See [10] on properties and some open problems related to countable equivalence relations. But in the nonstandard setting the structure of equivalence

---

[g] This condition can be weakened to $[x]_\mathsf{E} \cap \operatorname{dom} R \neq \varnothing$ for any $x \in X$ without any harm.
[h] Note that in this case any invariant pre-injection is a partial map that can be immediately extended to a reduction, and hence in fact $\mathsf{E} \leq_{\mathrm{eff}}^{+} \mathsf{D}_Y$ holds.

relations in a much wider class turns out to be considerably simpler: all of them admit effective transversals.

Recall that a *transversal* of an equivalence relation is any set having exactly one element in common in every equivalence class.

Theorem 9.1. *Any Δ_2^{ss} equivalence relation* E, *on an internal set H and with s. s. classes, has a Δ_2^{ss} transversal and hence is Δ_2^{ss}-smooth.*

Opposed to this, the Vitali equivalence on the reals is obviously countable but not smooth (via Borel maps), neither it admits a Borel transversal.

Proof. First of all, a Δ_2^{ss} transversal implies Δ_2^{ss}-smoothness: let $\rho(x)$ denote the only element of the transversal equivalent to x and apply 2.1 to show that ρ is still Δ_2^{ss}. Let us prove the existence of a Δ_2^{ss} transversal.

By definition $\mathsf{E} = \bigcup_{a \in A} \bigcap_{b \in B} E_{ab}$, where $E_{ab} \subseteq H \times H$ are internal sets while $A, B \in \mathsf{WF}$. Put $P"x = \{y : \langle x, y \rangle \in P\}$ for $P \subseteq H \times H$ and $x \in H$.

Lemma 9.2. *There exists a standard size family $\mathscr{F}$ of internal maps F : $H \to H$ such that $[x]_\mathsf{E} \subseteq \{F(x) : F \in \mathscr{F}\}$ for all $x \in H$.*

Proof. It suffices to prove the lemma for each "constituent" $E_a = \bigcap_{b \in B} E_{ab}$ of E. According to 1.3.6 in [17], the intersection $\bigcap \mathscr{X}$ of a s. s. family $\mathscr{X}$ of internal sets either is not a s. s. set or it is finite and there is a finite $\mathscr{X}' \subseteq \mathscr{X}$ such that $\bigcap \mathscr{X}' = \bigcap \mathscr{X}$. It follows that every set $E_a[x]$ is finite and moreover there is a finite set $\beta_{ax} \subseteq B$ such that $E_a"x = \bigcap_{b \in \beta_{ax}} E_{ab}"x$. Put, for any $n \in \mathsf{N}$ and any finite $\beta \subseteq B$,

$$E_{a\beta} = \bigcap_{b \in \beta} E_{ab} \quad \text{and} \quad P_{a\beta n} = \{\langle x, y \rangle \in E_{a\beta} : \operatorname{card} E_{a\beta}"x \le n\}.$$

All sets $P_{a\beta n}$ are internal. We define $F_{a\beta ni}(x) = $ "i- th element of $P_{a\beta n}$ in the sence of a fixed internal linear ordering of $P_{a\beta n}$" in the case when $1 \le i \le n$ and $P_{a\beta n}$ contains at least i elements, and $F_{a\beta ni}(x) = y_0$ otherwise, where y_0 is a once and for all fixed element of H. It remains to define $\mathscr{F}$ to be the family of all functions $F_{a\beta ni}$. $\qquad\square$

Let $\mathscr{F}$ be as in the lemma. The sets

$$D_F = \operatorname{dom}(\mathsf{E} \cap F) = \{x \in H : x \mathrel{\mathsf{E}} F(x)\} \qquad (F \in \mathscr{F})$$

belong to Δ_2^{ss} by Proposition 2.1. Let us fix an internal wellordering $\prec$ of the set H. Suppose that $F \in \mathscr{F}$. For any $x \in H$ we carry out the following construction called the *F-construction for x.* Define an internal $\prec$-decreasing sequence $\{x_{(a)}\}_{a \le a(x)}$ of length $a(x) + 1 \in {}^*\mathsf{N}$. Its terms $x_{(a)}$ are defined by induction on a. Put $x_{(0)} = x$. Assume that $x_{(a)}$ has been defined.

If $z = F(x_{(a)}) \prec x_{(a)}$ then put $x_{(a+1)} = z$, otherwise put $a(x) = a$ and stop the construction. Eventually the construction ends since $x_{(a+1)} \prec x_{(a)}$ for all a. Put $\nu_F(x) = 0$ if $a(x)$ is even and $\nu_F(x) = 1$ otherwise.

Define $\psi(x)(F) = \nu_F(x)$ for any $x \in H$, $F \in \mathscr{F}$; thus $\psi : H \to 2^{\mathscr{F}}$.

Lemma 9.3. *If $r \in 2^{\mathscr{F}}$ then $\Psi_r = \{x \in H : \psi(x) = r\}$ belongs to $\boldsymbol{\Delta}_2^{\mathbf{ss}}$.*

Proof. Note that $x \in \Psi_r$ iff $\nu_F(x) = r(F)$ for all $F \in \mathscr{F}$. On the other hand, all sets $X_F = \{x \in H : \nu_F(x) = 0\}$ $(F \in \mathscr{F})$ are internal because the F-construction is internal. It remains to apply Proposition 2.1. $\square$

According to the next lemma, any two different but E-equivalent elements $x \in H$ have different "profiles" $\psi(x)$.

Lemma 9.4. *If $x \neq y \in H$ and $x \mathsf{E} y$ then $\psi(x) \neq \psi(y)$.*

Proof. Suppose that $y \prec x$. There exists a function $F \in \mathscr{F}$ such that $y = F(x)$. Then $y = x_{(1)}$ in the sense of F-construction for x. It follows that the F-construction for y has exactly one step less than the F-construction for x. Thus $\nu_F(x) \neq \nu_F(y)$ and $\psi(x) \neq \psi(y)$. $\square$

We continue the proof of Theorem 9.1. Note that $2^{\mathscr{F}}$ and $\mathscr{P}(2^{\mathscr{F}})$ are sets of standard size together with $\mathscr{F}$ (1.3.3 in [17]). Thus by the axiom of Standard Size Choice there is a map $A \mapsto r_A$ such that $r_A \in A$ for any non-empty $A \subseteq 2^{\mathscr{F}}$. Its graph $C = \{\langle A, r \rangle : A \subseteq 2^{\mathscr{F}} \wedge r = r_A\}$ is a s.s. set together with $\mathscr{P}(2^{\mathscr{F}})$. For any $x \in H$ put $A(x) = \{\psi(y) : y \in [x]_{\mathsf{E}}\}$, a non-empty subset of $2^{\mathscr{F}}$. Now $X = \{x \in H : \psi(x) = r_{A(x)}\}$ is a transversal for E by Lemma 9.4.

Prove that X is a $\boldsymbol{\Delta}_2^{\mathbf{ss}}$ set. By definition $X = \bigcup_{\langle A, r \rangle \in C} Y_A \cap \Psi_r$, where $Y_A = \{x \in H : A(x) = A\}$. However $\Psi_r \in \boldsymbol{\Delta}_2^{\mathbf{ss}}$ by Lemma 9.3. It remains to check that $Y_A \in \boldsymbol{\Delta}_2^{\mathbf{ss}}$ for each $A \subseteq 2^{\mathscr{F}}$. Note that $A(x) = \{\psi(F(x)) : F \in \mathscr{F} \wedge x \in D_F\}$, and hence $A(x) = A$ is equivalent to

$$\forall r \in A \, \exists F \in \mathscr{F} \, (x \in D_F \wedge F(x) \in \Psi_r) \, \wedge$$

$$\wedge \, \forall F \in \mathscr{F} \, \exists r \in A \, (x \in D_F \implies F(x) \in \Psi_r).$$

Yet the sets Ψ_r and D_F are $\boldsymbol{\Delta}_2^{\mathbf{ss}}$ (see above), while the domains A and $\mathscr{F}$ are sets of standard size. Now apply Proposition 2.1.

$$\square \ (Thm \ 9.1)$$

10. Monadic partitions

A cut $U \subseteq {}^*\mathbb{N}$ is *additive* if $a \in U \implies 2a \in U$. Any such cut U induces an equivalence relation $x \, \mathsf{M}_U \, y$ iff $|x - y| \in U$ on ${}^*\mathbb{N}$. (The additivity implies that M_U is transitive.) Its equivalence classes $[x]_U = \{y : x \, \mathsf{M}_U \, y\} = \{y : |x-y| \in U\}$, are called *U-monads* and relations of the form M_U, accordingly, *monadic* equivalence relations or *monadic partitions*.

Monads of various kinds are considered in nonstandard analysis. As for those induced by additive cuts in $^*\mathbb{N}$, see [11, 20].

The following is an elementary corollary of Proposition 5.2:

Proposition 10.1. *If $\varnothing \neq U \subsetneqq {}^*\mathbb{N}$ is an additive $\mathbf{\Delta}_2^{\mathrm{ss}}$ cut then U is non-internal and either standard size cofinal or standard size coinitial.* $\square$

Any additive $\mathbf{\Delta}_2^{\mathrm{ss}}$ cut $U \subseteq {}^*\mathbb{N}$ defines a $\mathbf{\Delta}_2^{\mathrm{ss}}$ quotient $^*\mathbb{N}/U = {}^*\mathbb{N}/\mathsf{M}_U$, the set of all U-monads. According to the next theorem, effective cardinalities of those quotients are determined by two factors. The first of them is

$$\mathrm{wid}\, U = \bigcap\nolimits_{u\in U,\, u'\in {}^*\mathbb{N}\smallsetminus U}[0, \tfrac{u'}{u}) = \bigcap\nolimits_{u\in U}\bigcup\nolimits_{u'\in U,\, u'>u}[0, \tfrac{u'}{u})$$

the width of U.[i] The second one is the cofinality/coinitiality. The *cofinality* $\mathrm{cof}\, U$ of a standard size (s. s.) cofinal non-internal cut, is the least cardinal $\vartheta \in \mathsf{Card}$ such that U has an increasing cofinal sequence of type ϑ. The *coinitiality* $\mathrm{coi}\, U$ of a standard size coinitial cut is defined similarly, with a reference to coinitial sequences in $^*\mathsf{Card} \smallsetminus U$. Note that $\mathrm{cof}\, U$ and $\mathrm{coi}\, U$ are infinite regular cardinals.

Additive cuts of lowest possible width are obviously those of the form $U = c\mathbb{N}$, $c \in {}^*\mathbb{N}$ and $U = c/\mathbb{N}$, $c \in {}^*\mathbb{N} \smallsetminus \mathbb{N}$, which we call *slow*; they satisfy $\mathrm{wid}\, U = \mathbb{N}$. Other additive cuts will be called *fast*.

Theorem 10.2. *Suppose that U, V are additive $\mathbf{\Delta}_2^{\mathrm{ss}}$ cuts in $^*\mathbb{N}$ other than $\varnothing$ and $^*\mathbb{N}$. Then (i) $|^*\mathbb{N}|^{\mathrm{eff}} \leq |^*\mathbb{N}/U|^{\mathrm{eff}}$. In addition,*

(ii) *$^*\mathbb{N}/U$ is $\mathbf{\Delta}_2^{\mathrm{ss}}$-smooth iff $^*\mathbb{N}/U$ has a $\mathbf{\Delta}_2^{\mathrm{ss}}$ transversal iff U is slow;*

(iii) *if U is slow then $|^*\mathbb{N}/U|^{\mathrm{eff}} \leq |^*\mathbb{N}/V|^{\mathrm{eff}}$;*

(iv) *if both U, V are s. s. cofinal cuts and U is fast then $|^*\mathbb{N}/U|^{\mathrm{eff}} \leq |^*\mathbb{N}/V|^{\mathrm{eff}}$ iff: $\mathrm{cof}\, U = \mathrm{cof}\, V$ and $\mathrm{wid}\, U \subseteq \mathrm{wid}\, V$;*

(v) *if both U, V are s. s. coinitial cuts and U is fast then $|^*\mathbb{N}/U|^{\mathrm{eff}} \leq |^*\mathbb{N}/V|^{\mathrm{eff}}$ iff: $\mathrm{coi}\, U = \mathrm{coi}\, V$ and $\mathrm{wid}\, U \subseteq \mathrm{wid}\, V$;*

(vi) *if U, V are fast cuts, U is s. s. cofinal and V is s. s. coinitial then $|^*\mathbb{N}/U|^{\mathrm{eff}}$ and $|^*\mathbb{N}/V|^{\mathrm{eff}}$ are incomparable.*

---

[i]Also called *the thickness* of U in some papers on **AST**.

Thus either of the two classes of monadic partitions (s. s. cofinal and s. s. coinitial) is linearly $\leq_{\mathtt{eff}}$-(pre)ordered in each subclass of the same cofinality (coinitiality), slow partitions of both classes form the $\leq_{\mathtt{eff}}$-least type, and there is no other $\leq_{\mathtt{eff}}$-connection between the two classes and their same-cofinality/coinitiality subclasess.

See [16] for earlier results of countably determined and Borel reducubility of monadic partitions for *countably* cofinal/coinitial cuts.

11. The proof of the reducibility theorem

We begin the proof of Theorem 10.2 with the following observation.

Remark 11.1. Call a set $X \subseteq {}^*\mathbb{N}$ *scattered* iff there is a number $c \in {}^*\mathbb{N} \setminus \mathbb{N}$ such that $\frac{|X \cap I|^{\mathtt{int}}}{c}$ is infinitesimal for any interval I in ${}^*\mathbb{N}$ of length c. It is quite clear that ${}^*\mathbb{N}$ is not a finite union of scattered sets, and hence, by Saturation, ${}^*\mathbb{N}$ *is not a standard size union of internal scattered sets.* $\qquad\square$

Proof of Theorem 10.2. (i) Choose a number $h \in {}^*\mathbb{N} \setminus U$. The map $x \mapsto [xh]_U$ is an injection of ${}^*\mathbb{N}$ into ${}^*\mathbb{N}/U$.

(ii) If ${}^*\mathbb{N}/U$ admits a $\mathbf{\Delta}_2^{\mathtt{ss}}$ transversal then it is $\mathbf{\Delta}_2^{\mathtt{ss}}$-smooth. (Let, for $x \in {}^*\mathbb{N}$, $\rho(x)$ be the only element of the transversal equivalent to x.) Suppose that ${}^*\mathbb{N}/U$ is smooth, *i.e.* $\mathsf{M}_U \leq_{\mathtt{eff}} \mathsf{D}_R$ for a suitable $\mathbf{\Delta}_2^{\mathtt{ss}}$ set Z. This is witnessed by a $\mathbf{\Delta}_2^{\mathtt{ss}}$ reduction $\rho : {}^*\mathbb{N} \to Z$ By Theorem 3.1(i) the set $\mathbf{ran}\,\rho$ can be covered by an internal set Y with $|Y|^{\mathtt{int}} \leq |{}^*\mathbb{N}|^{\mathtt{int}}$. Thus $|{}^*\mathbb{N}/U|^{\mathtt{eff}} \leq |{}^*\mathbb{N}|^{\mathtt{eff}}$. Then $|{}^*\mathbb{N}/U|^{\mathtt{eff}} \leq |{}^*\mathbb{N}/V|^{\mathtt{eff}}$ for any other additive $\mathbf{\Delta}_2^{\mathtt{ss}}$ cut V by (i), thus U must be slow by (vi). Finally, if U is slow then ${}^*\mathbb{N}/U$ has a $\mathbf{\Delta}_2^{\mathtt{ss}}$ transversal by Theorem 1.4.7 in [17].[j]

(iii) If U is slow then ${}^*\mathbb{N}/U$ is $\mathbf{\Delta}_2^{\mathtt{ss}}$-smooth, and in fact $|{}^*\mathbb{N}/U|^{\mathtt{eff}} \leq |{}^*\mathbb{N}|^{\mathtt{eff}}$, see the proof of (ii). It remains to apply (i).

(iv) Thus let U, V be additive s. s. cofinal cuts. Choose increasing sequences $\{u_\xi\}_{\xi < \vartheta}$ and $\{v_\eta\}_{\eta < \tau}$ cofinal in resp. U and V; $\vartheta = \mathsf{cof}\,U$ and $\tau = \mathsf{cof}\,V$ being infinite regular cardinals in $\mathbb{WF}$. As U is supposed to be fast, we can assume that $\frac{u_{\xi+1}}{u_\xi}$ is infinitely large for all ξ.

Part 1: assuming $|{}^*\mathbb{N}/U|^{\mathtt{eff}} \leq |{}^*\mathbb{N}/V|^{\mathtt{eff}}$, we prove that $\mathtt{wid}\,U \subseteq \mathtt{wid}\,V$. Let, by 8.2, $R \subseteq {}^*\mathbb{N} \times {}^*\mathbb{N}$ be a (U, V)-invariant pre-injection, thus $\mathbf{dom}\,R = {}^*\mathbb{N}$, and $|x - x'| \in U \iff |y - y'| \in V$ for all pairs $\langle x, y \rangle$ and $\langle x', y' \rangle$ in R.

---

[j] Theorem 9.1 yields a $\mathbf{\Delta}_2^{\mathtt{ss}}$ transversal for ${}^*\mathbb{N}/\mathbb{N}$, and hence for any ${}^*\mathbb{N}/(h\mathbb{N})$ by multiplication. Transversals defined this way are countably determined but not Borel. Yet partitions of the form ${}^*\mathbb{N}/(h/\mathbb{N})$ have no countably determined transversals by 9.7.14 in [17]. These theorems were obtained in [17] on the base of earlier results in [11].

Since R is $\mathbf{\Delta}_2^{\mathrm{ss}}$, we have, by definition, $R = \bigcup_{a \in A} \bigcap_{b \in B} R_{ab}$, where $A, B \in \mathsf{WF}$ and the sets $R_{ab} \subseteq {}^*\mathbb{N} \times {}^*\mathbb{N}$ are internal.

Let us fix $a \in A$.

Then $R_a = \bigcap_{b \in B} R_{ab} \subseteq R$, hence for any $\eta < \tau$ we have

$$\forall b\,(x\,R_{ab}\,y \wedge x'\,R_{ab}\,y') \wedge |y - y'| < v_\eta \implies \exists \xi < \vartheta\,(|x - x'| < u_\xi)$$

for all $x, x', y, y' \in {}^*\mathbb{N}$. We obtain, by **Saturation**,

$$\forall \eta < \tau \quad \exists \text{ finite } F \subseteq B \; \exists \xi < \vartheta \quad \forall x, x', y, y' \in {}^*\mathbb{N}:$$
$$x\,R_{aF}\,y \wedge x'\,R_{aF}\,y' \wedge |y - y'| < v_\eta \implies |x - x'| < u_\xi, \qquad (7)$$

where $R_{aF} = \bigcap_{b \in F} R_{ab}$. A similar (symmetric) argument yields:

$$\forall \xi < \vartheta \quad \exists \text{ finite } F' \subseteq B \; \exists \eta < \tau \quad \forall x, x', y, y' \in {}^*\mathbb{N}:$$
$$x\,R_{aF'}\,y \wedge x'\,R_{aF'}\,y' \wedge |x - x'| < u_\xi \implies |y - y'| < v_\eta. \qquad (8)$$

Suppose, towards the contrary, that $\mathtt{wid}\,U \not\subseteq \mathtt{wid}\,V$. Then there exists $\eta < \tau$ such that the sequence $\{\frac{v_{\eta'}}{v_\eta}\}_{\eta < \eta' < \tau}$ is not cofinal in $\mathtt{wid}\,U$.

Keeping $a \in A$ still fixed, we let F and ξ satisfy (7) for this η. By the choice of η, there exists an ordinal $\xi' > \xi$ such that $\frac{u_{\xi'}}{u_\xi} > \frac{v_{\eta'}}{v_\eta}$ for any $\eta' > \eta$, hence in fact $\frac{u_{\xi'}}{u_\xi} > \ell \cdot \frac{v_{\eta'}}{v_\eta}$ for any $\eta' > \eta$ and any $\ell \in \mathbb{N}$. We now let F' and η' satisfy (8) (as F and η) for the ξ' considered. We may assume that $F \subseteq F'$ and $\eta' \geq \eta$ — otherwise take, resp., the union and the maximum of the two. Then we have, for all $\langle x, y \rangle$, $\langle x', y' \rangle$ in the set $R(a) = R_{aF'}$:

$$\left. \begin{array}{l} |y - y'| < v_\eta \implies |x - x'| < u_\xi \\ |x - x'| < u_{\xi'} \implies |y - y'| < v_{\eta'} \end{array} \right\} ; \quad \xi, \xi', \eta, \eta' \text{ depend on } a. \qquad (9)$$

Put $D(a) = \mathrm{dom}\,R(a)$, an internal subset of ${}^*\mathbb{N}$ together with $R(a)$.

Note that any interval of length $v_{\eta'}$ in ${}^*\mathbb{N}$ consists of approximately $s = \frac{v'_\eta}{v_\eta}$ subintervals of length v_η. Accordingly any interval of length $v_{\xi'}$ consists of approximately $t = \frac{u'_\xi}{u_\xi}$ subintervals of length u_ξ, while $\frac{s}{t}$ is infinitesimal by the above. It follows by (9) that $\frac{|I \cap D(a)|^{\mathrm{int}}}{|I|^{\mathrm{int}}}$ is infinitesimal for any interval I in ${}^*\mathbb{N}$ of length u'_ξ, hence $D(a)$ is scattered in the sense of 11.1.

On the other hand ${}^*\mathbb{N} = \mathrm{dom}\,R = \bigcup_{a \in A} D_a = \bigcup_{a \in A} D(a)$, where $D_a = \mathrm{dom}\,R_a$, simply because $R_a \subseteq R(a)$, which is a contradiction with 11.1.

Part 2: in the same assumptions and notation as in Part 1, we prove that $\mathtt{cof}\,U = \mathtt{cof}\,V$. This means to prove $\vartheta = \tau$. Suppose $\vartheta \neq \tau$. Let say $\vartheta < \tau$. (The other case is similar.) Then, for a fixed $a \in A$, there is an ordinal $\eta < \tau$, one and the same for all $\xi < \vartheta$, such that (8) takes the form:

$$\forall \xi < \vartheta \quad \exists \text{ finite } F' \subseteq B \quad \forall x, x', y, y' \in {}^*\mathbb{N} :$$
$$x \, R_{aF'} \, y \wedge x' \, R_{aF'} \, y' \wedge |x - x'| < u_\xi \implies |y - y'| < v_\eta. \qquad (10)$$

Take an ordinal $\xi < \vartheta$ for this η by (7), and then apply (10) for $\xi + 1$. We obtain a finite set $F \subseteq B$ such that, for all $x, x' \in D(a) = \operatorname{dom} R_{aF}$:

$$|x - x'| < u_{\xi+1} \implies |x - x'| < u_\xi. \qquad (11)$$

However, as U is fast, the cofinal sequence $\{u_\xi\}$ can be chosen so that $\frac{u_\xi}{u_{\xi+1}}$ is infinitesimal for all ξ. Then the set $D(a)$ is scattered by (11), and so on towards the contradiction as in Part 1.

Part 3. Suppose that $\operatorname{cof} U = \operatorname{cof} V = \vartheta$ (an infinite regular cardinal in Card) and $\operatorname{wid} U \subseteq \operatorname{wid} V$. To prove $|{}^*\mathbb{N}/U|^{\mathsf{eff}} \leq |{}^*\mathbb{N}/V|^{\mathsf{eff}}$ it suffices, by 8.2, to define a reduction of ${}^*\mathbb{N}/U$ to ${}^*\mathbb{N}/V$. Let $\{u_\xi\}_{\xi<\vartheta}$, $\{v_\xi\}_{\xi<\vartheta}$ be increasing cofinal sequences in the cuts resp. U, V. Due to additivity of the cuts, we may w.l.o.g. assume that all terms u_ξ, v_ξ are powers of 2.

We first define subsequences of the cofinal sequences satisfying a certain term-to-term inequality. Note that $\operatorname{wid} U \subseteq \operatorname{wid} V$ basically means

$$\forall v \in V \quad \exists u \in U \quad \forall u' \in U, u' > u \quad \exists v' \in V, v' > v \left(\frac{u'}{u} \leq \frac{v'}{v} \right).$$

This allows us to define an unbounded subsequence of $\{u_\xi\}_{\xi<\vartheta}$ such that, after the renumeration, the following holds (ξ, η, ζ are ordinals $< \vartheta$):

$$\forall \zeta \quad \forall \xi > \zeta \quad \exists \eta > \zeta \left(\frac{u_\xi}{u_\zeta} \leq \frac{v_\eta}{v_\zeta}, \quad \text{that is,} \quad \frac{v_\zeta}{u_\zeta} \leq \frac{v_\eta}{u_\xi} \right),$$

and then to once again define an unbounded subsection of, now, $\{v_\eta\}_{\eta<\vartheta}$ to satisfy, after the renumeration, the following:

$$\forall \xi < \eta < \vartheta \left(\frac{v_\xi}{u_\xi} \leq \frac{v_\eta}{u_\eta}, \quad \text{that is,} \quad \frac{u_\eta}{u_\xi} \leq \frac{v_\eta}{v_\xi} \right). \qquad (12)$$

Finally, we may assume that $u_0 = 1$. (Replace each u_ξ by $u'_\xi = \frac{u_\xi}{u_0}$. As all u_ξ are powers of 2, these fractions belong to ${}^*\mathbb{N}$. The sequence $\{u'_\xi\}$ is then cofinal in the cut $U' = U/u_0 = \{u : uu_0 \in U\}$. The inequality $|{}^*\mathbb{N}/U|^{\mathsf{eff}} \leq |{}^*\mathbb{N}/U'|^{\mathsf{eff}}$ is witnessed by the map $[x]_U \mapsto [\text{entire part of } \frac{x}{u_0}]_{U'}$.)

Note that the map f sending each u_ξ to v_ξ satisfies the following: $\operatorname{dom} f = \{u_\xi : \xi < \vartheta\}$ is a s.s. set, $\operatorname{dom} f$ and $\operatorname{ran} f$ consist of powers of 2, and $\frac{f(u)}{u} \leq \frac{f(u')}{u'}$ for all $u < u'$ in $\operatorname{dom} f$ by (12). By Saturation there is an internal function F with $D = \operatorname{dom} F$ a hyperfinite subset of ${}^*\mathbb{N} \smallsetminus \{0\}$, such that $\operatorname{dom} f \subseteq \operatorname{dom} F$, $F(u_\xi) = v_\xi$ for all ξ, and still $D = \operatorname{dom} F$ and $Z = \operatorname{ran} F$ consist of powers of 2 and $\frac{F(d)}{d} \leq \frac{F(d')}{d'}$ for all $d < d'$ in D.

Let $h = |D|^{\mathrm{int}} = |Z|^{\mathrm{int}}$ and $D = \{d_1, d_2, \ldots, d_h\}$, $Z = \{z_1, z_2, \ldots, z_h\}$, in the increasing order of ${}^*\mathbb{N}$ in $\mathbb{I}$. Then $z_\nu = F(d_\nu)$ for all $\nu = 1, \ldots, h$. As all d_ν, z_ν are powers of 2, the fractions $j_\nu = \frac{d_{\nu+1}}{d_\nu}$ and $k_\nu = \frac{z_{\nu+1}}{z_\nu}$ belong to ${}^*\mathbb{N}$ and $j_\nu \leq k_\nu$ by the above. Note also that $d_1 = u_0 = 1$.

Any number $x \in {}^*\mathbb{N}$ admits, in $\mathbb{I}$, a unique representation in the form $x = \sum_{\nu=1}^{h} \alpha_\nu d_\nu$, where $\alpha_\nu \in {}^*\mathbb{N}$ and $0 \leq \alpha_\nu < j_\nu$ for all $\nu = 1, \ldots, h-1$ (but α_h is not restricted, of course). The first idea that comes to mind is to try $\sigma(x) = \sum_{\nu=1}^{h} \alpha_\nu z_\nu$ as a reduction of ${}^*\mathbb{N}/U$ to ${}^*\mathbb{N}/V$. However this does not work. Indeed let $x = \sum_{\nu=1}^{h} d_\nu$ and $x' = \sum_{\nu=1}^{h-1}(j_\nu - 1)d_\nu$, so that $x - x' = 1$ but $|\sigma(x) - \sigma(x')|$ can be very big in the case when, say, $k_\nu > j_\nu$ for all ν. However there is a useful modification.

Suppose that $x = \sum_{\nu=1}^{h} \alpha_\nu d_\nu \in {}^*\mathbb{N}$, and $0 \leq \alpha_\nu < j_\nu$ for $\nu = 1, \ldots, h-1$, as above. Say that x is *type-1* if there exist indices $1 \leq \nu' < \nu'' \leq h - 1$ such that $d_{\nu'} \in U$, $d_{\nu''} \notin U$, and $\alpha_\nu = j_\nu - 1$ for all ν such that $\nu' \leq \nu \leq \nu''$. Then take the largest ν'' and the least ν' such that the pair ν', ν'' has this property, and put $\bar\alpha_\nu = a_\nu$ for all $\nu < \nu'$ and $\nu > \nu''$, $\bar\alpha_\nu = 0$ for $\nu' \leq \nu \leq \nu''$, and $\bar\alpha_{\nu''+1} = \alpha_{\nu''+1}+1$, and define $\bar x = \sum_{\nu=1}^{h} \bar\alpha_\nu d_\nu$. Otherwise ($x$ is *type-2*) put $\bar x = x$. Easily $\bar x - x = d_{\nu'} \in U$ in the type-1 case.

Prove that the map $\rho(x) = \sigma(\bar x)$ is a reduction of ${}^*\mathbb{N}/U$ to ${}^*\mathbb{N}/V$, that is, $|x - x'| \in U \iff |\sigma(\bar x) - \sigma(\bar y)| \in V$ holds for all $x, x' \in {}^*\mathbb{N}$.

Assume that $x = \sum_{\nu=1}^{h} \alpha_\nu d_\nu$ and $y = \sum_{\nu=1}^{h} \gamma_\nu d_\nu$, where $\alpha_\nu, \gamma_\nu < j_\nu$, and $|x - y| \in U$, hence $|x - y| < u_\xi = d_\nu$ for some $\xi < \vartheta$, $\nu < h$. Let $x < y$. Assume w.l.o.g. that x, y are of type-2. (Otherwise change x, y to $\bar x, \bar y$.) There exist infinitely (but $\mathbb{I}$-finitely) many indices $\nu' > \nu$ such that $\alpha_{\nu'} \neq j_{\nu'} - 1$. In this case $\alpha_{\nu'} = \gamma_{\nu'}$ for all $\nu' \geq \nu$ by the assumption $|x - y| < d_\nu$. Thus $|\sigma(x) - \sigma(y)| \in V$ (since $j_\nu \leq k_\nu$ for all ν), as required.

Now suppose that $x < y$ are as above, in particular, of type-2, but $|x - y| \notin U$, hence $|x - y| > u_\xi$ for all $\xi < \vartheta$. Then $D' = \{d_\nu \in D : \alpha_\nu \neq \gamma_\nu\}$ is an internal set, hence it has the largest element, say $d_{\nu''} = \max D'$. Note that $d_{\nu''} \notin U$. (Use the assumption $|x - y| > u$ for all $u \in U$.) We have $\alpha_{\nu''} < \gamma_{\nu''}$ (as $x < y$). Then the only opportunity for $|\sigma(x) - \sigma(y)|$ to belong to V is obviously the existence of an index $\nu' < \nu''$ such that $z_{\nu'} \in V$ and $\gamma_\nu = 0$, $\alpha_\nu = j_\nu - 1 = k_\nu - 1$ for all ν between ν' and ν''. But this contradicts the assumption that x is of type-2. Thus $|\sigma(x) - \sigma(y)| \notin V$, as required.

(v) The proof of this item follows the same line as the proof of (iv), but with appropriate changes, of course. It will appear elsewhere.

(vi) Suppose that U, V are resp. s.s. cofinal, s.s. coinitial additive fast cuts. Prove that $|{}^*\mathbb{N}/U|^{\mathtt{eff}} \not\lesssim |{}^*\mathbb{N}/V|^{\mathtt{eff}}$; the proof of $|{}^*\mathbb{N}/V|^{\mathtt{eff}} \not\lesssim |{}^*\mathbb{N}/U|^{\mathtt{eff}}$ is similar. Choose an increasing sequence $\{u_\xi\}_{\xi<\vartheta}$ and a decreasing sequence $\{v_\eta\}_{\eta<\tau}$ resp. cofinal in U and coinitial in ${}^*\mathbb{N} \setminus V$; $\vartheta = \mathtt{cof}\, U$ and $\tau = \mathtt{coi}\, V$ being infinite regular cardinals in $\mathbb{WF}$.

Suppose on the contrary that $R \subseteq {}^*\mathbb{N} \times {}^*\mathbb{N}$ is an invariant pre-injection of ${}^*\mathbb{N}/U$ to ${}^*\mathbb{N}/V$, that is, $|x - x'| \in U \iff |y - y'| \in V$ for any pairs $\langle x, y \rangle$ and $\langle x', y' \rangle$ in R, and $\mathtt{dom}\, R = {}^*\mathbb{N}$. Then $R = \bigcup_{a \in A} \bigcap_{b \in B} R_{ab}$, where $A, B \in \mathbb{WF}$ and R_{ab} are internal sets. Arguing as above in the proof of (iv) (parts 1,2), we obtain by **Saturation** for any fixed $a \in A$:

$$\exists \text{ finite } F \subseteq B \;\; \exists \xi < \vartheta \;\; \exists \eta < \tau \quad \forall x, x', y, y' \in {}^*\mathbb{N}:$$
$$x \, R_{aF} \, y \wedge x' \, R_{aF} \, y' \wedge |y - y'| < v_\eta \implies |x - x'| < u_\xi, \qquad (13)$$

where $R_{aF} = \bigcap_{b \in F} R_{ab}$, and, in the opposite direction,

$$\forall \xi < \vartheta \;\; \forall \eta < \tau \;\; \exists \text{ finite } F' \subseteq B \quad \forall x, x', y, y' \in {}^*\mathbb{N}:$$
$$x \, R_{aF'} \, y \wedge x' \, R_{aF'} \, y' \wedge |x - x'| < u_\xi \implies |y - y'| < v_\eta. \qquad (14)$$

Let $a \in A$. Take ξ, η, F as in (13). Take then F' as in (14) for $\xi + 1$ and η. We may assume that $F \subseteq F'$. Then for all x, x' in the set $D(a) = \mathtt{dom}\, R(a)$, where $R(a) = R_{aF'}$, we have $|x - x'| < u_{\xi+1} \implies |x - x'| < u_\xi$. Assuming w. l. o. g. that $\frac{u_{\xi+1}}{u_\xi}$ is infinitely large for all ξ, we conclude that each $D(a)$ is an internal scattered set in the sense of 11.1, and so on towards contradiction as above.

$$\square \; (Thm\ 10.2)$$

12. On small and large effective sets

Here we prove a "small–large" type theorem related to $\boldsymbol{\Delta}_2^{\mathtt{ss}}$ quotients. The notions of smallness and largeness will be connected with a cut $U \subseteq {}^*\mathtt{Card}$, as in Corollary 5.6. By necessity there also will be a gap between the largeness and smallness, but we don't know whether its size can be reduced.

Recall that a cut (initial segment) $U \subseteq {}^*\mathtt{Card}$ is called *exponential* iff $\kappa \in U \implies 2^\kappa \in U$, or, equivalently, $U = 2^U$ holds, where $2^U = \{\vartheta \in {}^*\mathtt{Card} : \exists \kappa \in U \, (\vartheta \le 2^\kappa)\}$. ($2^\kappa$ is understood as the cardinal exponentiation in $\mathbb{I}$.)

We write $\lambda \ge 2^U$ to mean $\lambda \ge 2^\kappa$ for all $\kappa \in U$.

Theorem 12.1. *Suppose that* E *is a* $\boldsymbol{\Delta}_2^{\mathtt{ss}}$ *equivalence relation on an internal set* H *and* $U \subseteq {}^*\mathtt{Card}$ *is a* $\boldsymbol{\Delta}_2^{\mathtt{ss}}$ *cut such that* $\mathbb{N} \subseteq U$. *Then at least one of the following conditions holds:*

(A) *for any $\lambda \in {}^*\mathrm{Card}$ with $\lambda \geq 2^U$ and any $m \in {}^*\mathbb{N} \smallsetminus \mathbb{N}$ there is an internal map ρ defined on H such that $|\operatorname{ran}\rho|^{\mathrm{int}} \leq \lambda^m$ ($=\lambda$ whenever $\lambda \notin {}^*\mathbb{N}$) and $\rho(x) = \rho(y) \Longrightarrow x \mathrel{\mathsf{E}} y$ for all $x, y \in H$;*

(B) *there exists an internal set $Y \subseteq H$ of pairwise E-inequivalent elements such that $|Y|^{\mathrm{int}} \notin U$.*

If U is an exponential non-internal cut then (A) *and* (B) *are incompatible even in the case when $\boldsymbol{\Delta}_2^{\mathrm{ss}}$ maps ρ are allowed in* (A).

In terms of effective cardinals (B) means $\kappa \leq |H/\mathsf{E}|^{\mathrm{eff}}$ (and even by means of an internal reduction) for some $\kappa = |Y|^{\mathrm{int}} \in {}^*\mathbb{N} \smallsetminus U$, that is a restriction of the cardinality of the quotient H/E from below. Accordingly (A) means that for all $\lambda \geq 2^U$ and $m \in {}^*\mathbb{N} \smallsetminus \mathbb{N}$ and any internal Z with $|Z|^{\mathrm{int}} = \lambda^m$ there is an equivalence relation F on Z (in terms of (A), $\rho(x) \mathrel{\mathsf{F}} \rho(y)$ iff $x \mathrel{\mathsf{E}} y$) such that $|H/\mathsf{E}|^{\mathrm{eff}} \leq |Z/\mathsf{F}|^{\mathrm{eff}}$ (still by means of an internal reduction), a restriction of the cardinality of H/E from above.

Some theorems of this form are known from descriptive set theory, for instance Silver's theorem on $\boldsymbol{\Pi}_1^1$ equivalence relations in [28], in which "small" means at most countably many equivalence classes while "large" means that there exists a pairwise E-inequivalent perfect set.

Note that the implication $\rho(x) = \rho(y) \Longrightarrow x \mathrel{\mathsf{E}} y$ in (A) cannot be replaced by the equivalence $\rho(x) = \rho(y) \Longleftrightarrow x \mathrel{\mathsf{E}} y$: indeed the latter would imply the $\boldsymbol{\Delta}_2^{\mathrm{ss}}$ smoothness of E, which, generally speaking, is not the case even for equivalence relations of the form M_U by Theorem 10.2.

Proof (Theorem 12.1). *Case 1*: U is standard size cofinal, including internal cuts. In this case we prove a stronger disjunction $(\mathrm{A}') \vee (\mathrm{B})$, where

(A') there exist a set $D \in \mathbb{WF}$, and for each $d \in D$ an internal set R_d and an internal map $f_d : H \to R_d$ such that $|R_d|^{\mathrm{int}} \in 2^U$ and $f(x) = f(y) \Longrightarrow x \mathrel{\mathsf{E}} y$ for all $x, y \in H$, where $f(x) = \{f_d(x)\}_{d \in D}$.

We first show that (A') implies (A). Suppose that $\lambda \geq 2^U$, $m \in {}^*\mathbb{N} \smallsetminus \mathbb{N}$. Recall that the map $d \mapsto {}^*d$ is an injection $D \to {}^*D$. Its image $D' = \{{}^*d : d \in D\} \subseteq {}^*D$ is a set of standard size together with D. By 4.1(iii), D' can be covered by an internal set $S \subseteq {}^*D$ such that $|S|^{\mathrm{int}} \leq m$. The Extension principle (1.3.13 in [17]) yields an internal function F defined on $S \times H$ so that $F({}^*d, x) = f_d(x)$ for all $d \in D$, $x \in H$. By the same reasons there is an internal map r defined on S so that $r({}^*d) = R_d$ for all $d \in D$. We can assume that for any $s \in S$, $r(s)$ is an internal set with $|r(s)|^{\mathrm{int}} < \lambda$, and $F(s, x) \in r(s)$ for all $x \in H$. (Otherwise redefine r and F by $r(s) = \{0\}$

and $F(s,x) = 0$ for all "bad" s — but none of $s = {}^*d$, $d \in D$, is "bad" in the assumptions of (A').) Put $\rho(x)(s) = F(s,x)$ for $x \in H$, $s \in S$.

We begin the proof of (A') $\vee$ (B). By definition $\mathsf{E} = \bigcup_{a \in A} \bigcap_{b \in B} E_b^a$, where all sets $E_b^a \subseteq H \times H$ are internal while $A, B \in \mathsf{WF}$. We may w. l. o. g. assume that every set E_b^a is symmetric (similarly to E itself), that is, $E_b^a = (E_b^a)^{-1}$, where $E^{-1} = \{\langle y,x \rangle : x \, E \, y\}$: indeed

$$\mathsf{E} = \mathsf{E} \cap \mathsf{E}^{-1} = \bigcup_{a \in A} \bigcap_{b,b' \in B} E_b^a \cup (E_{b'}^a)^{-1} = \bigcup_{a \in A} \bigcap_{b,b' \in B} C_{bb'}^a \,,$$

where the sets $C_{bb'}^a = (E_b^a \cup (E_{b'}^a)^{-1}) \cap (E_{b'}^a \cup (E_b^a)^{-1})$ are symmetric. (We write $x \, E \, y$ for $\langle x,y \rangle \in E$ whenever E is a binary relation.)

It follows from the transitivity of E that for any $x, y \in H$

$$\exists\, a \in A \,\exists\, z \in H \,\forall\, b \in B \;(x \, E_b^a \, z \wedge y \, E_b^a \, z) \implies x \, \mathsf{E} \, y\,.$$

The axiom of Saturation transforms this to

$$\exists\, a \in A \,\forall\, B' \in \mathscr{P}_{\mathtt{fin}}(B) \,\exists\, z \in H \;(x \, E_{B'}^a \, z \wedge y \, E_{B'}^a \, z) \implies x \, \mathsf{E} \, y\,,$$

where $E_{B'}^a = \bigcap_{b \in B'} E_b^a$. As the two leftmost quantifiers are restricted to the sets A and $\mathscr{P}_{\mathtt{fin}}(B)$ in WF, the last formula is equivalent to

$$\forall\, \varphi \in \Phi \,\exists\, a \in A \,\exists\, z \in H \;(x \, E_{\varphi(a)}^a \, z \wedge y \, E_{\varphi(a)}^a \, z) \implies x \, \mathsf{E} \, y\,, \tag{15}$$

where $\Phi \in \mathsf{WF}$ is the set of all functions $\varphi : A \to \mathscr{P}_{\mathtt{fin}}(B)$.

As U is standard size cofinal, there is an increasing sequence $\{\nu_\xi\}_{\xi < \vartheta}$ of elements $\nu_\xi \in U$, cofinal in U, with ϑ being an infinite cardinal in Card, or simply U is internal, $\vartheta = 1 = \{0\}$, and ν_0 is the least element in $^*\mathsf{Card} \smallsetminus U$.

Suppose that (B) of the theorem fails, *i.e.* there is no pairwise E-inequivalent sets Y with $|Y|^{\mathtt{int}} \notin U$. More formally,

$$\forall\, Y \in P \left(\forall\, \xi < \vartheta \,(|Y|^{\mathtt{int}} \geq \nu_\xi) \implies \exists\, x \neq y \in Y \,\exists\, a \in A \,\forall\, b \in B \,(x \, E_b^a \, y)\right),$$

where $P = \mathscr{P}^{\mathtt{I}}(H) = \{Y \subseteq H : Y \text{ is internal}\}$. Saturation converts the expression to the right of $\implies$ to

$$\exists\, a \in A \,\forall\, B' \in \mathscr{P}_{\mathtt{fin}}(B) \,\exists\, x \neq y \in Y \,(x \, E_{B'}^a \, y)\,,$$

and then to $\forall\, \varphi \in \Phi \,\exists\, a \in A \,\exists\, x \neq y \in Y \,(x \, E_{\varphi(a)}^a \, y)$. We conclude that for any function $\varphi \in \Phi$

$$\forall\, Y \in P \left(\forall\, \xi < \vartheta \,(|Y|^{\mathtt{int}} \geq \nu_\xi) \implies \exists\, a \in A \,\exists\, x \neq y \in Y \,(x \, E_{\varphi(a)}^a \, y)\right).$$

Saturation yields an ordinal $\xi(\varphi) < \vartheta$ and a finite set $A_\varphi \subseteq A$ such that

$$\forall\, Y \in P \left(|Y|^{\mathtt{int}} \geq \nu_{\xi(\varphi)} \implies \exists\, a \in A_\varphi \,\exists\, x \neq y \in Y \,(x \, E_{\varphi(a)}^a \, y)\right). \tag{16}$$

Let Y_φ be any maximal (internal) subset of H such that $\neg\, x\, E^a_{\varphi(a)}\, y$ for all $a \in A_\varphi$ and $x \neq y \in Y_\varphi$. Then (16) implies $|Y_\varphi|^{\text{int}} < \nu_{\xi(\varphi)}$, while the properties of maximality of Y_φ and symmetricity of E^a_b imply

$$\forall\, x \in H\ \exists\, y \in Y_\varphi\ \exists\, a \in A_\varphi\ (x\, E^a_{\varphi(a)}\, y)\,. \tag{17}$$

Put $\zeta_x(\varphi, a) = \{y \in Y_\varphi : x\, E^a_{\varphi(a)}\, y\}$ for $x \in H$, $\varphi \in \Phi$, $a \in A_\varphi$. Thus ζ_x belongs to the set Z of all functions ζ defined on the set $D = \{\langle \varphi, a\rangle :$ $\varphi \in \Phi \wedge a \in A_\varphi\} \in \mathsf{WF}$ and satisfying $\zeta_x(\varphi, a) \in R_\varphi = \mathscr{P}^{\|}(Y_\varphi)$. The sets R_φ are internal and satisfy $|R_\varphi|^{\text{int}} \in 2^U$ (because $|Y_\varphi|^{\text{int}} \in U$).

We claim that $\zeta_x = \zeta_y$ implies $x\, \mathsf{E}\, y$. It suffices, by (15), to prove that for every $\varphi \in \Phi$ there exist $a \in A$, $z \in H$ such that $x\, E^a_{\varphi(a)}\, z$ and $y\, E^a_{\varphi(a)}\, z$. Note that $\zeta_x(\varphi, a) = \zeta_y(\varphi, a) \neq \varnothing$ for some $a \in A_\varphi$ by (17). Take any $z \in \zeta_x(\varphi, a)$. Then $z \in Y_\varphi$, thus $\langle x, z\rangle$ and $\langle y, z\rangle$ belong to $E^a_{\varphi(a)}$, as required.

To accomplish the proof of (A$'$) in the assumption $\neg(B)$, we put $f_d(x) = \zeta_x(\varphi, a)$ and $R_d = R_\varphi$ for all $x \in H$ and $d = \langle \varphi, a\rangle \in D$.

Case 2: U is standard size coinitial, but non-internal. Suppose that (B) fails, and consider any $m \in {}^*\mathbb{N} \smallsetminus \mathbb{N}$ and $\lambda \geq 2^U$. Then (B) fails also for the internal, hence, s.s. cofinal, cut $U' = \{\kappa \in {}^*\mathtt{Card} : 2^\kappa \leq \lambda\}$: indeed, $U \subseteq U'$ by the choice of λ. Therefore (A) holds for U'. Thus there is an internal map ρ, $\operatorname{dom}\rho = H$, such that $|\operatorname{ran}\rho|^{\text{int}} \leq \lambda^m$ and $\rho(x) = \rho(y) \Longrightarrow x\, \mathsf{E}\, y$.

Incompatibility. Assume that $Y \subseteq H$ witnesses (B), in particular, $\kappa = |Y|^{\text{int}} \notin U = 2^U$. Then $U' = \{\lambda \in {}^*\mathtt{Card} : 2^\lambda < \kappa\}$ is an internal cut with $U \subseteq U'$. Thus $U \subsetneq U'$ since U is non-internal. Therefore there is $\vartheta \notin U$ such that $2^\vartheta < \kappa$. Applying this trick once again, we find $\vartheta \notin U$ with $2^{2^\vartheta} < \kappa$. Suppose on the contrary that ρ witnesses (A) for $\lambda = 2^\vartheta$ and some $m \in {}^*\mathbb{N} \smallsetminus \mathbb{N}$, $m < \vartheta$. Then $\rho\restriction Y$ is an internal injection of Y into an internal set $Z = \rho''Y$ satisfying $|Z|^{\text{int}} \leq 2^{\vartheta m}$. But this contradicts Theorem 3.1, since by definition $2^{\vartheta m} \cdot n < 2^{\vartheta \cdot \vartheta} < 2^{2^\vartheta} < \kappa = |Y|^{\text{int}}$ for any $n \in \mathbb{N}$. $\qquad\square$

The case $U = \mathbb{N}$ deserves special attention. Since $\mathbb{N}$ is a s.s. cofinal cut, a stronger dichotomy holds: (A$'$) $\vee$ (B). Clearly (B) claims the existence of an infinite internal set of pairwise E-inequivalent elements in this case. On the other hand, the sets R_d in (A$'$) are finite, hence $P = \prod_{d \in D} R_d$ is a set of standard size, and so is any quotient of the form P/F, where F is an equivalence relation on P. Thus (A$'$) implies that H/E itself is a set of standard size. Such a dichotomy (*i.e.* standard size of H/E or an infinite internal pairwise inequivalent set) is contained in Theorem 1.4.11 in [17]. Similar dichotomies appeared in [16] for countably determined equivalence relations. P. Zlatoš informed us that a close result for $U = \mathbb{N}$ was earlier obtained by Vencovská (unpublished) in the frameworks of **AST**.

13. Nonstandard version of the finite Ramsey theorem

The following corollary of Theorem 12.1 is a Ramsey–like result. Recall that $[A]^n = \{X \subseteq A : \operatorname{card} X = n\}$. By a *partition* of $[A]^n$ we understand any equivalence relation E on $[A]^n$, and a *homogeneous set* for E is any $H \subseteq A$ such that the sets $X \in [H]^n$ are pairwise E-equivalent.

The finite Ramsey theorem claims (in **ZFC**) that

$(*)$ *for any natural numbers ℓ, n, s there is $k \in \mathbb{N}$ such that $k \to (\ell)^n_s$.*

Here $k \to (\ell)^n_s$ means that for any partition of $[k]^n$ into s-many parts there is an ℓ-element homogeneous set $H \subseteq k$. We refer to [25], and also to 3.3.7 in [1], §6 in [21], or [3] for a modern proof, details and related results.

Let $K(\ell, s, n)$ denote the least k satisfying $k \to (\ell)^n_s$. It is known that $K(\ell, s, n)$ is rapidly increasing as a function of ℓ for any fixed n, s, see [3]. But of course K is a recursive function.

It is an easy nonstandard corollary of $(*)$ that $\kappa \xrightarrow{\text{int}} (\ell)^n_s$ for all $n, s, \ell \in \mathbb{N}$ and $\kappa \in {}^*\mathbb{N} \setminus \mathbb{N}$ where int over the arrow means that the partition and the homogeneous set are assumed to be internal. A nicer nonstandard version, also well-known, is $\kappa \xrightarrow{\text{int}} (\infty)^n_s$ *for any $n, s \in \mathbb{N}$ and $\kappa \in {}^*\mathbb{N} \setminus \mathbb{N}$*, that is, any internal partition $[\kappa]^n$ into s parts admits an infinite internal homogeneous set. By the way, its quantifier structure is simpler than that of $(*)$:

$$\forall \kappa, \ell, n, s \; \forall \text{ partition } \exists A \; \forall u, v \in [A]^n.$$

The following theorem contains a much more general claim. In **HST**, define a function K in $\mathbb{WF}$ as above. Then ${}^*K$ is a standard function ${}^*\mathbb{N}^3 \to {}^*\mathbb{N}$ having in the internal universe $\mathbb{I}$ the same properties as K in $\mathbb{WF}$.

Theorem 13.1. *Suppose that $U \subsetneq {}^*\mathbb{N}$ is a $\mathbf{\Delta}^{\text{ss}}_2$ cut with $\mathbb{N} \subseteq U$, closed under ${}^*K$ and exponential, $n \in \mathbb{N}$, $\kappa \in {}^*\mathbb{N} \setminus U$, and E is a $\mathbf{\Delta}^{\text{ss}}_2$ equivalence relation on $[\kappa]^n$. If there is no internal pairwise E-inequivalent sets $Y \subseteq [\kappa]^n$ satisfying $|Y|^{\text{int}} \notin U$, then the partition E admits an internal homogeneous set $A \subseteq \kappa$ such that $|A|^{\text{int}} \notin U$.*

A similar result was obtained in [22] in the case $U = \mathbb{N}$ for countably determined equivalence relations. See Theorem 2.8 in [19] for a somewhat weaker result in the case when t in the proof of 13.1 is predefined.

Proof. Define, in $\mathbb{WF}$, $f(s) = K(s, s, n)$ for each $s \in \mathbb{N}$. Then $f : \mathbb{N} \to \mathbb{N}$ and $s \leq f(s)$, $\forall s$. The map ${}^*f$ has the same properties with respect to ${}^*\mathbb{N}$. As U is ${}^*K$-closed and exponential, there exist $s, \vartheta \in {}^*\mathbb{N} \setminus U$ and $m \in {}^*\mathbb{N} \setminus \mathbb{N}$ such that ${}^*f(s) = {}^*K(s, s, n) \leq \kappa$ and $2^{\vartheta m} \leq s$.

In our assumptions, (B) of Theorem 12.1 fails, hence (A) holds, that is, there exists an internal map ρ defined on $[\kappa]^n$ such that $|\operatorname{ran}\rho|^{\mathrm{int}} \leq 2^{\vartheta m} \leq s$ and $\rho(u) = \rho(u) \implies u \mathrel{\mathsf{E}} u$ for all $u, v \in [\kappa]^n$. On the other hand, we have $\kappa \to (s)^n_s$ by the choice of s, therefore the partition of $[\kappa]^n$ induced by ρ has an internal homogeneous set A such that $|A|^{\mathrm{int}} = s \notin U$. Thus $\rho(u) = \rho(v)$, and hence $u \mathrel{\mathsf{E}} v$, for all $u, v \in [A]^n$. $\qquad\square$

References

1. C. C. Chang and H. J. Keisler, *Model Theory*, 3rd edition. Amsterdam: North Holland, 1992, xiv + 650 pp.
2. K. Čuda and P. Vopenka, Real and imaginary classes in the AST, *Comment. Math. Univ. Carol.*, **20**, pp. 639–653 (1979).
3. P. Erdös, A. Hajnal, A. Máté, and R. Rado, *Combinatorial set theory: partition relations for cardinals*. Amsterdam: North Holland, 1977.
4. P. Frankl, Families of finite sets satisfying an intersection condition. *Bull. Austral. Math. Soc.* **15**, 1, 73–79 (1976).
5. E. I. Gordon, A. G. Kusraev, and S. S. Kutateladze, *Infinitesimal analysis*, Kluwer, Dordrecht, 2002. xiv+422 pp.
6. G. Hjorth, Orbit cardinals: on the effective cardinalities arising as quotient spaces of the form X/G where G acts on a Polish space X, *Israel J. Math.* **111**, pp. 221–261 (1999).
7. G. Hjorth, *Classification and Orbit Equivalence Relations* (Mathematical surveys and monographs, 75), AMS, 2000.
8. K. Hrbaček, Axiomatic foundations for nonstandard analysis, *Fund. Math.* **98**, pp. 1–19 (1978).
9. K. Hrbaček, Nonstandard set theory, *Amer. Math. Monthly* **86**, pp. 659–677 (1979).
10. S. Jackson, A. S. Kechris, and A. Louveau, Countable Borel equivalence relations, *J. Math. Logic*, **2**, 1, pp. 1–80 (2002).
11. R. Jin, Existence of some sparse sets of nonstandard natural numbers, *J. Symbolic Logic* **66**, 2, pp. 959–973 (2001).
12. M. Kalina and P. Zlatoš, Borel classes in AST, measurability, cuts, and equivalence, *Comment. Math. Univ. Carol.*, **30**, pp. 357–372 (1989).
13. V. Kanovei, Undecidable hypotheses in Edward Nelson's Internal Set Theory, *Russian Math. Surveys* **46**, 6, pp. 1–54 (1991).
14. V. Kanovei and M. Reeken, Internal approach to external sets and universes. *Studia Logica*, **55**, 2, pp. 229–257 (1995), **55**, 3, pp. 347–376 (1995), **56**, 3, pp. 293–322 (1996).
15. V. Kanovei and M. Reeken, Isomorphism property in nonstandard extensions of a **ZFC** universe, *Ann. Pure Appl. Logic*, **88**, pp. 1–25 (1997).
16. V. Kanovei and M. Reeken, Borel and countably determined reducibility in nonstandard domain. *Monats. für Math.*, **140**, 3, pp. 197–231 (2003).
17. V. Kanovei and M. Reeken, *Nonstandard Analysis: Axiomatically*, Springer, 2004.

18. A. S. Kechris, New directions in descriptive set theory, *Bull. Symbolic Logic*, **2**, pp. 161–174 (1999).

19. H. J. Keisler, K. Kunen, A. Miller, and S. Leth, Descriptive set theory over hyperfinite sets, *J. Symbolic Logic*, **54**, pp. 1167–1180 (1989).

20. H. J. Keisler and S. Leth, Meager sets on the hyperfinite time line, *J. Symbolic Logic* **56**, pp. 71–102 (1991).

21. K. Kunen, Combinatorics, in: *Handbook of mathematical logic*, Studies in Logic and Foundations of Math., 90, North-Holland, Amsterdam, 1977, pp. 371–401.

22. J. Mlček and P. Zlatoš, Some Ramsey-type theorems for countably determined sets, *Arch. Math. Logic* **41**, 7, pp. 619–630 (2002).

23. E. Nelson, Internal set theory; a new approach to nonstandard analysis, *Bull. Amer. Math. Soc.* **83**, 6, pp. 1165–1198 (1977).

24. R. L. Panetta, A finite intersection property and the measurability of ultrafilters on hyperfinite sets, *Ann. Math. Artif. Intell.* **6**, 1–3, pp. 267-270 (1992).

25. F. P. Ramsey, On a problem in formal logic, *Proc. London Math. Soc.*, **30**, pp. 264–286 (1930).

26. A. Robinson *Non-standard analysis*, North-Holland, Amsterdam, 1966, xi+293 pp.

27. K. Schilling, Vanishing Borel sets. *J. Symbolic Logic*, **63**, 1, pp. 262–268 (1998).

28. J. Silver, Counting the number of equivalence classes of Borel and coanalytic equivalence relations, *Ann. Math. Log.*, **18**, pp. 1–28 (1980).

29. V. A. Uspensky, What is nonstandard analysis? (Russian), Nauka, M., 1987.

MODEL-THEORETIC METHODS OF ANALYSIS OF COMPUTER ARITHMETIC

SERGE P. KOVALYOV

Institute of Computational Technologies,
6 Lavrentiev Ave,
Novosibirsk, 630090, Russia
E-mail: kovalyov@nsc.ru

Practical problems associated with engineering efficient robust algorithms for real world computers lay beyond the traditional scope of mathematical theory of algorithms. Special mathematical methods are required to formally specify and verify empirical approaches routinely used by technicians. Such methods based on model theory and multiple-valued logics are presented in this report. A construct of partial interpretation is elaborated for developing formal specifications of computer arithmetics taking resource limitations into account. Finite-valued Łukasiewicz logic is proven to be capable to express and verify operations used in computer implementations of integral arithmetic.

1. Introduction

One of the key problems arising at developing computing systems is caused by restrictions on available amount of resources. Computations performed on real devices are limited by finite amounts of time (performance) and space (memory). Due to memory limitations computer implementations of arithmetic fail to satisfy standard arithmetic axioms that have only infinite models. Nevertheless arithmetic devices are required to have supported numbers behaving similarly to their theoretical originals. Software engineers qualify this situation as conflict between functional and non-functional requirements to computation models. With regard to semiconductor computers this conflict is considered as de-facto resolved few decades ago (although, as shown in the report, not ideally). However, when developing novel non-traditional computing devices the problem arises again.

General mathematical methods are needed to solve it. Such methods based on model theory and multiple-valued logics are presented in this report. A construct of partial interpretation is elaborated for developing formal specifications of computer arithmetics taking resource limitations into account. Finite-valued Łukasiewicz logic is employed to express and verify operations used in computer implementations of integral arithmetic.

In fact, it is shown that Lukasiewicz logic is "right" (natural) abstraction of various finite approximations of arithmetic.

2. Specifications of computer arithmetics

Traditionally, the Abstract Data Type technique is used to specify computer implementations of real world objects [5]. However it has a serious limitation: it doesn't offer tools for abstract modeling of infinite entities by finite structures. The author of the report has suggested one such tool in [4]. It is special modification of standard model-theoretic approach called partial interpretation of first-order theory T. It lies in constructing algebraic system that must verify only those statements from T that contain only terms that can be substituted by constants from given finite subsignature of its signature. Thus the formal specification of resource available to represent objects described by T is constituted by the explicitly given set of constant symbols. Observe that even the usual properties of equality relation are allowed to fail beyond this set. Isomorph embedding of this set to universes of models of various theories allows formalizing the concept of polymorphism.

The precise definition of this construction follows.

Definition 1. We will consider first-order languages without equality. Let σ be a signature containing the equality sign $\equiv$, let T be a theory of σ containing the standard axioms of equality, and let σ_0 be a finite subsignature of σ. The *projection* $T \downarrow \sigma_0$ is the set of quantifier-free sentences φ of σ_0 such that $T \vdash \varphi$ and, for every term t occurring in φ, there exists a constant symbol $c \in \sigma_0$ such that $T \vdash t \equiv c$. A *partial interpretation* of T in signature σ_0 is a pair $\langle \mathfrak{A}, \sigma_0 \rangle$, where $\mathfrak{A}$ is an algebraic system of signature $\sigma(\mathfrak{A}) \supseteq \sigma_0$ such that $\mathfrak{A} \models T \downarrow \sigma_0$.

Let $C(\sigma_0) \rightleftharpoons \{c_1, \ldots, c_n\}$ be the set of constant symbols of σ_0. The *relativization* of a formula φ on σ_0 is the formula $\varphi \upharpoonright \sigma_0$ obtained from φ by replacing all subformulas of the form $\forall x_0 \psi(x_0, x_1, \ldots, x_k)$ with the conjunctions $\bigwedge_{i=1,\ldots,n} \psi(c_i, x_1, \ldots, x_k)$ and the subformulas of the form $\exists x_0 \psi(x_0, x_1, \ldots, x_k)$ with the disjunctions $\bigvee_{i=1,\ldots,n} \psi(c_i, x_1, \ldots, x_k)$. The pair $\langle \mathfrak{A}, \sigma_0 \rangle$ is said to *support* a sentence φ if $\mathfrak{A} \models \varphi \upharpoonright \sigma_0$.

Let $D\sigma_0(\mathfrak{A})$ be the set of quantifier-free sentences of σ_0 valid in $\mathfrak{A}$. The *polymorphic structure* $\mathrm{Pol}\langle \mathfrak{A}, \sigma_0 \rangle$ is the collection of pairs $\langle \mathfrak{B}, \sigma_0 \rangle$ such that $\sigma(\mathfrak{B}) \supseteq \sigma_0$ and $D\sigma_0(\mathfrak{A})$ coincides with $D\sigma_0(\mathfrak{B})$. A *homomorphism* of pairs $\langle \mathfrak{A}, \sigma_0 \rangle$ and $\langle \mathfrak{B}, \sigma_1 \rangle$ is a partial map $h : |\mathfrak{A}| \to |\mathfrak{B}|$ defined on the set of interpretations of constant symbols in $C(\sigma_0 \cap \sigma_1)$ which is a partial homomorphism from the reduct $\mathfrak{A} \restriction (\sigma_0 \cap \sigma_1)$ into the reduct $\mathfrak{B} \restriction (\sigma_0 \cap \sigma_1)$.

$\square$

Computer arithmetics are formally specified as partial interpretations of arithmetic theories over the universe

$$E_{n+1} \rightleftharpoons \{0, 1, \ldots, n\}.$$

In this report such partial interpretations of integral arithmetic are considered that show the high degree of similarity to their originals. The degree of similarity is determined by the capability to support (in the sense of Definition 1) different arithmetic axioms. As shown in [4], the following two partial interpretations are the most adequate.

Definition 2.

a) *Initial segment of nonnegative integers with overflow* is the algebraic system

$$\mathrm{OA}_{n+1} \rightleftharpoons \langle E_{n+1}, 0, 1, \ldots, n-1, =, |+|, |-|, |\times| \rangle,$$
$$x \mid + \mid y \rightleftharpoons \min(n, x+y),$$
$$x \mid - \mid y \rightleftharpoons \max(0, x-y),$$
$$x \mid \times \mid y \rightleftharpoons \min(n, xy);$$

b) *(Almost) symmetric modular segment of integers* is the algebraic system

$$\mathrm{MA}_{n+1} \rightleftharpoons \langle E_{n+1}, 0, 1, \ldots, [(n-1)/2], -[n/2], \ldots, -1,$$
$$(=), (+), (-), (\times), \mathrm{Carry} \rangle,$$
$$x \, (=) \, y \rightleftharpoons (x = y) \vee (x, y) \in \{(0, n), (n, 0)\},$$
$$x \, (+) \, y \rightleftharpoons (x + y) \bmod n,$$
$$(-)x \rightleftharpoons n - x,$$
$$x \, (\times) \, y \rightleftharpoons xy \bmod n,$$
$$\mathrm{Carry}(x) \rightleftharpoons (x = n).$$

3. Arithmetics design method

Designers of computation models and algorithms use specifications of computer arithmetics as input data. They particularly need them while performing mapping of computing algorithms to computer architecture, i.e. binding computation stream to functional capabilities of employed computing devices [8]. Abstract mathematical method of modeling computer architectures is needed here. It should offer verification technique based on formal proof. As a basis for such method the author of the report has suggested to employ multiple-valued Lukasiewicz logic [4]. Numerical data storage units (variables) are used as architecture elements of computer arithmetic. Their values (states) correspond to logical constants. Computation operations are described as compositions of base logic functions. Such setting traditionally disposes one to apply multiple-valued logics. However, Lukasiewicz logic and its enrichments weren't thoroughly employed earlier. They provide rich capabilities to evaluate efficiency of computing models against different criteria: functional power, performance, energy consumption etc.

For these purposes multiple-valued logic is represented as matrix – algebraic system with the universe E_{n+1}. The following matrix corresponds to Lukasiewicz logic [3]:

$$L_{n+1} \rightleftharpoons \langle E_{n+1}, \sim, \to, \{n\}\rangle,$$
$$\sim x \rightleftharpoons n - x,$$
$$x \to y \rightleftharpoons \min(n, n - x + y).$$

The connectives of this matrix can be used to express many-valued disjunction and conjunction:

$$x \vee y \rightleftharpoons \max(x, y) = (x \to y) \to y,$$
$$x \wedge y \rightleftharpoons \min(x, y) = \sim (\sim x \vee \sim y) = \sim (x \to \sim (x \to y)),$$

Let's consider properties of Lukasiewicz logic as a clone – class of functions on E_{n+1} closed with respect to function composition. Every clone is a subclass of the Post logic P_{n+1} which consists of all functions on E_{n+1}. For any nonempty set $X \subset E_{n+1}$ let

$$C_{n+1}^X \rightleftharpoons \{f \in P_{n+1} \mid f(X, \ldots, X) \subseteq X\},$$
$$D_{n+1}^X \rightleftharpoons \{f \in P_{n+1} \mid f(E_{n+1}, \ldots, E_{n+1}) \subseteq X\},$$
$$Q_{n+1}^X \rightleftharpoons \{f \in P_{n+1} \mid \text{ there exists } c \in X \text{ such that } f(X, \ldots, X) = c\},$$
$$T_{n+1} \rightleftharpoons C_{n+1}^{\{0,n\}}.$$

Class C_{n+1}^X is *precomplete* in P_{n+1}, i.e. it is closed and the closure of its union with an arbitrary function not expressible in it equals P_{n+1}. Regarding Lukasiewicz logic Evans and Schwartz have shown in [1] that it is *weakly complete* [7], i.e. the system of functions obtained by uniting it with the set of all constant functions on E_{n+1} is complete in P_{n+1}. Precisely due to functional incompleteness Lukasiewicz logic seems inadequate in modeling finite arithmetic. However, its incompleteness is overcome by addition numbers themselves (constant functions). It means that Łukasiewicz logic allows discovering structural properties of arithmetic operations that don't depend on their arguments values.

In [3], the following number-theoretic characterization is given for the lattice of clones that contain Łukasiewicz logic. We call *L-closed* a clone that contains connectives of L_{n+1}. Denote by $D(n)$ the set of all divisors of number n. A subset $Y \subseteq D(n)$ is said to be LCM-closed iff $1 \in Y$ and $\text{LCM}(x, y) \in Y$ whenever $x, y \in Y$. Henceforth, we denote by $F_{n+1}(Y)$ the class of functions f in P_{n+1} that satisfy the following set of divisibility conditions: for every $m \in Y$, if all $x_1, \ldots, x_k$ are divisible by m, then $f(x_1, \ldots, x_k)$ is also divisible by m. Then $F_{n+1}(Y) = \bigcap_{u \in Y} C_{n+1}^{V(u)}$, where $V(u) \rightleftharpoons \{v \mid v \in E_{n+1} \wedge u \in D(v)\}$ (thus $F_{n+1}(Y)$ is a clone). It is proven that class K_{n+1} is L-closed if and only if there exists such LCM-closed set $Y \subseteq E_{n+1}$ that $K_{n+1} = F_{n+1}(Y)$.

In particular, on the one hand L_{n+1} is contained in the precomplete class T_{n+1} and coincides with it if and only if n is prime. On the other hand, L_{n+1} contains class $D_{n+1}^{\{0,n\}}$ and doesn't coincide with it if $n > 1$ (hereinafter we assume that this condition holds). There exists a bijective correspondence between functions from $D_{n+1}^{\{0, n\}}$ and predicates on E_{n+1} that interpret arithmetic relations. They are constructed via disjunction and conjunction from Rosser–Turquette functions

$$J_i(x) \rightleftharpoons \begin{cases} n, & x = i, \\ 0, & x \neq i. \end{cases}$$

In the present report apparatus of Łukasiewicz logic is employed to analyze properties of operations of computer arithmetics described in Definition 2. The following results are obtained.

Proposition 1. *An expression whose computation is determined by a formula A does not cause overflow if and only if $\sim J_n(A)$ is a tautology in L_{n+1}.* $\qquad\square$

Lemma 1. *The following equalities hold.*
   a) $(x = y) = J_n((x \to y) \wedge (y \to x))$;
   b) $x \mid + \mid y = \sim x \to y$;
   c) $x \mid - \mid y = \sim (x \to y)$;
   d) $x \mid \times \mid y = \bigvee_{i \in E_{n+1}} (\sim x \to^i 0) \wedge J_i(y)$.

Theorem 1.
   a) *The functions $=, \mid + \mid, \mid - \mid, \mid \times \mid$ are expressible in L_{n+1}.*

b) *The system of functions $\{\ |+|, |-|\ \}$ forms a basis in the class $L_{n+1} \cap C_{n+1}^{\{0\}}$ which is precomplete in L_{n+1}.*

c) *The systems of functions $\{\ \sim, |+|\ \}$, $\{n, |-|\ \}$, and $OAL_{n+1} \rightleftharpoons \{=, |-|\}$ are bases in L_{n+1}.*

Corollary 1.1. *The set of constants and functions of the overflow arithmetic OA_{n+1} is complete in P_{n+1}.* $\qquad\square$

Lemma 2. *The following equalities hold.*

a) $x \bmod n = x \mid - \mid J_n(x)$;

b) $x\,(+)\,y = ((x \mid + \mid y) \bmod n) \mid + \mid \big((x \mid - \mid (n \mid - \mid y)) \bmod n\big)$;

c) $(-)x = n \mid - \mid x$;

d) $x\,(\times)\,y = \bigvee_{i \in E_{n+1}} (0\,(+)^i\,x) \wedge J_i(y)$;

e) $\mathrm{Carry}(x) = J_n(x)$;

f) $x\,(=)\,y = \mathrm{Carry}((-)[x\,(+)\,((-)y)])$.

Denote by M_{n+1} the class of functions that preserve the modular equality relation $(=)$ of the system MA_{n+1}. It is known [7] that the class M_{n+1} is precomplete but not weakly complete. Denote by M_{n+1}^Λ the class of functions that preserve the quaternary relation

$$\Lambda_{n+1} \rightleftharpoons \{\langle x, x, x, x \rangle \mid x \in E_{n+1}\}$$
$$\cup\,(\{0, n\}^4 \cap \{\langle x_1, x_2, x_3, x_4 \rangle \mid x_1 + x_2 = x_3 + x_4(\bmod 2n)\}).$$

Denote by U_{n+1}^1 class of all unary functions on E_{n+1}. Let

$$\mathrm{MP}_{n+1} \rightleftharpoons \{f \in \mathrm{P}_{n+1} \mid f(x_1, \ldots, x_{s-1}, 0, x_{s+1}, \ldots, x_k)$$
$$= f(x_1, \ldots, x_{s-1}, n, x_{s+1}, \ldots, x_k)\ \text{for all}\ s = 1, \ldots, k\},$$
$$\mathrm{J}_{n+1} \rightleftharpoons \{id, \sim, J_n, J_0, J_0 J_n, J_0 J_0\},$$
$$\mathrm{R}_{n+1} \rightleftharpoons (\mathrm{MP}_{n+1} \cap (\mathrm{D}_{n+1}^{\{0,n\}} \cup \mathrm{D}_{n+1}^{E_{n+1}\backslash\{n\}} \cup \mathrm{D}_{n+1}^{E_{n+1}\backslash\{0\}})) \cup \mathrm{J}_{n+1},$$
$$\mathrm{S}_{n+1} \rightleftharpoons (\mathrm{Q}_{n+1}^{\{0,n\}} \cap \mathrm{M}_{n+1}) \cup (\mathrm{T}_{n+1} \cap \mathrm{U}_{n+1}^1).$$

Theorem 2.

a) *The functions $(=)$, $(+)$, $(-)$, $(\times)$ are expressible in L_{n+1}.*

b) *The system of functions $\{(+), (-), (\times), Carry\}$ forms a base in the class*

$$MAC_{n+1} \subseteq L_{n+1} \cap R_{n+1}$$
$$\subset L_{n+1} \cap S_{n+1}$$
$$\subset L_{n+1} \cap M_{n+1}^{\Lambda}$$
$$\subset L_{n+1} \cap M_{n+1}$$
$$\subset L_{n+1},$$

where MAC_{n+1} coincides with $L_{n+1} \cap R_{n+1}$ if and only if n is prime.

   c) *The system of functions*

$$MAL_{n+1} \rightleftharpoons \begin{cases} \{(-), Carry, \vee\}, & n = 2, \\ \{(+), (-), Carry, \vee\}, & n > 2, \end{cases}$$

forms a base in L_{n+1}.

Corollary 2.1. *The set of constants and functions of the modular arithmetic MA_{n+1} is neither complete nor maximal in P_{n+1} but becomes complete after enrichment by function $\max(x,y)$, $(x,y) \in E_{n+1} \times E_{n+1}$.* $\qquad\square$

4. Digital number systems

Explicit expressions for arithmetic operations shown in Lemmas 1 and 2 demonstrate the following well-known fact: when n increases, computational complexity of machine implementations of operations increases. The traditional approach to decreasing complexity lays in employing digital number representation in positional system. Arithmetic over q-digit $(q > 1)$ number representation in b-ary positional system is interpreted over the universe $E_{b\uparrow q}$, where

$$b \uparrow q \rightleftharpoons b^q + 1.$$

The functional structure of digitwise operations is induced by decomposing a number x into summands of the form $b_j(x) = x_j b^j$, $j = 0, 1, \ldots, q - 1$, each of which is computed by the formula

$$b_j(x) \rightleftharpoons \begin{cases} x \bmod b^{j+1} - x \bmod b^j, & 0 \leq j < q - 1, \\ x - x \bmod b^{q-1}, & j = q - 1. \end{cases}$$

An arbitrary function $f : E_{b\uparrow q}^k \to E_{b\uparrow q}$ may be defined by a *quasi-Post representation (q. p. r.)* – a family of functions $\{f_j : E_{b+1}^{qk} \to E_{b+1} \mid j = 0, \ldots, q-1\}$ such that, if the decomposition of every x_i in base b has the form

152

$x_i = \sum_j x_{i,j} b^j$, then the decomposition of $f(x_1, \ldots, x_k)$ has the form

$$f(x_1, \ldots, x_k) = \sum_j f_j(x_{1,0}, \ldots, x_{1,q-1}, \ldots, x_{k,0}, \ldots, x_{k,q-1}) b^j.$$

This representation is called *regular* (r. q. p. r.) if it is done by means of a family that maps the set $E_b^q \cap (\{0\}^{q-1} \times \{b\})$ into itself. If the function has an r. q. p. r. in which each f_j depends only on $(x_{1,j}, \ldots, x_{k,j}) \in E_{b+1}^k$, then it is called *digital*. Further, if it has an r. q. p. r. in which each f_j depends only on $(x_{1,0}, \ldots, x_{1,j}, \ldots, x_{k,0}, \ldots, x_{k,j}) \in E_{b+1}^{(j+1)k}$, then it is called *progressive*. Progressive functions play an important role in machine arithmetics: functions $b_j(x)$, and all functions of the system MA_{n+1} are progressive.

Consider the r. q. p. r. as a mapping of classes of functions as follows: given a class K_{b+1} of functions on E_{b+1}, for every $q > 1$, denote by $K_{b+1} \uparrow q$ the class of functions on $E_{b\uparrow q}$ that have r. q. p. r. with all terms belonging to K_{b+1}. We also put $K_{b+1} \uparrow 1 \rightleftharpoons K_{b+1}$ because of the equality $b \uparrow 1 = b+1$. Since the r. q. p. r. is concordant with superposition (i. e., for every tuple of functions, there exists an r. q. p. r. of their superposition whose terms are superpositions of the terms of r. q. p. r.'s of these functions), it follows that closedness of the class K_{b+1} implies closedness of $K_{b+1} \uparrow q$. The class of all digital functions from $K_{b+1} \uparrow q$ will be denoted by $K_{b+1} \uparrow^- q$, and the class of all progressive functions from $K_{b+1} \uparrow q$ – by $K_{b+1} \uparrow^+ q$.

Analysis of functional properties of digital computer arithmetics yield the following results.

Theorem 3.

a) *The class K_{b+1} coincides with T_{b+1} if and only if the class $K_{b+1} \uparrow q$ coincides with $T_{b\uparrow q}$ for every $q \geq 1$.*

b) *For every $q > 1$, the class $L[L_{b+1} \uparrow q]$ coincides with $T_{b\uparrow q}$.*

c) *For every $q > 1$, if $b = p^s$, with p being a prime and $s > 0$, then all progressive functions in $L_{b+1} \uparrow^+ q$ (in particular, all the functions b_j, $j = 0, \ldots, q - 1$) are expressible in $L_{b\uparrow q}$; otherwise, none of the b_j's is expressible in $L_{b\uparrow q}$ and the set of all b_j's enriches $L_{b\uparrow q}$ to the class*

$$L_{b\uparrow q}^+ \rightleftharpoons L[L_{b+1} \uparrow^+ q]$$
$$= L[L_{b+1} \uparrow^- q]$$
$$= F_{b\uparrow q}(\{db^j \mid d \in D(b), \ j = 0, \ldots, q-1\}).$$

d) *If b is not a prime power then the system of functions*

$$QPL_{b\uparrow q} \rightleftharpoons \{\rightarrow, b_0, b_0 \ (+) \ b_1, \ldots, b_0 \ (+) \ldots (+) \ b_{q-2}\}$$

forms a base in $L_{b\uparrow q}^+$ for every $q > 1$.

Corollary 3.1. *The following statements are equivalent:*

a) *b is a prime;*

b) *the class $L_{b+1} \uparrow q$ coincides with $T_{b\uparrow q}$ for every $q \geq 1$;*

c) *for every $q > 1$ and every function expressible in $L_{b\uparrow q}$, there exists a q. p. r. all of whose members are expressible in L_{b+1}.*

5. Application: dataflow computations verification

Deep formal performance analysis and optimization of arithmetic algorithms is required in control engineering. Control applications, such as sound mixers or digital electricity meters, must satisfy hard real time and reliability requirements while running on devices with limited resource capacities. One of widely used approaches to their design is based on the dataflow paradigm [2]. Here data values are represented as flows – numeric sequences $\{x_i \mid i = 1, 2, \ldots\}$ indexed by ticks of infinite discrete clock. At clock tick i, i-th values of output flows (reactions) are computed from i-th values of input flows (stimuli) and possibly l-th values of certain flows for some $l < i$ (memory). The computation is:

a) directed (each x_i is assigned exactly once, at tick i),

b) bounded (the total number of operations performed at tick i doesn't exceed fixed constant that doesn't depend on i),

c) finite (all data domains are finite sets).

At implementation phase it must be verified that computation always ends before the next tick and an overflow is signaled if occurred. Presented results are used for this purpose as follows. Computation and overflow detection rules are specified as finite superpositions of base functions of suitable clone enriching Łukasiewicz logic. Requirements a)-c) guarantee that such specification exists. When mapping it to the target microprocessor, arithmetic unit commands are expressed using the same base. Then subexpression matching is performed, resulting in formal definition of the algorithm in terms of target hardware capabilities. It is optimized using various techniques, e.g. caching multiply used results in intermediate flows. The resulting expressions' computation time is determined using hardware performance characteristics.

Such analysis requires large amount of routine Łukasiewicz logic reasoning when applied to real-world algorithms. For brief illustration here we draw very simple example. Consider sound filter described by the following finite-difference equation:

$$y_i = \sum_{l=0,\ldots,N} b_l x_{i-l} - \sum_{l=1,\ldots,M} a_l y_{i-l}.$$

In order to implement it using overflow and modular arithmetics (e.g. on TMS digital signal processor from Texas Instruments Corp.), it is specified according to Lemmas 1 and 2 as follows:

$$y(z_0, \ldots, z_{N+M}) =$$
$$[(+)_{l=0,\ldots,N}(0\,(+)^{b_l}\,z_l)]\,(+)\,[(-)\,(+)_{l=1,\ldots,M}\,(0\,(+)^{a_l}\,z_{N+l})],$$
$$o(z_0, \ldots, z_{N+M}) =$$
$$\mathrm{Carry}(|+|_{l=0,\ldots,N}\,(0\,|+|^{b_l}\,z_l)) \vee \mathrm{Carry}(|+|_{l=1,\ldots,M}\,(0\,|+|^{a_l}\,z_{N+l})),$$

where o is a two-valued (Boolean) overflow signal. Mapping of this specification is straightforward.

6. Conclusion

To summarize presented results we quickly review the traditional approach to computer arithmetics implementation. In general-purpose hardware platforms and programming environments modular arithmetic $MA_{2\uparrow q}$ is implemented with binary digital number representation (usually q equals 16, 32 or 64). At the hardware level, the choice of number 2 for the base is justified by Corollary 3.1 and item (c) of Theorem 2. In addition, the q. p. rúsed for constructing arithmetic operations is implemented "for free" (without extra transformations) by means of a simple commutation of the input and output lines.

However, the Carry predicate (flag) doesn't have software implementation, which complicates integer overflow control. For example, the predicate $\mathrm{Carry}(x\,|+|\,y)$ is implemented by the following rather verbose procedure on the C programming language:

```
int carry_sum(unsigned int x, unsigned int y)
{
    return (y != 0) && (x >= -y);
}
```

Moreover, functional incompleteness of MA_{n+1} leads to necessity of such inefficient operations as conditional branch (cf Corollary 2.1). Important alternative is offered by multiple-valued arithmetic-logical units (MVALU) such as hardware implementation of overflow subtraction $|-|$ by polymer conductors [6]. According to Theorem 1 it is enough to implement this operation in order to build full-functional MVALU.

References

1. Evans T., Schwartz P.B., On Słupecki T-functions. *J. Symbolic Logic* **23** (1958), 267–270.
2. Halbwachs N., Caspi P., Raymond P., Pilaud D., The synchronous data flow programming language LUSTRE. *Proc. IEEE* **79** (1991), 1305–1320.
3. Karpenko A.S., *Lukasiewicz's Logics and Prime Numbers* (Moscow, Nauka, 2000). (Russian)
4. Kovalyov S.P., Mathematical foundations of computer arithmetics. *Siberian Advances in Mathematics* **15(4)** (2005), 34–70.
5. Liskov B.H., Zilles S.N., Specification techniques for data abstractions. *IEEE Trans. on Software Engineering* **SE-1(1)** (1975), 7–19.
6. Mills J.W., *Polymer Processors.* Indiana University, Computer Science Dept, Technical Report TR580 (Indiana University, 2003). http://www.cs.indiana.edu/pub/techreports/TR580.pdf.
7. Rosenberg I.G., Completeness properties of multiple-valued logic algebras. *Computer Science and Multiple-Valued Logic*, (Amsterdam, North Holland, 1977), 144–186.
8. Voevodin V.V., Mapping the problems of computational mathematics onto computer systems architecture. *Russian J. Numer. Anal. Math. Modelling* **15(3–4)** (2000), 349–359.

THE FUNCTIONAL COMPLETENESS OF LEŚNIEWSKI'S SYSTEMS

FRANÇOIS LEPAGE

Département de philosophie
Université de Montréal
C.P. 6128, succ. Centre-Ville
Montréal (Québec)
H3C 3J7, Canada
E-mail: francois.lepage@umontreal.ca

After a brief presentation of Leśniewski's system of Protothetics and Ontology, we introduce the λ–operator and then show that the language of Protothetics is functionally complete. Furthermore, after having introduce a very simple algebra over names, we show that language of Ontology is also functionally complete.

1. Introduction

In the first half of the last century, the Polish logician Stanisław Leśniewski introduced three new systems of logic — Protothetic, Ontology and Mereology — that were free from logical contradictions and strong enough to be used as a foundation of mathematics. His work did not receive all the attention it deserves, most likely because it arrived too late.

In this paper, I will first present a rough sketch of the first two of Leśniewski systems. Secondly, I will expand upon these systems by adding the λ abstractor. Lastly, I will introduce two special functors, which will engender a new system that is functionally complete in a sense to be specified in the second section.

2. Leśniewski's Systems

The simplest of the three systems is Protothetic, which is incorporated into the two others. Protothetic is a generalized calculus of propositions containing variables of arbitrary syntactic categories which are defined starting with the basic category S of sentences.

Definition 2.1.

(1) S is a syntactic category.
(2) If X and Y are syntactic categories, (X/Y) is a syntactic category[a].
(3) Nothing else is a syntactic category.

The wff's of Protothetic are

(1) A variable of type S;
(2) Identity statements $\lceil \equiv (AB) \rceil$ where A and B are wff's;
(3) Generalization: $\lfloor v_1 \ldots v_n \rfloor \lceil A \rceil$ where A is a wff;
(4) All the expressions $N(v_1 \ldots v_n)$ where N is introduced by a definition. The general form of these definitions is

$$\lfloor v_1 \ldots v_n \rfloor \lceil \equiv (N(v_1 \ldots v_n)A\{v_1 \ldots v_n\}) \rceil$$

where N is a new constant and $N(v_1 \ldots v_n)$ is of category S and $A\{v_1 \ldots v_n\}$ is an already defined wff containing the variables $v_1 \ldots v_n$.

Example 2.1.

(1) p where p is a propositional variable
(2) $\lfloor p \rfloor \lceil p \rceil$ (which will be used to designate falsity $\bot$)
(3) $\lfloor pq \rfloor \lceil \equiv (pq) \rceil$
(4) $\lfloor p \rfloor \lceil \equiv (pp) \rceil$ (which will be used as truth $\top$)
(5) $\lfloor p \rfloor \lceil \equiv (\neg(p) \equiv (p\bot)) \rceil$ (this definition introduces negation)
(6) $\lfloor pq \rfloor \lceil \equiv (\wedge(pq)\lfloor f \rfloor \lceil \equiv (p \equiv (\lfloor r \rfloor \lceil \equiv (pf(r)) \rceil \lfloor r \rfloor \lceil \equiv (qf(r)) \rceil)) \rceil) \rceil$ (this definition introduces conjunction)
In the last example, f is of category S/S.
(7) $\lfloor pq \rfloor \lceil \equiv (\vee(pq)\neg(\wedge(\neg(p)\neg(q)))) \rceil$
(8) $\lfloor p \rfloor \lceil \equiv (\supset (pq) \vee (\neg(p)q)) \rceil$

The Protothetic system can be axiomatized in many similar ways. For example, here is a system taken from Słupecki 1950:

A1. $\lfloor pqr \rfloor \lceil \equiv (\equiv (pq)(\equiv (\equiv (rq) \equiv (pr)))) \rceil$
A2. $\lfloor pq \rfloor \lceil \equiv (\equiv (pq)\lfloor f \rfloor] \equiv (f(p)f(q))]) \rceil$
A3. $\lfloor pq \rfloor \lceil \equiv (\equiv (pq) \equiv (\lfloor f \rfloor \lceil \equiv (f(p)f(q)) \equiv (pq) \rceil))) \rceil$

---

[a]The (2) of this definition is much simpler than the one we can find in the literature on Protothetic: (2) If X and $X_1, \ldots, X_n$ are syntactic categories, $X/X_1 \ldots X_n$ is a syntactic category. We can easily show that $(\ldots (X/X_i) \ldots X_n)$ can accomplish the same task. We will write $A(BC)$ instead of $(A)(B)(C)$

A4. $\lfloor f \rfloor \lceil \equiv (f(\lfloor p \rfloor \lceil p \rceil)$
$\equiv (\equiv (f(\lfloor p \rfloor \lceil p \rceil) \lfloor p \rfloor \lceil p \rceil) \equiv (\lfloor q \rfloor \lceil f(\lfloor p \rfloor \lceil p \rceil) f(q) \rceil)))\rceil$.

If we add the following 5 rules:

R1. Substitution
R2. Detachment
R3. Distribution of quantifiers
R4. Extensionality (of any expression of any category)
R5. Rule of definition (every definition above is a theorem)

we have a complete system in the following sense: Every closed wff of category S is either a theorem or its negation is a theorem.

It is worth saying few words about the distinction between Protothetic and the theory of propositional types. First, Protothetic is purely nominalistic: There is no formal semantics in terms of a hierarchy of functions built on truth values. Indeed, we have an implicit semantic: The theorems are taken as the true statements and expressions of the type S/S can be seen as denoting one place propositional functions *etc.*, but these considerations play no role in the theory. Protothetic is a syntactic theory and thus purely inscriptional: It deals with uninterpreted strings of symbols. This brings me to a second remark. A Protothetic system is never completely developed. One can always introduce new constants and thus produce new theorems. Each system is complete but this completeness is relative to the constant functors already introduced. As a result, each system is a work in progress. This is the major difference with propositional type theory in which the hierarchy of propositional functions is given once and for all.

The second system is Ontology. It expands upon the Protothetic by adding a second basic category, the category N of names.

Definition 2.2.

(1) S and N are syntactic categories;
(2) If X and Y are syntactic categories, (X/Y) is a syntactic category;
(3) Nothing else is a syntactic category.

We will omit parentheses when unnecessary, the default association being on the left: We will thus write X/Y for (X/Y), $X/Y/Z$ for $((X/Y)/Z)$ and $X/(Y/Z)$ for $(X/(Y/Z))$.

Ontology contains a new constant "ε" of the category $S/N/N$. $\varepsilon(AB)$ should read "A is a part of B". "ε" should not be confused with the "$\in$" of set theory. For example, "ε" is transitive whereas "$\in$" is not.

Let us introduce equality in Protothetic. For expression of type S, equality is identity. Let us consider an expression A of type X. Categories $X_1, \ldots, X_n$ such that for any expressions $A_1, \ldots, A_n$ respectively of category $X_1, \ldots, X_n$, $A(A_1 \ldots An)$ is of type S. We define identity as follows: $\lfloor A_1...A_n \rfloor \lceil \equiv (= (AB) \equiv (A(A_1...A_n)B(A_1...A_n)))\rceil$. We will assume the rule of substitution for $=$: From $= (AB)$ and $\Psi(...A...)$ we can write $\Psi(...B...)$. A particular case would be the detachment rule: From $= (AB)$ and B, we can write A.

A system for Ontology is obtained by adding

A5. $\lfloor Ab \rfloor \lceil \equiv (\varepsilon\{Ab\} \wedge (\neg(\lfloor B \rfloor \lceil \neg(\varepsilon\{BA\})\rceil))$
$\wedge (\lfloor DC \rfloor \lceil \supset (\wedge(\varepsilon\{DA\}\varepsilon\{CA\})\varepsilon\{DC\})\lfloor D \rfloor \lceil \supset (\varepsilon\{DA\}\varepsilon\{Db\})\rceil))))\rceil$

Identity between names is introduced by the following definition: $\lfloor AB \rfloor \lceil \equiv (= (AB) \wedge (\varepsilon\{AB\}\varepsilon\{BA\}))\rceil$

Equality between arbitrary functions is defined in the same way it was in Protothetic.

Finally, we introduce λ–abstractor (which is not in the original Protothetic) with the following axiom:

$$= (\lfloor \lambda x \rfloor \lceil A(B) \rceil A[x|B]) \quad (\ \lambda \text{ reduction })$$

where $A[x|B]$ is the formula obtained from A by substituting B to each free occurrence of x in A.

The introduction of λ–abstractor in Protothetic authorizes structural definitions while they are always contextual in the original system. We have already introduced negation by the contextual definition

$$\lfloor p \rfloor \lceil \equiv (\neg(p) \equiv (p\bot))\rceil$$

This definition is contextual because it does not provide a complex expression X of the category S/S, which gives the negation of the argument. Given the λ–abstractor we can provide such an expression:

$$(\lfloor \lambda p \rfloor \lceil \equiv (p\bot)\rceil$$

It is clear that both definitions are equivalent. Replacing $\wedge$ by $(\lfloor \lambda p \rfloor \lceil \equiv (q\bot)\rceil$ in the contextual definition we get (changing the first p for q for clarity's sake)

$$\lfloor q \rfloor \lceil \equiv (\lfloor \lambda p \rfloor \lceil \equiv (p\bot)\rceil(q) \equiv (q\bot)\rceil$$

by λ reduction

$$\lfloor q \rfloor \lceil \equiv (\equiv (q\bot) \equiv (q\bot)\rceil$$

which is a theorem.

In fact, we can prove that:

$$\lceil \equiv (\lfloor v_1 \ldots v_n \rfloor \lceil \equiv (C(v_1 \ldots v_n)A\{v_1 \ldots v_n\})\rceil$$
$$\equiv (\lfloor \lambda v_1 \ldots v_n \rfloor \lceil C(v_1 \ldots v_n)\rceil \lfloor \lambda v_1 \ldots v_n \rfloor \lceil A\{v_1 \ldots v_n\}\rceil))\rceil$$

3. The Functional Completeness of λ–Protothetic

In classical propositional type theory the notion of functional completeness
is very simple: Does any function in the hierarchy of propositional types
have a name, i.e., is there an expression of type theory that has this function
as semantic value? The answer is yes. Now, the question becomes whether
Protothetic is functionally complete. As we don't have an explicit semantic,
we must specify what we mean by functional completeness. For example,
we already have introduced $\bot$, $\top$, $\neg$ and $\wedge$. $\neg$ is a name for negation and
$\wedge$ for conjunction because the following are theorems of Protothetic:

$$\lfloor p \rfloor \lceil \equiv (\neg(p) \equiv (p\lfloor q \rfloor \lceil q \rceil))\rceil$$
$$\lfloor p \rfloor \lceil \equiv (\neg(p) \equiv (p\bot))\rceil$$
$$\lceil \equiv (\neg(\top) \equiv (\top\bot))\rceil \text{ and } \lceil \equiv (\neg(\bot) \equiv (\bot\bot))\rceil$$

For conjunction, it is much more difficult to show but the following are also
theorems

$$\lceil \equiv (\wedge(\top\top)\top)\rceil$$
$$\lceil \equiv (\wedge(\top\bot)\bot)\rceil$$
$$\lceil \equiv (\wedge(\bot\top)\bot)\rceil$$
$$\lceil \equiv (\wedge(\bot\bot)\bot)\rceil$$

One can check that $\vee$ and $\supset$ have the expected properties.

In this line of thought, the question of the functional completeness of
Protothetic can be formulated as follows: Given the collection of expressions of category C_1 already introduced, and the collection of expressions
of category C_2 already introduced, can we introduce a new constant X of
category C_1/C_2 such that, for any arbitrary chosen A and B respectively
of category C_1 and C_2, $= (X(A)B)$ is a theorem? The answer is yes.

The recursive formula is quite complicated even if the idea is quite
simple (I take it from van Benthem 1992).

We will need the notion of *projector*. Let X be an expression of category
C. There exists a sequence (possibly empty if C is S) of categories such
that for any sequence of expressions $x_1, \ldots, x_n$ respectively of categories
$C_1, \ldots, C_n$, $X(x_1, \ldots, x_n)$ is of category S. $x_1, \ldots, x_n$ is called a projector
of X and $X(x_1, \ldots, x_n)$ is a projection. For simplicity's sake, we will use
$X(\overrightarrow{x})$ instead of $X(x_1, \ldots, x_n)$.

Any functor can be represented by a generalized truth table:

x_1	x_2	$\ldots$	x_n	$X(x_1, \ldots, x_n)$
A_1^1	A_2^1	$\ldots$	A_n^1	$+$
$\vdots$	$\vdots$	$\ldots$	$\vdots$	$\vdots$
A_1^j	A_2^j	$\ldots$	A_2^j	$+$
$\vdots$	$\vdots$	$\ldots$	$\vdots$	$\vdots$
A_1^m	A_2^m	$\ldots$	A_n^m	$+$

where $+$ stands for $\top$ or $\bot$.

There are two cases. First case, the right column contains only $\bot$s. In that case, $\lfloor \lambda x_1 \ldots x_n \rfloor \lceil \bot \rceil$ is an expression having this generalized truth table.

In the second case, the right column contains at least one $\top$. Let $pr(A_i^j)$ be the set of projectors of A_i^j. We will write $\vec{p} \in pr(A_i^j)$ for $\vec{p}$ being a projector of A_i^j and $\bigwedge_{\vec{p} \in X_{A_i}}$ for a conjunction over a set $X_{A_i} \subseteq p(A_i)$ of projectors of A_i. Similarly, we will use general disjunctions.

Let $\delta_{A_i^j}[x_i(\vec{p})]$ stand for $x_i(\vec{p})$ if $\lceil \equiv (A_i^j(\vec{p})\top) \rceil$ is a theorem and stand for $x_i(\vec{p})$ if $\lceil \equiv (A_i^j(\vec{p})\bot) \rceil$ is a theorem.

Let $\top(X)$ be the subset of the set of projectors of X for which X has a $\top$ in the generalized truth table.

Theorem 3.1.

$$\lfloor \lambda x_1 \ldots \lambda x_n \rfloor \lceil (\bigvee_{\vec{A}^j \in \top(X)} (\bigwedge_{\vec{p} \in pr(A_1^j)} \delta_{A_1^j}[x_1(\vec{p})] \wedge \cdots \wedge \bigwedge_{\vec{p} \in pr(A_n^j)} \delta_{A_n^j}[x_n(\vec{p})]))) \rceil$$

is an expression which has the generalized truth table of X.

This formula is recursive: Starting from $\top$ and $\bot$ it provides a structural name for any functor given by a generalized truth table. Without the λ–abstractor, we can only provide a contextual definition

$$\lfloor x_1 \ldots x_n \rfloor \lceil \equiv (X(\bigvee_{\vec{A}^j \in \top(X)} (\bigwedge_{\vec{p} \in pr(A_1^j)} \delta_{A_1^j}[x_1(\vec{p})] \wedge \cdots \wedge \bigwedge_{\vec{p} \in pr(A_n^j)} \delta_{A_n^j}[x_n(\vec{p})])))) \rceil$$

which is equivalent but less elegant.

4. The Functional Completeness of λ–Ontology

The question now becomes whether we can do a similar exercise in ontology without adding new resources? The answer is *no*. To see why, let's go back to the generalized truth table.

162

x_1	x_2	...	x_n	$X(x_1,\ldots,x_n)$
A_1^1	A_2^1	...	A_n^1	$+$
$\vdots$	$\vdots$	...	$\vdots$	$\vdots$
A_1^j	A_2^j	...	A_2^j	$+$
$\vdots$	$\vdots$	...	$\vdots$	$\vdots$
A_1^m	A_2^m	...	A_n^m	$+$

What permits us to use the technique of generalized truth tables is the fact that the behaviour of any A_i^j rests on the behaviour of its projections which are of category S and for that category we have an algebra, namely the Boolean algebra which needs only $\neg$, $\wedge$, $\bot$ and $\top$ and all these expressions are definable using quantification and $\equiv$. We just do not have a similar tool or the categories built on N.

For the moment, I will restrict myself to a hierarchy based on the category N and I will treat mixed categories (e.g. $(S/N)/(S/N)$) at the end of the paper.

Let us introduce two new constants of category N, $\curlywedge$ (up) and $\curlyvee$ (down) and two new functors $\otimes$ and $\oplus$ of category N/N defined as follows:

$$\lfloor \lambda x_1 \lambda x_2 \lambda x_3 \rfloor \lceil \equiv (= (\otimes(x_1 x_2)x_3) \vee (\wedge (= (x_1 x_2) = (x_3 \curlywedge))$$
$$\wedge (\neg(= (x_1 x_2)) = (x_3 \curlyvee)))) \rceil$$

The "meaning" is clear: $\otimes(xy)$ reduces to $\curlywedge$ when x and y are equal, reduces to $\curlyvee$ when x and y are not. For all other cases, the function is the value $\bot$. Let us check the first case

$$\lfloor \lambda x_1 \lambda x_2 \lambda x_3 \rfloor \lceil \equiv (= (\otimes(x_1 x_2)x_3) \vee (\wedge (= (x_1 x_2) = (x_3 \curlywedge))$$
$$\wedge (\neg(= (x_1 x_2)) = (x_3 \curlyvee)))) \rceil (aa \curlywedge)$$

By λ reduction

$$\equiv (= (\otimes(aa)\curlywedge) \vee (\wedge(= (aa) = (\curlywedge \curlywedge)) \wedge (\neg(= (aa)) = (\curlywedge \curlyvee))))$$

By the usual rules, the completeness and detachment

$$= (\otimes(aa)\curlywedge)$$

The second functor is a little bit more complex. In order to simplify the formula, we will assume that $\oplus$ is commutative and we will write

$\bigvee(ABCD\dots)$ instead of $\vee(A \vee (B \vee (C \vee (D\dots))))$ as above.

$$\lfloor \lambda x_1 \lambda x_2 \lambda x_3 \rfloor \lceil \equiv (= (\oplus(x_1 x_2)x_3)$$
$$\bigvee((\bigwedge(=(x_2 \curlywedge) = (x_3 x_1)\neg(= x_1 \curlywedge)\neg(= x_1 \Upsilon))$$
$$(\bigwedge(=(x_2 \Upsilon) = (x_3 \curlywedge)\neg(= (x_1 \curlywedge))\neg(= (x_1 \Upsilon)))$$
$$(\bigwedge(=(x_1 \curlywedge) = (x_2 \Upsilon) = (x_3 \Upsilon))$$
$$(\bigwedge(=(x_1 \Upsilon) = (x_2 \Upsilon) = (x_3 \Upsilon))$$
$$(\bigwedge(=(x_1 \curlywedge) = (x_2 \curlywedge) = (x_3 \curlywedge))))\rceil$$

This large formula is justified because it gives the following theorems:

$$= (\oplus(a\curlywedge)a)$$
$$= (\oplus(a\Upsilon)\curlywedge)$$
$$= (\oplus(\curlywedge\Upsilon)\Upsilon)$$
$$= (\oplus(\Upsilon\Upsilon)\Upsilon)$$
$$= (\oplus(\curlywedge\curlywedge)\curlywedge)$$

Let us consider in detail a very simple example: That of an arbitrary functor X of category N/N.

It is entirely described by the following table

x	$X(x)$
a_1	b_1
a_2	b_2
$\vdots$	$\vdots$
a_n	b_n

The following formula F is such that $= (F(a_i)b_i)$ is a theorem for any $i \leq n$. F is

$$\lfloor \lambda x \rfloor \lceil [\oplus[\dots [\oplus[\oplus(b_1(\otimes(xa_1))) \oplus (b_2(\otimes(xa_2)))]\dots](\oplus(b_n(\otimes(xa_n))))]\rceil$$

Let us check by applying F to a_i. By λ reduction, we get

$$\lceil [\oplus[\dots [\oplus[\oplus(b_1(\otimes(a_i a_1))) \oplus (b_2(\otimes(a_i a_2)))]\dots](\oplus(b_n(\otimes(a_i a_n))))]\rceil$$

There are two kinds of terms $\oplus(b_j(\otimes(a_i a_j)))$, namely those where $i \neq j$ and the one where $i = j$. In the first case, we have

$$(\oplus(b_j(\otimes(a_i a_j))))$$

which reduces to

$$(\oplus(b_j \Upsilon))$$

164

which reduces to

$$\curlywedge.$$

In the second case, we have

$$(\oplus(b_j(\otimes(a_i a_i))))$$

which reduces to

$$(\oplus(b_i \curlywedge))$$

which reduces to

$$b_i.$$

So the whole expression reduces to

$$\lceil [\oplus[\ldots [\oplus[\cdots \oplus [\curlywedge \curlywedge]\ldots]b_i]\ldots]\curlywedge]\rceil$$

which reduces to

$$b_i.$$

This is the elementary case for a functor of category N/N. For the general case of any functor based on the category N, we use the same strategy that was used for the functors based on S. Let us consider the following arbitrary table of category $((\ldots((N/X_n)/X_{n-1})\ldots)/N_1)$:

x_1	x_2	$\ldots$	x_n	$X(x_1,\ldots,x_n)$
A_1^1	A_2^1	$\ldots$	A_n^1	a_1
$\vdots$	$\vdots$	$\ldots$	$\vdots$	$\vdots$
A_1^j	A_2^j	$\ldots$	A_2^j	a_j
$\vdots$	$\vdots$	$\ldots$	$\vdots$	$\vdots$
A_1^m	A_2^m	$\ldots$	A_n^m	a_n

Theorem 4.1. *The following expression corresponds exactly to the table above:*

$$\lfloor \lambda x_1 \ldots \lambda x_n \rfloor \lceil \bigoplus_{\vec{A^j} \in pr(X)} [(a_j \bigoplus_{i=0}^{m} [\bigoplus_{\vec{g} \in pr(A_i^j)} (\otimes(x_i(\vec{g})A_i^j(\vec{g})))))]]] \rceil$$

where $\bigoplus_{\vec{A^j} \in P(X)} [Z\{\vec{A^j}\}]$ *stands for* $\oplus(Z\{\vec{A^1}\}(\ldots(\oplus(Z\{\vec{A^{m-1}}\}Z\{\vec{A^m}\}))\ldots))$

and $\bigoplus_{i=0}^{m} [Z\{A_i\}]$ *stands for* $\oplus(Z\{A_1\}(\ldots(\oplus(Z\{A_{m-1}\}Z\{A_m\}))\ldots)).$

5. Conclusion

The two constructions above deal with only one basic category: S for the first one and N for the second one. What would happen in mixed cases? For example, functors of the category $((S/(N/S))/((S/N)/(N/S)))$. It is not clear there is a simple way to merge both approaches. Fortunately, we do not need to do so because the second approach is universal and, moreover, can be generalized to any language having a finite number of basic categories.

Let us suppose we have a finite number of categories $N_1, N_2, \ldots, N_n$. (We can think of N_1 as S and N_2 as N.) We then generalize the definitions of $\oplus$ and $\otimes$ as being category free (or having $\oplus_{N_i}$ and $\otimes_{N_i}$ for each category N_i) and the second formula above can be used to provide a name for any functor.

Ironically, this brings us back to the philosophical starting point: Nominalism. In particular, the category S is the category of the name of truth values.

References

1. ANDREW, P. B., "A Reduction of the Axioms for the Theory of Propositional Types", *Fundamenta Mathematicæ* LII, 1963, 345–350.
2. VAN BENTHEM, J., *Language In Action*, North-Holland, Amsterdam/MIT Press, Cambridge, 1995.
3. HENKIN, L., "A Theory Of Propositional Types", *Fundamenta Mathematicæ*, LII, 1963, 323–344.
4. LEPAGE, F., "Partial Monotonic Protothetic", *Partiality And Modality*, E. Thijsse, F. Lepage & H. Wansing (eds.) Special Issue Of *Studia Logica*, Vol. 66, No. 1, 2000, 147–163.
5. MONTAGUE, R., "Universal Grammar", *in Formal Philosophy*, Yale University Press, New Haven, 1974, 222–246.
6. POST, E., "Introduction To A General Theory Of Elementary Propositions", *in From Frege To Gödel*, J. van Heijenoort (ed.), Harvard University Press, Cambridge (Mass.), 1967, 264–283.
7. RICKEY, F., "A Survey Of Leśniewski's Logic", *Leśniewski's System: Protothetic*, J. Srzednicki, Z. Stachniak (eds.), 1998, 23–42.
8. SŁUPECKI, J., "St. Leśniewski Protothetics", *Studia Logica*, 1, 1953, 44–112.
9. TARSKI, A., "Sur le terme primitif de la logistique", *Fundamenta Mathematicæ* IV, 1923, 59–74.

Acknowledgement

This work is supported by a *Social Science and Humanities Research Counsil of Canada* research grant.

ANALYSIS OF A NEW REDUCTION CALCULUS FOR THE SATISFIABILITY PROBLEM

S. NOUREDDINE

Department of Mathematics and Computer Science
University of Lethbridge, 4401 University Drive
Lethbridge, T1K 3M4, AB, Canada
E-mail: soufiane.noureddine@uleth.ca

This paper proposes a new reduction calculus for propositional formulas. We prove that the presented calculus is refutation-complete. Hence, it is theoretically as powerful as resolution. Resolution-based algorithms for propositional tautology/satisfiability testing are in general optimized by reduction rules, but the main message of the paper is that by this mere optimization we are able to solve the satisfiability problem even if this solution is still of exponential worst-case complexity. We outline an algorithm for hybrid reduction/resolution and prove its partial correctness if only reduction is used. The hybrid approach promises more efficiency in practice, however, we will not analyze its complexity in the present paper.

1. Introduction

The ubiquitous tautology and its sibling the satisfiability problem are still open problems with extreme impact on theory and applications of computer science and related disciplines. The problem is so needed in practice that unceasingly papers for approximating it are published every year. No one knows whether the problem is efficiently solvable or not. The origin of the problem is in mathematical logic. However, hundreds of other problems stemming from other disciplines and equivalent to it have been discovered over the last decades. All these problems have in common that no efficient solution is known for them and if at least one of them is efficiently solvable then all of them are too. Theorists hence have a name for this kind of problems, namely, NP-complete problems.

There are many approaches to tackle the tautology/satisfiability problem. The principal ones are:

(i) Manipulation of formulas.

(ii) Methods of optimization theory.

(iii) Methods of graph theory.

(*iv*) Probabilistic approximation methods.

Manipulation of propositional formulas is the immediate approach for the tautology/satisfiability problem as these are originally problems of propositional logic. Prominent approaches in this respect are the method of resolution and the Davis-Putnam (DPLL) procedure. Resolution is a calculus and hence a method of (dis-)proof [2]. DPLL is an manipulation algorithm that is based on Boolean-algebraic identities [1]. Both are well used in practice. Our approach in this paper is akin to the method of resolution and falls in this category.

Methods of optimization theory convert the problem to a functional optimization one. They approximate the solution by the search for the global minimum of a well-selected function under specific constraints [3]. Of course, they rather see the problem as an optimization problem and not as a decision problem. In [4], we described a method that uses exterior, exact penalty optimization with a coercive objective function. The method is partially heuristic and delivers sub-optimal results. In [5], we extended the theory of geometric programming [10] to cope with the satisfiability problem. In both methods we focused on the so-called exact satisfiability problem [6], which is known to be NP-complete [7].

Finally, graph-theoretical methods [8] can be used to solve special cases of the satisfiability problem (*e.g.*, 2SAT). Graph theory is in this respect extremely useful for analytical purposes. The implementation of these methods, however, is often more efficient by way of formula manipulation methods. Last but not least, probabilistic algorithms try to circumvent the hardness of the problem by stochastic reasoning [9]. These methods proved to be very successful in practice and they are very promising in theory.

2. Notation and Problem Statement

Let $\mathfrak{F}$ be the set of all propositional formulas over a set of n logic variables (*i.e.*, atomic formulas) $V = \{x_1, \ldots, x_n\}$. Let B be the set of all valuations b:

$$b : V \mapsto \{0, 1\}$$

The domain of b is inductively extended to $\mathfrak{F}$.

A formula $F \in \mathfrak{F}$ in disjunctive normal form (DNF) will be represented as a finite set of terms, where each term t is either the empty set $\{\}$ or a set of literals over V. The empty term $\{\}$ is equivalent to the valid terminal *true*. Without loss of generality, nonempty terms are assumed to be satisfiable (thus, excluding terms like $\{false\}, \{x, \neg x, y\}$). The empty formula $\{\}$ is equivalent to the unsatisfiable terminal false. Occasionally, we will write $V(F)$ for the set of variables occurring in F.

Throughout the paper, we will use the following symbols to specify formulas in $\mathfrak{F}$:

- The letter x with/without subscript is used for logic variables.
- The letter t with/without subscript is used for conjunctive terms.
- Capital letters like F, G, and X with/without subscripts are used for arbitrary formulas in DNF.

The problem we shall address is the following:

Given a formula $F \in \mathfrak{F}$ in DNF, decide whether or not F is a *tautology* (*i.e.*, a *theorem* of propositional logic). More precisely, we want to decide whether or not:
$$\forall b \in B : \qquad b(F) = 1$$
Since $|B| = 2^n$ (or in other words, F admits 2^n possible valuations), this problem is known to be in Co-NP. It is in a sense the hardest problem in Co-NP, since it is also Co-NP-complete.

3. Reduction Calculus

The *reduction calculus* (*RC*) has a single rule called *reduction rule*, which operates in a purely syntactic manner.

The Reduction Rule

Two disjunctively connected conjunctive terms of the form $\{x\} \cup t, \{\neg x\} \cup t$ are reduced to the term t. Formally:

$$\{\{x\} \cup t, \{\neg x\} \cup t\} \Vdash \{t\}$$

Lemma 3.1. *The reduction rule preserves the validity and the satisfiability of terms.*

Proof. This can be immediately seen since by boolean algebra both sides of the reduction rule are logically equivalent. $\qquad\square$

Corollary 3.1. *The reduction rule is valid in both directions.*

We shall however use the specified direction only in order to reduce the original formula.

Definition 3.1. Two terms are said to be *reducible* to t if they are of the form $\{x\} \cup t$, $\{\neg x\} \cup t$.

Lemma 3.2. *Let $F \in \mathfrak{F}$ be in DNF and represented as a set of terms. Let t_1, $t_2 \in F$ be reducible to t. Then:*

$$F \equiv F \backslash \{t_1, t_2\} \cup \{t\}$$

Proof. The (semantic) equivalence of the formulas immediately follows from the basic identity:

$$x \wedge t \vee \neg x \wedge t \equiv t \qquad \qquad \square$$

Definition 3.2. Let $F \in \mathfrak{F}$ be represented as a set of terms. We define the following term sets:

1. $\quad R(F) = \begin{cases} F \backslash \{t_1, t_2\} \bigcup \{t\} & \text{if } t_1, t_2 \in F \text{ reducible to } t \text{ exist} \\ F & \text{otherwise} \end{cases}$

2. $\quad R^0(F) = F, \ R^{n+1}(F) = R(R^n(F)) \quad \text{for } n \geq 0$

3. $\quad R^*(F) = \bigcup_{n \geq 0} R^n(F)$

The set $R^*(F)$ defined in equation (3) is called the *closure set* of F *w.r.t.* reduction.

Lemma 3.3. *Let $F \in \mathfrak{F}$ be represented as a set of terms. Then:*

$$|R^{n+1}(F)| \leq |R^n(F)| \quad \forall n \geq 0$$

Proof. We know: $\quad R^{n+1}(F) = R(R^n(F))$.
Hence, by definition if $t_1, t_2 \in R^n(F)$ reducible to t exist, then:
$\quad |R^{n+1}(F)| = |R^n(F)| - 1$ otherwise $|R^{n+1}(F)| = |R^n(F)|$. $\qquad \square$

Theorem 3.1. *Soundness of Reduction Calculus*
If $\{\} \in R^*(F)$ then F is valid.

Proof.
Since $\{\} \in R^*(F)$, there exists an $n \geq 0$ with $\{\} \in R^n(F)$. Thus, $R^n(F)$ represents a valid formula. By Lemma 3.2 we know:

$$F \equiv R^n(F)$$

Thus, F is valid, too. $\qquad \square$

The next technical Lemma is obvious and is needed in the proof of Theorem 3.2.

Lemma 3.4. *If F consists of only positive or only negative literals, then F is not valid.*

Theorem 3.2. *Completeness of Reduction Calculus*
If F is valid then $\{\} \in R^*(F)$.

Proof.
We proceed by *induction* over $|V(F)| = n$, the number of variables in F.
Base. $n = 1$.
Since F is valid, $F = \{\{x\}, \{\neg x\}\}$. Applying the reduction rule yields $\{\}$.
Induction hypothesis.
If F is valid and $|V(F)| < n$ then $\{\} \in R^*(F)$.
Induction step.
Let F be valid with $|V(F)| = n$. Choose a variable x of $V(F)$ such that $\neg x$ occurs in F. Since F is valid, such a variable exits by Lemma 3.4. Let F_0 and F_1 be the formulas that result from fixing $b(x) = 0$ and $b(x) = 1$, respectively. Thus, by construction, F can be put in the following form without impairing its validity:

$$F = x * F_1 \cup \neg x * F_0 \tag{1}$$

Here, the multiplication symbol $(*)$ of a literal x by a term set S means adding x to each term of S. We assert that both F_1 and F_0 are valid. Suppose F_0 is not valid. Thus, there would exist a b such that $b(F_0) = 0$. Consider the valuation b' over $V(F)$ defined by:

1. $b'(y) = b(y)$ if $y \in V(F_0)$
2. $b'(x) = 0$
3. $b'(y)$ arbitrary otherwise.

Obviously, by (1) , $b'(F) = 0$ contradicting the validity of F. Thus, F_0 is valid. By a dual argument we can infer that F_1 is valid, too. Since both F_0 and F_1 are valid and additionally $|V(F_0)| < n$ and $|V(F_1)| < n$, we can apply the induction hypothesis on both formulas. This yields:

$$\{\} \in R^*(F_0) \tag{2}$$
$$\{\} \in R^*(F_1) \tag{3}$$

Let n_0 and n_1 be the smallest integers such that $\{\} \in R^{n_0}(F_0)$ and $\{\} \in R^{n_1}(F_1)$, respectively. Build a reduction of F to $\{\}$ by the following procedure:

1. Add in each reduction step taken in building $R^{n_0}(F_0)$ the literal $\neg x$ in each term.
2. Add in each reduction step taken in building $R^{n_1}(F_1)$ the literal x in each term.

3. By (2) and (3):

   a. After n_0 steps the term $\{\neg x\}$ is generated.
   b. After $n_0 + n_1$ steps the term $\{x\}$ is generated.

4. Generate the empty term $\{\}$ in an extra step by applying the reduction rule on the subset $\{\{x\}, \{\neg x\}\}$.

Thus, $\{\} \in R^{n_0+n_1+1}(F)$, which ends the proof. $\square$

With Theorem 3.1 and Theorem 3.2, we have the following main theorem.

Theorem 3.3. *Fundamental Theorem of Reduction Calculus*
F is valid if and only if $\{\} \in R^*(F)$.

4. A Reduction Algorithm for Tautology Test

We now develop and analyze an algorithm for tautology (and thus for satisfiability) test based on the fundamental theorem of reduction calculus. The following is the reduction algorithm for tautology test. **Reduction Algorithm**

```
Reduction (Formula F) {
Input: A DNF formula F.
Output: {} if F is a tautology.
Convert F to a term set;
Eliminate unsatisfiable terms in F;
repeat
    G := F;
    F := R(F);
until F = G          // or ({} ∈ F)
if {} ∈ F then
    return {};
else
    return F;
}
```

Theorem 4.1. *Partial Correctness of the Reduction Algorithm*
If the output of the reduction algorithm is $\{\}$ then F is a tautology.

Proof. Observe first that the algorithm does not mimic the proof of the completeness theorem (Theorem 3.2). By Lemma 3.2, we have:

$$R_1(F) \equiv F \tag{4}$$

Hence:

$$\{\} \in R_1(F) \implies R_1(F) \quad \text{is a tautology (by soundness theorem)}$$
$$\equiv F \quad \text{is a tautology by (4)} \qquad \square$$

Observe that Theorem 4.1 could be made stronger, if the algorithm is slightly modified so as to really mimic the proof of the completeness theorem. Though the proof of the completeness theorem is constructive per se, it yields exponential worst-case complexity in practice. It is for this reason that we tried to avoid it in the previous algorithm though sacrificing strong correctness.

Theorem 4.2. *The reduction algorithm is of polynomial-time complexity.*

Proof. The algorithm just applies the reduction rule successively until no more reductions are possible. It is more efficient to quit the loop as soon as the empty term $\{\}$ is generated. The concrete complexity bound is not important in our context but it polynomially depends on the number of the variables and the size of terms in the formula F. $\qquad \square$

The algorithm we presented is akin to the well-known resolution algorithm. In resolution, however, the term set is expanded and not reduced as in our algorithm. Resolution algorithms have exponential complexity if implemented to achieve total correctness. The immediate question that arises now, what to do when our algorithm delivers a non-empty set? It is clear that in this case the original formula F may still be a tautology. It is here where a combination of our algorithm with resolution is convenient. The idea is depicted in Figure 1.

We propose to combine the two algorithms in series. First reduction is used. If reduction detects that F is a tautology, the algorithm is exited. If not, resolution takes over and expands the formula. This expansion should be restricted in size, however, to prevent exponential growth of the term set. If resolution can detect the tautology property the algorithm is exited. If not, reduction is visited again to lower the size of the term set. This loop is repeated forever. Clearly, resolution can detect non-satisfiability, too, in which case the algorithm is terminated. The above described reduction algorithm cannot handle this case.

Though reduction seems to be inferior to resolution in practice, in theory it is not. In fact, we proved in last section that reduction calculus is refutation-complete. It is well-known that resolution is refutation-complete, too. Thus, theoretically the two methods are on par in some sense. This

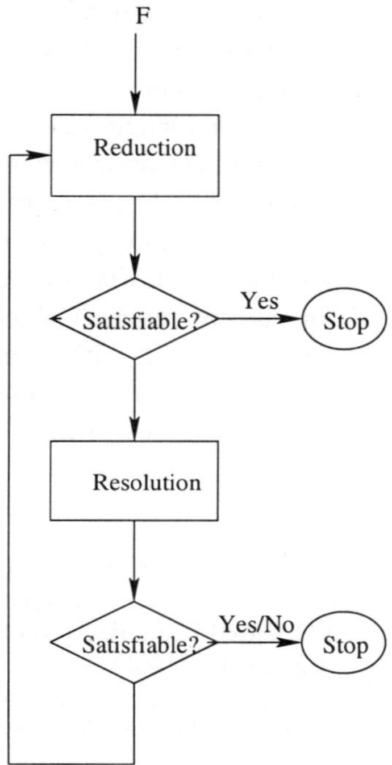

Figure 1.   Interplay between Reduction and Resolution.

is a very interesting theoretical insight. Some kind of formula simplification (akin to reduction) is often used in algorithms for resolution as in the algorithm we proposed above. The major contribution of the paper is rather to show that such an apparently helper reduction procedure is theoretically not inferior to the main resolution procedure.

5. Conclusion

The paper presented a new reduction calculus for propositional formulas. The calculus is proved to be refutation complete. We also proposed a partially correct algorithm for tautology test based on reduction calculus. The algorithm is of polynomial complexity. We tried to avoid exponential complexity even at the price of weaker correctness results. We also discussed the interplay between reduction and resolution.

There is no doubt that both reduction calculus and resolution calculus serve as an algorithmic approach for checking the satisfiability of logic formulas. The unexpected result that simple reduction is sound as well as complete with respect to refutation is exciting. This implies that resolution is not stronger than simple reduction from a theoretical point of view. On the other hand, the mechanics of resolution is more powerful in general, as reduction calculus has a simple and single rule of inference. Algorithmically, it is recommended to combine both approaches as outlined above. The combination of reduction and resolution is a straightforward approach. Apparently the process of repeated reduction(s) after each expansion step(s) is beneficial. However, a rigorous complexity analysis of this approach is needed and this is one of our current research objectives. The analysis is expected to be complicated as the two algorithms may influence the term set in opposite manners. Also, the algorithm we presented has the potential to be parallelized. This will be investigated in future.

References

1. F. Bacchus: Enhancing Davis Putnam with Extended Binary Clause Reasoning, 18th *National Conference on Artificial Intelligence*, (2002).
2. J. A. Robinson: A Machine-Oriented Logic Based on the Resolution Principle, *J. Assoc. Comput. Mach.*, **12**, 23–41, (1965).
3. S. R. Fletcher: Practical Methods of Optimization, 2nd Edition, Wiley, (1987).
4. S. Noureddine: An Approach for the Satisfiability Problem via Exterior Penalty Optimization, *Journal of Computer Science*, (2005).
5. S. Noureddine: A Geometric Programming Approach for the Satisfiability Problem, *Technical Report*, Dept. of Mathematics & Computer Science, University of Lethbridge, Lethbridge, Canada, (2005).
6. S. Porschen, et al.: Exact 3-satisfiability is Decidable in Time $O(2^{0.16254})$, *Annals of Mathematics and Artificial Intelligence*, (2002).
7. T. J. Schaefer: The Complexity of Satisfiability Problems, *Proceedings of the 10th annual ACM Symposium on Theory of Computing*, San Diego, California, U.S.A., 216–226, (1978).
8. B. Aspvall et al.: A Linear-time Algorithm for Testing the Truth of Certain Quantified Boolean Formulas, *Information Processing Letters*, **8(3)**: 121–123, (1979)
9. U. Schoening: A Probabilistic Algorithm for k-SAT and Constraint Satisfaction Problems, *Proceedings, 40th Annu. IEEE Symposium on Foundations of Computer Science*, FOCS'99, 410–414, (1999).
10. R.J. Duffin et al.: Geometric Programming: Theory and Applications, Wiley, New York, (1967).

ELEMENTARY TYPE SEMIGROUP FOR BOOLEAN ALGEBRAS WITH DISTINGUISHED IDEALS*

DMITRY PAL'CHUNOV

1. Introduction

We study model theoretical properties of Boolean algebras with distinguished ideals, further called as I-algebras. We consider I-algebras in the signature $\sigma_l \rightleftharpoons \langle \cup, \cap, {}^-, I_1, \ldots, I_l \rangle$, where I_1, ..., I_l are unary predicate symbols for the ideals. The number l of distinguished ideals is fixed.

Different model theoretical and algorithmic properties of I-algebras were studied in [1–12]. Some of results of this paper were announced in [8].

Definition 1. Denote

$$E \rightleftharpoons \{\mathfrak{A} \mid \mathfrak{A} \text{ is an } I\text{-algebra}\}/_\equiv = \{[\mathfrak{A}]_\equiv \mid \mathfrak{A} \text{ is an } I\text{-algebra}\},$$

where

$$[\mathfrak{A}]_\equiv = \{\mathfrak{B} \mid \mathfrak{B} \text{ is an } I\text{-algebra and } \mathfrak{B} \equiv \mathfrak{A}\}.$$

Denote $\mathbb{E} \rightleftharpoons \langle E; \times, e \rangle$, where $[\mathfrak{A}]_\equiv \times [\mathfrak{B}]_\equiv \rightleftharpoons [\mathfrak{A} \times \mathfrak{B}]_\equiv$ and $e \rightleftharpoons [\langle \{0\}, \sigma \rangle]_\equiv$ ($\langle \{0\}, \sigma \rangle$ is the degenerate Boolean algebra (in which $0 = 1$) with l distinguished ideals $I_j = \{0\}$).

Remark 1. If for I-algebras $\mathfrak{A}_1, \mathfrak{A}_2, \mathfrak{B}_1, \mathfrak{B}_2$ we have $\mathfrak{A}_1 \equiv \mathfrak{A}_2$ and $\mathfrak{B}_1 \equiv \mathfrak{B}_2$, then $\mathfrak{A}_1 \times \mathfrak{B}_1 \equiv \mathfrak{A}_2 \times \mathfrak{B}_2$. Therefore, Definition 1 is correct.

Proposition 1. *Let* $\mathfrak{A}, \mathfrak{B}, \mathfrak{C}$ *be I-algebras. Then*
   a) $\mathfrak{A} \times \mathfrak{B} \equiv \mathfrak{B} \times \mathfrak{A}$;
   b) $(\mathfrak{A} \times \mathfrak{B}) \times \mathfrak{C} \equiv \mathfrak{A} \times (\mathfrak{B} \times \mathfrak{C})$;
   c) $\mathfrak{A} \times \langle \{0\}; \sigma \rangle \equiv \mathfrak{A}$.

Hence $\mathbb{E}$ is a commutative semigroup with identity $e[\langle \{0\}; \sigma \rangle]_\equiv$.
The main purpose of the paper is to prove the following:

*Supported by RFBR grant N 05-01-04003-NNIO-a (DFG project COMO, GZ: 436 RUS 113/829/0-1), and by the Council for grants under RF President, project NSh-4413.2006.1.

Theorem 1. *Each of properties of I-algebras to be non-vanishing, local, finitely axiomatizable and ω-categorical may be represented in the semigroup $\mathbb{E}$ by a first order formula, i.e., there are first order formulas $N(x), L(x), F(x)$ and $C(x)$ of the signature $\{\times\}$ such that*
  1) *I-algebra $\mathfrak{A}$ is non-vanishing if and only if $\mathbb{E} \models N([\mathfrak{A}]_{\equiv})$;*
  2) *I-algebra $\mathfrak{A}$ is local if and only if $\mathbb{E} \models L([\mathfrak{A}]_{\equiv})$;*
  3) *I-algebra $\mathfrak{A}$ is finitely axiomatizable if and only if $\mathbb{E} \models F([\mathfrak{A}]_{\equiv})$;*
  4) *I-algebra $\mathfrak{A}$ is ω-categorical if and only if $\mathbb{E} \models C([\mathfrak{A}]_{\equiv})$.*

2. Non-vanishing *I*-algebras

For an I-algebra $\mathfrak{A}$ and $a \in \mathfrak{A}$ we denote $\widehat{a} \rightleftharpoons \{b \in \mathfrak{A} \mid b \leq a\}$, for $b \in \mathfrak{A}$ denote $\overline{b}^{a} \rightleftharpoons \overline{b} \cap a$. Then $\langle \widehat{a}; \cup, \cap, ^{-a}, I_1 \cap \widehat{a}, \ldots, I_l \cap \widehat{a} \rangle$ is an I-algebra, it is denoted as (a).

Remark 2. If $\mathfrak{A}$ is an I-algebra, then $\mathfrak{A} \cong (a) \times (\overline{a})$.

We write $\mathfrak{A} = (a) \times (\overline{a})$.

Definition 2. [7] An I-algebra $\mathfrak{A}$ is called non-vanishing if for any I-algebras $\mathfrak{B}$ and $\mathfrak{C}$, the statement $\mathfrak{A} = \mathfrak{B} \times \mathfrak{C}$ implies $\mathfrak{A} \equiv \mathfrak{B}$ or $\mathfrak{A} \equiv \mathfrak{C}$.

The meaning of this concept is the following: for any direct decomposition of a non-vanishing I-algebra its elementary theory does not vanish, and remains at least in one direct summund.

Remark 3. An I-algebra $\mathfrak{A}$ is non-vanishing if and only if for any $a \in \mathfrak{A}$, we have $\mathfrak{A} \equiv (a)$ or $\mathfrak{A} \equiv (\overline{a})$.

In [7] it was formulated the following

Question. Whether it is true, that if I-algebra $\mathfrak{A}$ is non-vanishing and $\mathfrak{B} \equiv \mathfrak{A}$, then $\mathfrak{B}$ is non-vanishing also.

To answer this question we prove the following

Proposition 2. *An I-algebra $\mathfrak{A}$ is non-vanishing if and only if for any I-algebras $\mathfrak{B}$ and $\mathfrak{C}$, $\mathfrak{A} \equiv \mathfrak{B} \times \mathfrak{C}$ implies $\mathfrak{A} \equiv \mathfrak{B}$ or $\mathfrak{A} \equiv \mathfrak{C}$.*

Proof. ($\Rightarrow$) Let an I-algebra $\mathfrak{A}$ be non-vanishing and $\mathfrak{A} \equiv \mathfrak{B} \times \mathfrak{C}$. Assume that $\mathfrak{A} \not\equiv \mathfrak{B}$ and $\mathfrak{A} \not\equiv \mathfrak{C}$. Then there exist sentences φ and ψ, such that $\mathfrak{B} \models \varphi$, $\mathfrak{C} \models \psi$, $\mathfrak{A} \not\models \varphi$ and $\mathfrak{A} \not\models \psi$. For a sentence ξ by a formula $\xi^R(x)$ we denote a relativization of the sentence ξ. Namely, $\xi^R(x)$ is a formula obtained from ξ by replacing of all occurrences of 1 by x, of $\exists y \varphi(y)$ by

$\exists y(y \leq x \,\&\, \varphi(y))$, and of $\forall y \varphi(y)$ by $\forall y(y \leq x \rightarrow \varphi(y))$ (we suppose that the variable x does not appear in the sentence ξ).

Denote $\mathfrak{M} \rightleftharpoons \mathfrak{B} \times \mathfrak{C}$, $b \rightleftharpoons (1^{\mathfrak{B}}, 0)$ and $c \rightleftharpoons (0, 1^{\mathfrak{C}})$. Then $c = \bar{b}$ and $\mathfrak{M} \cong (b) \times (c)$. We may assume that $\mathfrak{M} = (b) \times (\bar{b})$. Notice that $(b) \models \varphi$ and $(\bar{b}) \models \psi$, so $\mathfrak{M} \models \varphi^R(b)$ and $\mathfrak{M} \models \psi^R(\bar{b})$. Hence, $\mathfrak{M} \models \exists x(\varphi^R(x) \& \psi^R(\bar{x}))$. Moreover, $\mathfrak{A} \equiv \mathfrak{M}$, therefore $\mathfrak{A} \models \exists x(\varphi^R(x) \& \psi^R(\bar{x}))$. So there is an element $a \in \mathfrak{A}$ such that $\mathfrak{A} \models \varphi^R(a)$ and $\mathfrak{A} \models \psi^R(\bar{a})$. Denote $\mathfrak{A}_1 \rightleftharpoons (a)$ and $\mathfrak{A}_2 \rightleftharpoons (\bar{a})$. We may suppose that $\mathfrak{A} = \mathfrak{A}_1 \times \mathfrak{A}_2$, therefore $\mathfrak{A} \equiv \mathfrak{A}_1$ or $\mathfrak{A} \equiv \mathfrak{A}_2$. However, $\mathfrak{A}_1 \models \varphi$ and $\mathfrak{A}_2 \models \psi$, so $\mathfrak{A} \models \varphi$ or $\mathfrak{A} \models \psi$. But, by the hypothesis $\mathfrak{A} \not\models \varphi$ and $\mathfrak{A} \not\models \psi$ — a contradiction.

Hence, $\mathfrak{A} \equiv \mathfrak{B} \times \mathfrak{C}$ implies $\mathfrak{A} \equiv \mathfrak{B}$ or $\mathfrak{A} \equiv \mathfrak{C}$.

($\Leftarrow$) Let $\mathfrak{A} \equiv \mathfrak{B} \times \mathfrak{C}$ imply $\mathfrak{A} \equiv \mathfrak{B}$ or $\mathfrak{A} \equiv \mathfrak{C}$. If $\mathfrak{A} = \mathfrak{B} \times \mathfrak{C}$, then $\mathfrak{A} \equiv \mathfrak{B} \times \mathfrak{C}$, therefore $\mathfrak{A} \equiv \mathfrak{B}$ or $\mathfrak{A} \equiv \mathfrak{C}$.

Corollary 1. *If $\mathfrak{A} \equiv \mathfrak{B}$ and $\mathfrak{A}$ is non-vanishing, then $\mathfrak{B}$ is non-vanishing also.*

Proof. In fact, let $\mathfrak{B} = \mathfrak{M} \times \mathfrak{N}$ and $\mathfrak{A} \equiv \mathfrak{B}$. Hence $\mathfrak{A} \equiv \mathfrak{M} \times \mathfrak{N}$, so $\mathfrak{A} \equiv \mathfrak{M}$ or $\mathfrak{A} \equiv \mathfrak{N}$, therefore $\mathfrak{B} \equiv \mathfrak{M}$ or $\mathfrak{B} \equiv \mathfrak{N}$. Thus, $\mathfrak{B}$ is non-vanishing.

Corollary 2. *An I-algebra $\mathfrak{A}$ is non-vanishing if and only if $\mathbb{E} \models N([\mathfrak{A}]_{\equiv})$, where $N(x) \rightleftharpoons \forall y \forall z(x = y \times z \rightarrow (x = y \lor x = z))$.*

3. Ordered semigroup

Definition 3. For I-algebras $\mathfrak{A}$ and $\mathfrak{B}$ we denote $\mathfrak{A} \leq \mathfrak{B}$ if for some I-algebra $\mathfrak{C}$ we have $\mathfrak{B} \equiv \mathfrak{A} \times \mathfrak{C}$, denote $\mathfrak{A} < \mathfrak{B}$ if $\mathfrak{A} \leq \mathfrak{B}$ and $\mathfrak{A} \not\equiv \mathfrak{B}$. For elements $a, b \in \mathbb{E}$, we denote $a \leq b$ if $b = a \times c$ for some $c \in \mathbb{E}$, and $a < b$ if $a \leq b$ and $a \neq b$.

Remark 4. Consider I-algebras $\mathfrak{A}$ and $\mathfrak{B}$. We have $\mathfrak{A} \leq \mathfrak{B}$ if and only if $[\mathfrak{A}]_{\equiv} \leq [\mathfrak{B}]_{\equiv}$.

Proof. $\mathfrak{A} \leq \mathfrak{B} \Leftrightarrow \mathfrak{B} \equiv \mathfrak{A} \times \mathfrak{C}$ for some I-algebra $\mathfrak{C} \Leftrightarrow [\mathfrak{B}]_{\equiv} = [\mathfrak{A}]_{\equiv} \times [\mathfrak{C}]_{\equiv}$ for some I-algebra $\mathfrak{C} \Leftrightarrow [\mathfrak{B}]_{\equiv} = [\mathfrak{A}]_{\equiv} \times c$ for some $c \in \mathbb{E} \Leftrightarrow [\mathfrak{A}]_{\equiv} \leq [\mathfrak{B}]_{\equiv}$.

Corollary 3. *Suppose that for I-algebras $\mathfrak{A}, \mathfrak{B}, \mathfrak{M}$ and $\mathfrak{N}$ we have $\mathfrak{A} \equiv \mathfrak{M}$ and $\mathfrak{B} \equiv \mathfrak{N}$. Then $\mathfrak{A} \leq \mathfrak{B}$ if and only if $\mathfrak{M} \leq \mathfrak{N}$.*

Proposition 3. *The relation $\leq$ is a partial order in the semigroup $\mathbb{E}$.*

Proof. Reflexive and transitive properties are obvious. Antisymmetry property follows from [11, Lemma 3]

Proposition 4. *If $a, b, c, d \in \mathbb{E}$, $a \leq b$ and $c \leq d$, then $a \times c \leq b \times d$. It means that $\langle \mathbb{E}; \leq \rangle$ is an ordered semigroup.*

Proof. Let $a = [\mathfrak{A}]_{\equiv}$, $b[\mathfrak{B}]_{\equiv}$, $c = [\mathfrak{C}]_{\equiv}$ and $d = [\mathfrak{M}]_{\equiv}$. Then $\mathfrak{A} \leq \mathfrak{B}$ and $\mathfrak{C} \leq \mathfrak{M}$, therefore there exist $\mathfrak{N}$ and $\mathfrak{L}$ such that $\mathfrak{B} \equiv \mathfrak{A} \times \mathfrak{N}$ and $\mathfrak{M} \equiv \mathfrak{C} \times \mathfrak{L}$. Hence, $\mathfrak{B} \times \mathfrak{M} \equiv \mathfrak{A} \times \mathfrak{C} \times \mathfrak{N} \times \mathfrak{L}$, i.e., $\mathfrak{A} \times \mathfrak{C} \leq \mathfrak{B} \times \mathfrak{M}$ and $a \times c \leq b \times d$.

4. Local I-algebras

Definition 4. [7] An I-algebra is called basic if it is finitely axiomatizable and non-vanishing.

In [3] a sequence of formulas $V_n(x)$, $n \in \mathbb{N}$, of the I-algebra signature, was introduced.

Theorem 2. [7, Theorem 1] *An I-algebra $\mathfrak{A}$ is basic if and only if $\mathfrak{A} \models V_n(1)$ for some $n \in N$.*

Proposition 5. [3, 7] *For any $n \in \mathbb{N}$*
    1) *$V_n(1)$ is a complete sentence in the theory of the class of I-algebras.*
    2) *For any I-algebras $\mathfrak{A}, \mathfrak{B}$ and $a \in \mathfrak{A}$, we have:*
    a) *$\mathfrak{A} \models V_n(a) \Leftrightarrow (a) \models V_n(1)$,*
    b) *$\mathfrak{A} \models V_n(a) \Leftrightarrow \mathfrak{A} \times \mathfrak{B} \models V_n(a)$, where we identify $(a, 0)$ with a.*

Definition 5. For $n \in \mathbb{N}$ let $\mathfrak{A}_n$ be a countable I-algebra such that $\mathfrak{A}_n \models V_n(1)$. For $m \in \mathbb{N}$ we denote $V_n \leq V_m$ if $\mathfrak{A}_n \leq \mathfrak{A}_m$, and $V_n < V_m$ if $\mathfrak{A}_n < \mathfrak{A}_m$.

By the construction, the sequence of formulas V_n, $n \in \mathbb{N}$, possesses the following properties (see [3]):

Proposition 6. a) *For any m, n, if $V_m < V_n$, then $m \nmid n$.*
    b) *The number of minimal (up to the order $\leq$) formulas V_n is finite.*
    c) *For any n, the set $\{m \mid V_m < V_n\}$ is finite.*
    d) *For any n, the set $\{m \mid$ if $V_k < V_m$, then $k \leq n\}$ is finite.*

Definition 6. A basic I-algebra is called continuous if $\mathfrak{A} \equiv \mathfrak{A} \times \mathfrak{A}$, and pointwise if $\mathfrak{A} \not\equiv \mathfrak{A} \times \mathfrak{A}$.

Remark 5. If an I-algebra $\mathfrak{A}$ is continuous (pointwise) and $\mathfrak{B} \equiv \mathfrak{A}$, then $\mathfrak{B}$ is continuous (pointwise) also.

Proposition 7. [3, 7] 1) *If an I-algebra $\mathfrak{A}$ is continuous, then there exists an element $a \in \mathfrak{A}$ such that $\mathfrak{A} \equiv (a)$ and $\mathfrak{A} \equiv (\overline{a})$.*

2) If an I-algebra $\mathfrak{A}$ is pointwise and $\mathfrak{A} \equiv \mathfrak{B} \times \mathfrak{C}$ ($a \in \mathfrak{A}$), then either $\mathfrak{A} \not\equiv \mathfrak{B}$ or $\mathfrak{A} \not\equiv \mathfrak{C}$ (either $\mathfrak{A} \not\equiv (a)$ or $\mathfrak{A} \not\equiv (\overline{a})$ respectively).

Remark 6. A basic I-algebra $\mathfrak{A}$ is continuous if and only if $[a]_{\equiv}$ is an idempotent of the semigroup $\mathbb{E}$: $[a]_{\equiv} \times [a]_{\equiv} = [a]_{\equiv}$.

For an I-algebra $\mathfrak{A}$ we denote

$$M(\mathfrak{A}) \rightleftharpoons \{m \in \mathbb{N} \mid \mathfrak{A} \models \exists x V_m(x)\} = \{m \in \mathbb{N} \mid \mathfrak{A}_m \leq \mathfrak{A}\},$$

$$N(\mathfrak{A}) \rightleftharpoons \{m \in M(\mathfrak{A}) \mid \forall n \in M(\mathfrak{A}), \ V_m \not< V_n\} = \{m \in \mathbb{N} \mid \mathfrak{A}_m \text{ is a maximal}$$
$$\text{element in the set } \{\mathfrak{A}_n \mid n \in M(\mathfrak{A})\} \text{ w.r.t. the order relation } \leq\}.$$

Definition 7. An I-algebra $\mathfrak{A}$ is called local if the set $M(\mathfrak{A})$ is finite, i.e., the number of different, up to the elementary equivalence, basic direct summands of $\mathfrak{A}$ is finite.

Proposition 8. [3, 7] 1) *Every I-algebra $\mathfrak{A}$ has a basic direct summand, i.e., $M(\mathfrak{A}) \neq \emptyset$. If $\mathfrak{A}$ is local, then $N(\mathfrak{A}) \neq \emptyset$.*

2) Let an I-algebra $\mathfrak{A}$ be local and $N(\mathfrak{A}) = \{n_1, \ldots, n_k\}$. Then

a) $M(\mathfrak{A}) = \{m \mid V_m \leq V_{n_i} \text{ for some } i \leq k\}$.

b) There exists a decomposition $\mathfrak{A} = \mathfrak{B}_1 \times \ldots \times \mathfrak{B}_k$ such that $N(\mathfrak{B}_i) = \{n_i\}$.

3) Each basic I-algebra is local.

4) If an I-algebra $\mathfrak{A}$ is basic, then for some $n \in \mathbb{N}$, we have $\mathfrak{A} \models V_n(1)$ and $N(\mathfrak{A}) = \{n\}$.

5) If an I-algebra $\mathfrak{A}$ is local and $N(\mathfrak{A}) = \{n\}$, then either

a) $\mathfrak{A} \equiv (\mathfrak{A}_n)^k$, or

b) $(\mathfrak{A}_n)^k \leq \mathfrak{A}$ for any $k \in \mathbb{N}$.

6) A local I-algebra $\mathfrak{A}$ is non-vanishing if and only if $N(\mathfrak{A}) = \{n\}$ for some $n \in \mathbb{N}$ and either

a) $\mathfrak{A}$ is basic (in this case $\mathfrak{A} \models V_n(1)$), or

b) $\mathfrak{A}_n$ is pointwise and $(\mathfrak{A}_n)^k \leq \mathfrak{A}$ for any k. In this situation $\mathfrak{A} \times \mathfrak{A} \equiv \mathfrak{A}$.

7) Let I-algebras $\mathfrak{A}$ and $\mathfrak{B}$ be local. $\mathfrak{A} \equiv \mathfrak{B}$ if and only if $(\mathfrak{A}_n)^k \leq \mathfrak{A}$ is equivalent to $(\mathfrak{A}_n)^k \leq \mathfrak{B}$ for any $n, k \in \mathbb{N}$.

Definition 8. For any I-algebra $\mathfrak{A}$ we define a characteristic $r_{\mathfrak{A}} : \mathbb{N} \to \mathbb{N} \cup \{\infty\}$:

$$r_{\mathfrak{A}}(n) \rightleftharpoons \begin{cases} 0, & \text{if } \mathfrak{A}_n \not\leq \mathfrak{A}; \\ 1, & \text{if } \mathfrak{A}_n \leq \mathfrak{A} \text{ and } \mathfrak{A}_n \times \mathfrak{A}_n \equiv \mathfrak{A}_n; \\ k, & \text{if } (\mathfrak{A}_n)^k \leq \mathfrak{A} \text{ and } (\mathfrak{A}_n)^{k+1} \not\leq \mathfrak{A}; \\ \infty, & \text{if } \mathfrak{A}_n \times \mathfrak{A}_n \not\equiv \mathfrak{A}_n \text{ and } (\mathfrak{A}_n)^k \leq \mathfrak{A} \text{ for any } k \in \mathbb{N}. \end{cases}$$

For $a \in \mathfrak{A}$ we denote $r_a \rightleftharpoons r_{(a)}$.

Remark 7. An I-algebra $\mathfrak{A}$ is local if and only if the function $r_{\mathfrak{A}}$ has a finite support, i.e., there exists $n \in \mathbb{N}$ such that $r_{\mathfrak{A}}(k) = 0$ for any $k \geq n$.

Theorem 3. [3] *For local I-algebras $\mathfrak{A}$ and $\mathfrak{B}$ we have $\mathfrak{A} \equiv \mathfrak{B}$ if and only if $r_{\mathfrak{A}} = r_{\mathfrak{B}}$.*

Proposition 9. [3]
$$r_{\mathfrak{A} \times \mathfrak{B}}(n) \rightleftharpoons \begin{cases} \max\{r_{\mathfrak{A}}(n), r_{\mathfrak{B}}(n)\}, & \text{if } \mathfrak{A}_n \times \mathfrak{A}_n \equiv \mathfrak{A}_n; \\ r_{\mathfrak{A}}(n) + r_{\mathfrak{B}}(n), & \text{if } \mathfrak{A}_n \times \mathfrak{A}_n \not\equiv \mathfrak{A}_n. \end{cases}$$
(*Here* $\infty + x = \infty$).

Corollary 4. *If $\mathfrak{A} \leq \mathfrak{B}$, then $r_{\mathfrak{A}} \leq r_{\mathfrak{B}}$.*

Proof. Let $\mathfrak{A} \leq \mathfrak{B}$, then $\mathfrak{B} = \mathfrak{A} \times \mathfrak{C}$ for some I-algebra $\mathfrak{C}$. Therefore, $r_{\mathfrak{B}} = r_{\mathfrak{A} \times \mathfrak{C}}$, hence $r_{\mathfrak{A}} \leq r_{\mathfrak{B}}$.

Definition 9. A set $K \subseteq \mathbb{N}$ is called natural if for any $m, n \in \mathbb{N}$, $V_m \leq V_n$ and $n \in K$ imply $m \in K$.

Proposition 10. [3] *A set $K \subseteq \mathbb{N}$ is natural if and only if $K = M(\mathfrak{A})$ for some I-algebra $\mathfrak{A}$.*

Definition 10. An infinite natural set $K \subseteq \mathbb{N}$ is called stepwise if for any $n \in \mathbb{N}$ there exists $k \in \mathbb{N}$ such that if $m, j \in K$, $m \leq n$ and $j \geq k$ then $V_m \leq V_j$.

An I-algebra $\mathfrak{A}$ is called stepwise if the set $M(\mathfrak{A})$ is stepwise.

Theorem 1.1 [10] and the definition of the characteristic $r_{\mathfrak{A}}$ imply the following

Proposition 11. a) *A set $K \subseteq \mathbb{N}$ is stepwise if and only if it is minimal w.r.t. inclusion among infinite natural sets.*

b) *An I-algebra $\mathfrak{A}$ is stepwise if and only if for any non-local I-algebra $\mathfrak{B}$ (i.e., for any I-algebra $\mathfrak{B}$ with infinite $M(\mathfrak{B})$) if $M(\mathfrak{B}) \subseteq M(\mathfrak{A})$, then $M(\mathfrak{B}) = M(\mathfrak{A})$.*

Remind that every sentence $V_n(1)$ is complete, so it determines an I-algebra $\mathfrak{A} \models V_n(1)$ up to the elementary equivalence.

Definition 11. An I-algebra $\mathfrak{A}$ is called strictly stepwise if it is stepwise and for any $k \leq l$, $1^{\mathfrak{A}} \in I_k^{\mathfrak{A}}$ if and only if $1^{\mathfrak{A}_n} \in I_k^{\mathfrak{A}_n}$ for any $n \in M(\mathfrak{A})$ (i.e., a sentence $(V_n(1) \to (1 \in I_k)))$ is true on the class of I-algebras.

Thus, for any strictly stepwise I-algebra $\mathfrak{A}$ truth values of the sentences $I_k(1)$ are determined syntactically given the set $M(\mathfrak{A})$.

From Theorem 1.2 [10] and the definition of stepwise I-algebras it follows

Theorem 4. *For strictly stepwise I-algebras $\mathfrak{A}$ and $\mathfrak{B}$ we have $\mathfrak{A} \equiv \mathfrak{B}$ if and only if $M(\mathfrak{A}) = M(\mathfrak{B})$.*

Theorem 5. *For any non-local I-algebra $\mathfrak{A}$ there exists a strictly stepwise I-algebra $\mathfrak{B} \leq \mathfrak{A}$.*

Proof. First we prove

Proposition 12. *If an infinite set M is natural, then there exists $L \subseteq M$ which is minimal among infinite natural sets.*

Proof.

Let a set M be infinite and natural. We denote

$$M_0 \rightleftharpoons \{n \in M \mid V_n \text{ is minimal w.r.t. } \leq\},$$
$$M_{i+1} \rightleftharpoons \{n \in M \mid \text{ if } V_m < V_n, \text{ then } m \in M_i\}.$$

It is obvious that $M_0 \subseteq M_1 \subseteq \ldots$. Proposition 6 implies that for any $k \in \mathbb{N}$ the set M_k is finite, because M_0 is finite, and if M_i is finite, then M_{i+1} is finite also. Moreover, Proposition 6 implies that $M = \bigcup_{n \in \mathbb{N}} M_n$.

We choose a set $L_0 \subseteq M_0$ which is minimal w.r.t. inclusion and such that the set $\{n \in M \mid \text{ if } m \in M_0 \text{ and } V_m < V_n, \text{ than } m \in L_0\}$ is infinite.

Suppose that a set $L_i \subseteq M_i$ is constructed and the set $\{n \in M \mid \text{ if } m \in M_i \text{ and } V_m < V_n, \text{ than } m \in L_i\}$ is infinite.

Choose a set $L_{i+1} \subseteq M_{i+1}$ which is minimal w.r.t. inclusion and such that $L_{i+1} \cap M_i = L_i$, the set $\{n \in M \mid \text{ if } m \in M_{i+1} \text{ and } V_m < V_n, \text{ then } m \in L_{i+1}\}$ is infinite, and if $n \in L_{i+1}$ and $V_m < V_n$, then $m \in L_i$.

Put $L \rightleftharpoons \bigcup_{k \in \mathbb{N}} L_k$. It is easy to prove that the set L is infinite and $L \cap M_i L_i$ for any i.

Lemma 1. *The set L is natural.*

Proof. Let $n \in L$ and $V_m < V_n$. Then $n \in M$, hence $m \in M$. So, there exists $i \in \mathbb{N}$ such that $n \in M_{i+1}$. Then $m \in M_i$. Besides, $n \in M_{i+1} \cap L = L_{i+1}$ and $V_m < V_n$. Hence, $m \in L_i$, so $m \in L$. Therefore, the set L is natural.

The lemma is proved.

Lemma 2. *L is minimal among infinite natural sets.*

Proof. Suppose that L is not minimal among infinite natural sets. Hence, there is an infinite natural set $K \subseteq L$ with $K \neq L$. Therefore, $K \subseteq M$.

Denote $K_i \rightleftharpoons K \cap L_i$. Then $K \bigcup_{i \in \mathbb{N}} K_i$, and for any i, $K_i \subseteq L_i$. Hence, there exists $i \in \mathbb{N}$ with $K_i \neq L_i$. Let j be the least number such that $K_j \neq L_j$.

1) $j = 0$. Let $n \in K$, then $n \in M$. Assume that $m \in M_0$ and $V_m < V_n$. Then $n \in L$ and $m \in L$, since the set L is natural. However, $L \cap M_0 = L_0$, hence, $m \in L_0$, therefore $m \in K_0 = K \cap L_0$.

Thus, $K \subseteq \{n \in M \mid \text{ if } m \in M_0 \text{ and } V_m < V_n, \text{ then } m \in K_0\}$ and the set K is infinite. We arrive at a contradiction with L_0 being minimal w.r.t. inclusion, because $K_0 \subseteq L_0$ and $K_0 \neq L_0$.

2) $j > 0$. Then $j = i + 1$ for some i. Hence $K_m = L_m$ for any $m \leq i$. Therefore $K_{i+1} \subseteq L_{i+1} \subseteq M_{i+1}$, $L_i = K_i \subseteq K_{i+1}$ and $L_i \subseteq M_i$, so $L_i \subseteq K_{i+1} \cap M_i$ and $K_{i+1} \cap M_i \subseteq L_{i+1} \cap M_i = L_i$. Hence $K_{i+1} \cap M_i = L_i$. Let $n \in K_{i+1}$ and $V_m < V_n$, then $n \in L_{i+1}$, so $m \in L_i$.

Further, assume that $n \in K$, then $n \in L$ and $n \in M$. Let $m \in M_{i+1}$ and $V_m < V_n$. Then $m \in L_{i+1}$, and $m \in K$, because K is natural. Hence $m \in K \cap L_{i+1} = K_{i+1}$. Thus $K \subseteq \{n \in M \mid \text{ if } m \in M_{i+1} \text{ and } V_m < V_n, \text{ then } m \in K_{i+1}\}$ and these sets are infinite. Besides, $K_{i+1} \subseteq L_{i+1}$ and $K_{i+1} \neq L_{i+1}$. We arrive at a contradiction, because K_{i+1} is minimal w.r.t. inclusion.

Hence, an infinite natural set $K \subseteq L$, with $K \neq L$, does not exist. Therefore L is minimal.

Lemma 2, as well as Proposition 12, is proved.

Proposition 13. *If a set L is stepwise and $L \subseteq M(\mathfrak{A})$, then there exist a strictly stepwise I-algebra $\mathfrak{B} \leq \mathfrak{A}$ with $M(\mathfrak{B}) = L$.*

Proof. Let $L \subseteq M(\mathfrak{A})$ and L be stepwise. Consider a set

$$J \rightleftharpoons \{k \leq l \mid 1^{\mathfrak{A}_n} \in I_k^{\mathfrak{A}_n} \text{ for any } n \in L\}.$$

Consider a set of formulas

$$S(x) \rightleftharpoons \{\exists y(y \leq x \,\&\, V_n(y)) \mid n \in L\}$$
$$\cup \,\{\neg \exists y(y \leq x \,\&\, V_n(y)) \mid n \notin L\} \cup \{I_k(x) \mid k \in J\}.$$

Denote $T \rightleftharpoons Th\mathfrak{A}$. We show that the set of formulas $T \cup S(x)$ is consistent. First we show that it is locally consistent. Let $L' \subseteq L$ and L' be finite. We prove that

$$T \cup \{ \exists y(y \le x \,\&\, V_n(y)) \mid n \in L' \}$$
$$\cup \{\neg \exists y(y \le x \,\&\, V_n(y)) \mid n \notin L\} \cup \{I_k(x) \mid k \in J\}$$

is consistent.

Since $L \subseteq M(\mathfrak{A})$, we have $L' \subseteq M(\mathfrak{A})$, therefore, for any $n \in L'$, $\mathfrak{A} \models \exists y V_n(y)$. For any $n \in L'$ we find $a_n \in \mathfrak{A}$ such that $\mathfrak{A} \models V_n(a_n)$, put $a \rightleftharpoons \bigcup_{n \in L'} a_n$. Then $\mathfrak{A} \models \exists y(y \le a \,\&\, V_n(y))$ for any $n \in L'$. Moreover, if $k \in J$, then for any $n \in L$ a formula $(V_n(x) \to I_k(x))$ is true on the class of I-algebras. Therefore, $a_n \in I_k$ for any $n \in L'$. Hence $I_k(a)$ for any $k \in J$.

Suppose that $n \notin L$ and there exists $c \le a$ such that $\mathfrak{A} \models V_n(c)$. Let $L' = \{m_1, \ldots, m_k\}$. Denote $b_1 \rightleftharpoons a_{m_1}, b_2 \rightleftharpoons a_{m_2} \backslash a_{m_1}, \ldots, b_k \rightleftharpoons a_{m_k} \backslash (a_{m_1} \cup \ldots \cup a_{m_{k-1}})$. Then $a = b_1 \cup \ldots \cup b_k$ and $b_i \cap b_j = 0$ for $i \ne j$. We denote $c_1 \rightleftharpoons c \cap b_1, \ldots, c_k \rightleftharpoons c \cap b_k$. Then $(c) \cong (c_1) \times \ldots \times (c_k)$ and $(c) \models V_n(1)$. I-algebra (c) is non-vanishing, so $(c_i) \models V_n(1)$ for some $i \le k$. Moreover $c_i = c \cap b_i = c \cap (a_{m_i} \backslash (a_{m_1} \cup \ldots \cup a_{m_{i-1}})) \le a_{m_i}$.

Hence $(a_{m_i}) \models V_{m_i}(1)$, $(c_i) \models V_n(1)$ and $(c_i) \le (a_{m_i})$. Therefore, $V_n \le V_{m_i}$. Furthermore, $m_i \in L' \subseteq L$ and L is a natural set. So, $n \in L$ — a contradiction, because $n \notin L$ by our assumption.

Hence $\mathfrak{A} \models \neg \exists x((x \le a) \,\&\, V_n(x))$ for any $n \notin L$.

Thus we have proved that

$$\mathfrak{A} \models T \cup \{\exists y(y \le a \,\&\, V_n(y)) \mid n \in L'\}$$
$$\cup \{\neg \exists y(y \le a \,\&\, V_n(y)) \mid n \notin L\} \cup \{I_k(a) \mid k \in J\}.$$

Consequently the set $T \cup S(x)$ is locally consistent, so it is consistent and there exist an I-algebra $\mathfrak{C}$ and $b \in \mathfrak{C}$ such that $\mathfrak{C} \models T \cup S(b)$. Denote $\mathfrak{B} \rightleftharpoons (b)$ and $\mathfrak{L} \rightleftharpoons (\bar{b})$. Then $\mathfrak{C} \equiv \mathfrak{A}$, $\mathfrak{C} = \mathfrak{B} \times \mathfrak{L}$, $M(\mathfrak{B}) = L$ and $\mathfrak{B}$ is strictly stepwise. Hence, $\mathfrak{B} \le \mathfrak{A}$.

The proposition is proved.

Theorem 5 follows from Propositions 12 and 13.

Corollary 5. *Let $M(\mathfrak{C}) \subseteq M(\mathfrak{A})$ and I-algebra $\mathfrak{C}$ be strictly stepwise. Then $\mathfrak{A} \times \mathfrak{C} \equiv \mathfrak{A}$.*

Proof. Let I-algebra $\mathfrak{B}$ be strictly stepwise, $\mathfrak{B} \le \mathfrak{A}$ and $M(\mathfrak{B}) = M(\mathfrak{C})$. Then $\mathfrak{A} \equiv \mathfrak{B} \times \mathfrak{M}$ for some I-algebra $\mathfrak{M}$. Note that $M(\mathfrak{B} \times \mathfrak{C}) = M(\mathfrak{B})$ and $\mathfrak{B} \times \mathfrak{C}$ is strictly stepwise. Therefore $\mathfrak{B} \times \mathfrak{C} \equiv \mathfrak{B}$ and $\mathfrak{A} \times \mathfrak{C} \cong \mathfrak{B} \times \mathfrak{C} \times \mathfrak{M} \equiv \mathfrak{B} \times \mathfrak{M} \equiv \mathfrak{A}$.

Corollary 6. *If $\mathfrak{A} \equiv \mathfrak{B} \times \mathfrak{C}$, then $M(\mathfrak{A}) = M(\mathfrak{B}) \cup M(\mathfrak{C})$.*

Proof. It is obvious that $M(\mathfrak{B}) \cup M(\mathfrak{C}) \subseteq M(\mathfrak{A})$. Let $n \in M(\mathfrak{A})$ and $a \in \mathfrak{A}$. Then there exists $c \in \mathfrak{A}$ with $\mathfrak{A} \models V_n(c)$. Hence, as it was shown in the proof of Proposition 13, either $\mathfrak{A} \models V_n(c \cap a)$ or $\mathfrak{A} \models V_n(c \cap \bar{a})$. Therefore, either $n \in M(a)$, or $n \in M(\bar{a})$. Corollory is proved.

Theorem 6. *A strictly stepwise I-algebra is non-vanishing.*

Proof. Let an I-algebra $\mathfrak{A}$ be strictly stepwise and $\mathfrak{A} \equiv \mathfrak{B} \times \mathfrak{C}$. Then $M(\mathfrak{A}) = M(\mathfrak{B}) \cup M(\mathfrak{C})$. Therefore at least one of the sets $M(\mathfrak{B})$ or $M(\mathfrak{C})$ is infinite, suppose that $M(\mathfrak{B})$ is infinite. Hence the infinite natural set $M(\mathfrak{B}) \subseteq M(\mathfrak{A})$ and $M(\mathfrak{A})$ is stepwise, so $M(\mathfrak{B}) = M(\mathfrak{A})$. It is easy to prove that the I-algebra $\mathfrak{B}$ is strictly stepwise, hence $\mathfrak{A} \equiv \mathfrak{B}$. The theorem is proved.

Corollary 7. *If an I-algebra $\mathfrak{A}$ is strictly stepwise and $\mathfrak{B} \leq \mathfrak{A}$, then either $\mathfrak{B}$ is local, or $\mathfrak{A} \equiv \mathfrak{B}$.*

Corollary 8. *In the set $\{a \in \mathbb{E} \mid a$ is not local $\}$, under each element there is a minimal one. The minimal elements of this set are exactly the equivalence classes of strictly stepwise I-algebras.*

Corollary 9. *If an I-algebra $\mathfrak{A}$ is basic and $\mathfrak{A} \leq \mathfrak{B} \times \mathfrak{C}$, then $\mathfrak{A} \leq \mathfrak{B}$ or $\mathfrak{A} \leq \mathfrak{C}$.*

Proof. Let I-algebra $\mathfrak{A}$ be basic, $\mathfrak{A} \equiv \mathfrak{A}_n$ for some $n \in \mathbb{N}$, and $\mathfrak{A} \leq \mathfrak{B} \times \mathfrak{C}$. Then $\mathfrak{A}_n \leq \mathfrak{B} \times \mathfrak{C}$ and $\mathfrak{B} \times \mathfrak{C} \models \exists x V_n(x)$. Let $\mathfrak{M} = \mathfrak{B} \times \mathfrak{C}$, $b \in \mathfrak{M}$, $\mathfrak{B} \equiv (b)$ and $\mathfrak{C} \equiv (\bar{b})$. Then there exists $a \in \mathfrak{M}$ such that $\mathfrak{M} \models V_n(a)$. Hence, as shown above, $\mathfrak{M} \models V_n(a \cap b)$ or $\mathfrak{M} \models V_n(a \cap \bar{b})$. Therefore $(a \cap b) \models V_n(1)$ and $(a \cap b) \equiv \mathfrak{A}_n$, or $(a \cap \bar{b}) \models V_n(1)$ and $(a \cap \bar{b}) \equiv \mathfrak{A}_n$. However, $(a \cap b) \leq (b) \equiv \mathfrak{B}$ and $(a \cap \bar{b}) \leq (\bar{b}) \equiv \mathfrak{C}$. Thus $\mathfrak{A} \leq \mathfrak{B}$ or $\mathfrak{A} \leq \mathfrak{C}$.

The corollary is proved.

Corollary 10. *If an I-algebra $\mathfrak{A}$ is basic and $\mathfrak{B} < \mathfrak{A}$, then $\mathfrak{A} \times \mathfrak{B} \equiv \mathfrak{A}$.*

Proof. Let $\mathfrak{A}$ be basic and $\mathfrak{B} < \mathfrak{A}$. Then $\mathfrak{B} \leq \mathfrak{A}$, hence $\mathfrak{A} \equiv \mathfrak{B} \times \mathfrak{C}$. Moreover, $\mathfrak{A} \not\equiv \mathfrak{B}$ and $\mathfrak{A}$ is non-vanishing, therefore $\mathfrak{A} \equiv \mathfrak{C}$, i.e., $\mathfrak{A} \equiv \mathfrak{B} \times \mathfrak{A} \equiv \mathfrak{A} \times \mathfrak{B}$. The corollary is proved.

We denote

$$L'(x) \rightleftharpoons (\forall b \leq x)(\exists c < b)(\forall d < b)(((c \leq d) \ \& \ N(d)) \rightarrow c = d).$$

By $L(x)$ we denote a formula of the first order predicate logic of the signature $\{\times\}$, which is equivalent to $L'(x)$ on $\mathbb{E}$ (which is easily constructed).

Theorem 7. *An I-algebra $\mathfrak{A}$ is local if and only if $\mathbb{E} \models L([\mathfrak{A}]_{\equiv})$.*

Proof. ($\Rightarrow$) Let an I-algebra $\mathfrak{A}$ be local, $b \leq [\mathfrak{A}]_{\equiv}$ and $[\mathfrak{B}]_{\equiv} = b$. Then $\mathfrak{B} \leq \mathfrak{A}$.

Let $N(\mathfrak{B}) = \{n_1, \ldots, n_k\}$. By virtue of Proposition 8(2) there exists a decomposition $\mathfrak{B} = \mathfrak{B}_1 \times \ldots \times \mathfrak{B}_k$ with $N(\mathfrak{B}_i) = \{n_k\}$.Consider the following cases.

I. $k > 1$ and for some $i \leq k$, there exists $m \in \mathbb{N}$ such that $\mathfrak{B}_i \equiv (\mathfrak{A}_{n_i})^m$. Put $c \rightleftharpoons [\mathfrak{A}_{n_i}]_{\equiv}$. Then $N(\mathfrak{A}_{n_i}) = \{n_i\}$, therefore $c \neq b$ and $c \leq b$, i.e., $c < b$. Let $d < b$, $c \leq d$, $\mathbb{E} \models N(d)$ and $d = [\mathfrak{L}]_{\equiv}$ for some $\mathfrak{L}$. Then $\mathfrak{L} < \mathfrak{B}$, $\mathfrak{A}_{n_i} \leq \mathfrak{L}$ and $\mathfrak{L}$ is non-vanishing. Since $N(\mathfrak{L}) \subseteq \{n_1, \ldots, n_k\}$, then $N(\mathfrak{L}) = \{n_i\}$. Hence $\mathfrak{L} \leq (\mathfrak{A}_{n_i})^m$, and $\mathfrak{L} \equiv \mathfrak{A}_{n_i}$, because $\mathfrak{L}$ is non-vanishing. Thus $c = d$.

II. $k > 1$ and for any $i \leq k$ and $m \in \mathbb{N}$, $(\mathfrak{A}_{n_i})^m \leq \mathfrak{B}_i$. Put $c \rightleftharpoons [\mathfrak{B}_1]_{\equiv}$. By virtue of Proposition 8(6), $\mathfrak{B}_1$ is non-vanishing. Let $d < b, c \leq d$, $\mathbb{E} \models N(d)$, and $d[\mathfrak{L}]_{\equiv}$ for some $\mathfrak{L}$. Then $\mathfrak{B}_1 \leq \mathfrak{L}$ and $\mathfrak{L} < \mathfrak{B}$. Hence, $N(\mathfrak{L}) = \{n_1\}$ and for any $m \in \mathbb{N}$, $(\mathfrak{A}_{n_i})^m \leq \mathfrak{L}$. By the definition of the characteristic r, then $r_{\mathfrak{L}} = r_{\mathfrak{B}_1}$. Therefore $\mathfrak{L} \equiv \mathfrak{B}_1$, since $\mathfrak{L}$ and $\mathfrak{B}_1$ are local. Thus $c = d$.

III. $k = 1$ and $(\mathfrak{A}_{n_1})^2 \leq \mathfrak{B}$. Put $c \rightleftharpoons [\mathfrak{A}_{n_1}]_{\equiv}$, then $c < b$. Suppose, that $d = [\mathfrak{L}]_{\equiv}$, $\mathfrak{A}_{n_1} \leq \mathfrak{L}$, $\mathfrak{L} < \mathfrak{B}$ and $\mathfrak{L}$ is non-vanishing. $\mathfrak{L} \not\equiv \mathfrak{B}$, so $(\mathfrak{A}_{n_1})^m \not\leq \mathfrak{L}$ for some $m \in \mathbb{N}$. Hence, $\mathfrak{L} \equiv \mathfrak{A}_{n_1}$, i.e., $c = d$.

IV. Finally, let $k = 1$ and $\mathfrak{B} \equiv \mathfrak{A}_{n_1}$. Then there exists $m \in \mathbb{N}$ such that $V_m < V_{n_1}$, and for any k, if $V_m \leq V_k$ and $V_k < V_{n_1}$, then $m = k$.

a) If $\mathfrak{A}_m \times \mathfrak{A}_m \equiv \mathfrak{A}_m$, put $c \rightleftharpoons [\mathfrak{A}_m]_{\equiv}$. If $d = [\mathfrak{L}]_{\equiv}$, $\mathfrak{A}_m \leq \mathfrak{L}$ and $\mathfrak{L} < \mathfrak{B}$, then $N(\mathfrak{L}) \neq N(\mathfrak{B})$, therefore $N(\mathfrak{L}) = N(\mathfrak{A}_m) = \{m\}$. Hence $\mathfrak{L} \equiv \mathfrak{A}_m$ and $c = d$.

b) Let $\mathfrak{A}_m \times \mathfrak{A}_m \not\equiv \mathfrak{A}_m$. $\mathfrak{A}_m < \mathfrak{B}$, so $\mathfrak{B} \equiv \mathfrak{A}_m \times \mathfrak{M}$ for some I-algebra $\mathfrak{M}$. $\mathfrak{B}$ is non-vanishing and $\mathfrak{B} \not\equiv \mathfrak{A}_m$, therefore $\mathfrak{B} \equiv \mathfrak{M}$, i.e., $\mathfrak{B} \equiv \mathfrak{B} \times \mathfrak{A}_m \equiv \mathfrak{B} \times \mathfrak{A}_m \times \mathfrak{A}_m \equiv \ldots$. Hence, for any $k \in \mathbb{N}$, $\mathfrak{B} \equiv \mathfrak{B} \times (\mathfrak{A}_m)^k$ and $\mathfrak{B} \models \exists x \psi_k(x)$, where $\psi_k(x) = \exists y_1 \ldots \exists y_k((x = y_1 \cup \ldots \cup y_k) \& (\bigwedge_{i \neq j} y_i \cap y_j = 0) \& \bigwedge_i V_m(y_i))$.

Therefore the set $Th\mathfrak{B} \cup \{\psi_k(x) \mid k \in \mathbb{N}\} \cup \{\neg \exists y(y \leq x \ \& \ V_k(y)) \mid V_k \not< V_m\}$ is locally consistent and, so, it is consistent. Hence there exist $\mathfrak{M} \equiv \mathfrak{B}$ and $u \in \mathfrak{M}$ such that $N((u)) = \{m\}$ and $(\mathfrak{A}_m)^k \leq (u)$ for any $k \in \mathbb{N}$. We put $\mathfrak{C} \rightleftharpoons (u)$ and $\mathfrak{C}' \rightleftharpoons (\overline{u})$, then $\mathfrak{M} \cong \mathfrak{C} \times \mathfrak{C}'$ and $\mathfrak{B} \equiv \mathfrak{C} \times \mathfrak{C}'$. Put $c \rightleftharpoons [\mathfrak{C}]_{\equiv}$, then $\mathfrak{C} \not\equiv \mathfrak{B}$ and $c < d$.

Let $d = [\mathfrak{L}]_{\equiv}$, $\mathfrak{C} \leq \mathfrak{L}$ and $\mathfrak{L} < \mathfrak{B}$. Then $N(\mathfrak{L}) \neq N(\mathfrak{B})$, hence $N(\mathfrak{L}) = \{m\}$, and $(\mathfrak{A}_m)^k \leq \mathfrak{L}$ for any k. Hence, $\mathfrak{L} \equiv \mathfrak{C}$ and $c = d$.

($\Leftarrow$) Assume that $\mathfrak{A}$ is not local, i.e., $M(\mathfrak{A})$ is infinite.

Suppose that $\mathbb{E} \models L([\mathfrak{A}]_{\equiv})$. By virtue of Theorem 5 there exists a strictly stepwise I-algebra $\mathfrak{B} \leq \mathfrak{A}$. Put $b \rightleftharpoons [\mathfrak{B}]_{\equiv}$. Then $b \leq [\mathfrak{A}]_{\equiv}$ and

$\mathbb{E} \models (\exists c < b)(\forall d < b)(((c \leq d) \;\&\; N(d)) \to c = d)$. Let $c < b$ and $c = [\mathfrak{C}]_{\equiv}$. Then $\mathfrak{C} < \mathfrak{B}$, i.e., $\mathfrak{C} \leq \mathfrak{B}$ and $\mathfrak{C} \neq \mathfrak{B}$. Therefore, by virtue of Corollary 7, $\mathfrak{C}$ is local and $M(\mathfrak{C})$ is finite.

Consider $n = \max M(\mathfrak{C})$. $M(\mathfrak{B})$ is stepwise, so there exists $k \in \mathbb{N}$ such that if $m, j \in M(\mathfrak{B})$, $m \leq n$ and $k \leq j$, then $V_m < V_j$. The set $M(\mathfrak{B})$ is infinite, so there exists $j \in M(\mathfrak{B})$ such that $j \geq k$. Hence $V_m < V_j$ for any $m \in M(\mathfrak{C})$. Since $j \in M(\mathfrak{B})$, $\mathfrak{A}_j \leq \mathfrak{B}$. So $\mathfrak{A}_j \neq \mathfrak{B}$, because $\mathfrak{B}$ is not local. Put $d \rightleftharpoons [\mathfrak{A}_j]_{\equiv}$. Then $d < b$ and $\mathbb{E} \models (((c \leq d) \;\&\; N(D)) \to c = d)$.

By virtue of Theorem 2 $\mathfrak{A}_j$ is non-vanishing, therefore, $\mathbb{E} \models N(d)$. Prove that $c \leq d$. For any $m \in M(\mathfrak{C})$, $V_m < V_j$, hence $M(\mathfrak{C}) \subseteq M(\mathfrak{A}_j)$, $M(\mathfrak{C} \times \mathfrak{A}_j) = M(\mathfrak{C}) \cup M(\mathfrak{A}_j) = M(\mathfrak{A}_j)$ and $N(\mathfrak{C} \times \mathfrak{A}_j) = \{j\}$. By the definition of the characteristic r, then $r_{\mathfrak{C} \times \mathfrak{A}_j} = r_{\mathfrak{A}_j}$. Therefore $\mathfrak{C} \times \mathfrak{A}_j \equiv \mathfrak{A}_j$, i.e., $\mathfrak{C} \leq \mathfrak{A}_j$ and $c \leq d$. Hence $c = d$ and $\mathfrak{C} \equiv \mathfrak{A}_j$. Moreover, $j \in M(\mathfrak{C})$ and $V_n < V_j$, so $n < j$. However, $n = \max M(\mathfrak{C})$. We arrive at a contradiction, assuming $\mathbb{E} \models L([\mathfrak{A}]_{\equiv})$. Thus, $\mathbb{E} \models \neg L([\mathfrak{A}]_{\equiv})$.

The theorem is proved.

5. Finitely axiomatizable I-algebras

Denote

$$F'(x) \rightleftharpoons L(x) \;\&\; (\forall b \,\forall d \,\forall e)(((x = d \times e)$$
$$\&\; (x \neq e) \;\&\; (d = d \times d) \;\&\; (b \leq d) \;\&\; (b \neq b \times b)$$
$$\&\; N(b) \;\&\; N(d)) \to (\exists c)(N(c) \;\&\; (b < c) \;\&\; (c < d))).$$

By $F(x)$ we denote a formula of the first order predicate logic of the signature $\{\times\}$, which is equivalent to $F'(x)$ on the semigroup $\mathbb{E}$.

Theorem 3 from [7] immediately implies

Theorem 8. *An I-algebra $\mathfrak{A}$ is finitely axiomatizable if and only if it is a direct product of a finite number of basic I-algebras, i.e.,*

$$\mathfrak{A} \equiv (\mathfrak{A}_{n_1})^{m_1} \times \ldots \times (\mathfrak{A}_{n_k})^{m_k}$$

for some $m_1, \ldots, m_k \in \mathbb{N}$. In this case $\mathfrak{A}$ is local and $N(\mathfrak{A}) = \{n_1, \ldots, n_k\}$.

Theorem 9. *An I-algebra $\mathfrak{A}$ is finitely axiomatizable if and only if* $\mathbb{E} \models F([\mathfrak{A}]_{\equiv})$.

Proof. $(\Rightarrow)$ Let $\mathfrak{A}$ be finitely axiomatizable. Denote $a \rightleftharpoons [\mathfrak{A}]_{\equiv}$. We prove that $\mathbb{E} \models F(a)$, i.e., $\mathbb{E} \models F'(a)$.

$\mathfrak{A}$ is local, therefore, by virtue of Theorem 7, $\mathbb{E} \models L(a)$. Consider $b, d, e \in \mathbb{E}$. Let $b = [\mathfrak{B}]_{\equiv}$, $d = [\mathfrak{M}]_{\equiv}$, $e = [\mathfrak{N}]_{\equiv}$, $\mathfrak{A} \equiv \mathfrak{M} \times \mathfrak{N}$, $\mathfrak{A} \neq \mathfrak{N}$, $\mathfrak{M} \times \mathfrak{M} \equiv \mathfrak{M}$,

$\mathfrak{B} \leq \mathfrak{M}$, $\mathfrak{B} \neq \mathfrak{B} \times \mathfrak{B}$ and I-algebras $\mathfrak{B}$ and $\mathfrak{M}$ are non-vanishing. Then $\mathfrak{B}$ and $\mathfrak{M}$ are local. Let $N(\mathfrak{A}) = \{n_1, \ldots, n_k\}$, $m_1, \ldots, m_k \in \mathbb{N}$ and $\mathfrak{A} \equiv (\mathfrak{A}_{n_1})^{m_1} \times \ldots (\mathfrak{A}_{n_k})^{m_k}$.

I-algebras $\mathfrak{A}$ and $\mathfrak{N}$ are local, and $\mathfrak{A} \not\equiv \mathfrak{N}$, so $r_{\mathfrak{A}} \not\equiv r_{\mathfrak{N}}$. If for any $i \leq k$, $r_{\mathfrak{N}}(n_i) r_{\mathfrak{A}}(n_i)$, then $r_{\mathfrak{A}} = r_{\mathfrak{N}}$ by the definition of the characteristics r. Hence $r_{\mathfrak{N}}(n_i) < r_{\mathfrak{A}}(n_i)$ for some $i \leq k$.

Therefore there exists $i \leq k$ such that $r_{\mathfrak{M}}(n_i) > 0$. Since $\mathfrak{M} \leq \mathfrak{A}$, $r_{\mathfrak{M}} \leq r_{\mathfrak{A}}$ and $r_{\mathfrak{M}}(n_i) \leq m_i$. So $M(\mathfrak{M}) \subseteq M(\mathfrak{A})$ and V_{n_i} is maximal among V_j, $j \in M(\mathfrak{M})$, hence $n_i \in N(\mathfrak{M})$.

By virtue of Proposition 8 (6), $N(\mathfrak{M}) = \{n_i\}$ and $\mathfrak{M} \models V_{n_i}(1)$. Hence, $\mathfrak{M} \equiv \mathfrak{A}_{n_i}$.

$\mathfrak{B} \leq \mathfrak{M}$, $\mathfrak{M} \times \mathfrak{M} \equiv \mathfrak{M}$ and $\mathfrak{B} \neq \mathfrak{B} \times \mathfrak{B}$, therefore $\mathfrak{B} \neq \mathfrak{M}$ and $\mathfrak{B} < \mathfrak{M}$. Beside8(6), there exists $n \in \mathbb{N}$ such that $N(\mathfrak{B}) = \{n\}$. If $\mathfrak{A}_n$ is continuous, then $\mathfrak{B} \equiv \mathfrak{A}_n$ by Proposition 8(6 a). However, $\mathfrak{B} \neq \mathfrak{B} \times \mathfrak{B}$ — a contradiction. So $\mathfrak{A}_n$ is pointwise and $\mathfrak{A}_n \neq \mathfrak{A}_n \times \mathfrak{A}_n$. Then $\mathfrak{A}_n \leq \mathfrak{M}$ and, obviously, $\mathfrak{A}_n \not\equiv \mathfrak{M}$. Therefore $\mathfrak{A}_n < \mathfrak{M}$ and $V_n < V_{n_i}$. Then, by Corollary 10, $\mathfrak{M} \times \mathfrak{A}_n \equiv \mathfrak{M}$, so $\mathfrak{M} \times (\mathfrak{A}_n)^p \equiv \mathfrak{M}$ for any $p \in \mathbb{N}$. Thus, the set of sentences

$$Th(\mathfrak{M}) \cup \{\exists y_1 \ldots \exists y_p((x = y_1 \cup \ldots \cup y_p)$$
$$\& \, (\bigwedge_{i \neq j} y_i \cap y_j = 0) \, \& \, (\bigwedge_i V_n(y_i))) \mid p \in \mathbb{N}\}$$
$$\cup \{\neg \exists y((y \leq x) \, \& \, V_p(y)) \mid V_p \not< V_n\}$$

is locally consistent and, hence, is consistent.

Repeating the argumentation of the proof of Theorem 7, item IV, we prove that there exists an I-algebra $\mathfrak{C} \leq \mathfrak{M}$ such that $N(\mathfrak{C}) = \{n\}$ and $(\mathfrak{A}_n)^p \leq \mathfrak{C}$ for any $p \in \mathbb{N}$. Then $\mathfrak{C}$ is non-vanishing. $\mathfrak{B}$ is non-vanishing also, so $N(\mathfrak{B}) = \{n\}$. Moreover, $\mathfrak{B} \times \mathfrak{B} \neq \mathfrak{B}$, hence $\mathfrak{B} \equiv \mathfrak{A}_n$ and $\mathfrak{B} < \mathfrak{C}$, by Proposition 8 (6 a).

Put $c \rightleftharpoons [\mathfrak{C}]_{\equiv}$. Then $\mathbb{E} \models N(c)$, $b < c$ and $c \leq d$. Since $V_n < V_{n_i}$ and $\mathfrak{M} \equiv \mathfrak{A}_{n_i}$, $\mathfrak{C} \not\equiv \mathfrak{M}$. Thus $c < d$.

The part $(\Rightarrow)$ of the theorem is proved.

$(\Leftarrow)$ Assume that $\mathfrak{A}$ is not finitely axiomatizable. If $\mathfrak{A}$ is not local, then, by Theorem 7, $\mathbb{E} \models \neg L([\mathfrak{A}]_{\equiv})$, so $\mathbb{E} \models \neg F([\mathfrak{A}]_{\equiv})$.

Let $\mathfrak{A}$ be local. Suppose that $a = [\mathfrak{A}]_{\equiv}$ and $\mathbb{E} \models F'(a)$. The set $N(\mathfrak{A}) \neq \emptyset$ is finite, let $N(\mathfrak{A}) \models \{n_1, \ldots, n_k\}$. By Proposition 8(2) there exists a decomposition $\mathfrak{A} = \mathfrak{N}_1 \times \ldots \mathfrak{N}_k$ such that $N(\mathfrak{N}_i)\{n_i\}$. $\mathfrak{A}$ is not finitely axiomatizable, so there exists $i \leq k$ with $r_{\mathfrak{A}}(n_i) = \infty$. Hence $r_{\mathfrak{N}_i}(n_i) = \infty$ and for any $m \in \mathbb{N}$, $(\mathfrak{A}_{n_i})^m \leq \mathfrak{N}_i$. Then $\mathfrak{A}_{n_i}$ is pointwise, so $\mathfrak{N}_i$ is non-vanishing by virtue of Proposition 8 (6).

Put $\mathfrak{M} \rightleftharpoons \mathfrak{N}_1 \times \ldots \times \mathfrak{N}_{i-1} \times \mathfrak{N}_{i+1} \times \ldots \times \mathfrak{N}_k$, $d = [\mathfrak{N}_i]_\equiv$ and $e = [\mathfrak{M}]_\equiv$. Then $N(\mathfrak{M}) = N(\mathfrak{A}) \backslash \{n_i\}$, therefore $\mathfrak{M} \not\equiv \mathfrak{A}$. Hence $a = d \times e$ and $a \neq e$.

Let $b = [\mathfrak{A}_{n_i}]_\equiv$. Then $b \leq d$ and $\mathbb{E} \models N(b)$, because $\mathfrak{A}_{n_i}$ is non-vanishing. $\mathfrak{A}_{n_i} \times \mathfrak{A}_{n_i} \not\equiv \mathfrak{A}_{n_i}$ and $\mathfrak{N}_i \times \mathfrak{N}_i \equiv \mathfrak{N}_i$, so $b \neq b \times b$ and $d = d \times d$.

Hence $\mathbb{E} \models (\exists c)(N(c) \,\&\, (b < c) \,\&\, (c < d))$. Let $c \in \mathbb{E}$ and $c = [\mathfrak{C}]_\equiv$. Then $\mathfrak{C}$ is non-vanishing, $\mathfrak{A}_{n_i} < \mathfrak{C}$ and $\mathfrak{C} < \mathfrak{N}_i$ which implies $M(\mathfrak{C}) \subseteq M(\mathfrak{N}_i)$. Moreover, $\mathfrak{A}_{n_i} \leq \mathfrak{C}$ implies $n_i \in M(\mathfrak{C})$. Hence, $N(\mathfrak{C}) = \{n_i\}$. Since $\mathfrak{C}$ is non-vanishing and $\mathfrak{A}_{n_i} \neq \mathfrak{C}$, Proposition 8 (6) implies $(\mathfrak{A}_{n_i})^m \leq \mathfrak{C}$ for any $m \in \mathbb{N}$. Then $r_\mathfrak{C}(n_i) = \infty = r_{\mathfrak{N}_i}(n_i)$. Therefore $r_\mathfrak{C} = r_{\mathfrak{N}_i}$ and $\mathfrak{C} \equiv \mathfrak{N}_i$, i.e., $c = d$ — a contradiction. Thus $\mathbb{E} \not\models F'(a)$.

The theorem is proved.

6. ω-categorical I-algebras

Remind that an I-algebra $\mathfrak{A}$ is called ω-categorical if the elemenary theory $Th(\mathfrak{A})$ is categorical in the countable cardinality, i.e., for any countable $\mathfrak{B}$ and $\mathfrak{C}$ from $\mathfrak{A} \equiv \mathfrak{B}$ and $\mathfrak{A} \equiv \mathfrak{C}$ it follows that $\mathfrak{B} \cong \mathfrak{C}$.

Denote $C'(x) \rightleftharpoons \forall b((b \leq x) \to F(b))$ and

$$C(x) \rightleftharpoons \forall b \forall c \,((x = b \times c) \to F(b)).$$

Remark 8. $\mathbb{E} \models \forall x(C'(x) \leftrightarrow C(x))$.

Theorem 10. ([7], Theorem 9, (1) and (3)) *An I-algebra $\mathfrak{A}$ is ω-categorical if and only if for any decomposition $\mathfrak{A} \equiv \mathfrak{B} \times \mathfrak{C}$, I-algebra $\mathfrak{B}$ is finitely axiomatizable.*

Corollary 11. *I-algebra $\mathfrak{A}$ is ω-categorical if and only if $\mathbb{E} \models C([\mathfrak{A}]_\equiv)$.*

Corollary 2, Theorem 7, Theorem 9 and Corollary 11 imply Theorem 1.

7. Nonaxiomatizability of local, finitely axiomatizable and ω-categorical I-algebras

Theorem 11. *Classes of local, finitely axiomatizable and ω-categorical I-algebras are not axiomatizable.*

Proof. Let K be a class local I-algebras (a class finitely axiomatizable or a class of ω-categorical I-algebras). Suppose that K is axiomatizable, then $K = K(\Gamma)$ for some set of sentences Γ.

From [3] it follows that the set of ω-categorical basic I-algebras $\mathfrak{A}_n$ is infinite. Denote

$$C \rightleftharpoons \{n \mid \mathfrak{A}_n \text{ is } \omega\text{-categorical }\}.$$

Notice that $\mathfrak{A}_n$ is local and finitely axiomatizable for any $n \in C$. By the construction of the sequence of formulas $V_n(x)$ [3], for any finite $L \subseteq C$ there exists $n \in C$ such that $\mathfrak{A}_m < \mathfrak{A}_n$ for any $m \in L$. In this case $\mathfrak{A}_n \models \exists x V_m(x)$ for any $m \in L$, and $\mathfrak{A}_n \models \Gamma$. Therefore $\Gamma \cup \{\exists x V_n(x) \mid n \in C\}$ is locally consistent, so it is consistent. Hence there exists $\mathfrak{B} \models \Gamma \cup \{\exists x V_n(x) \mid n \in C\}$. Then $\mathfrak{B} \models \Gamma$ and $C \subseteq M(\mathfrak{B})$, so $M(\mathfrak{B})$ is infinite. Consequently $\mathfrak{B}$ is not local, and hence it is neither ω-categorical nor finitely axiomatizable — a contradiction.

The theorem is proved.

Question. Whether classes of non-local, not finitely axiomatizable and non-ω-categorical I-algebras are axiomatizable?

Thus, each of these classes of I-algebras, which are not even axiomatizable in the language of I-algebras, is described by one sentence of the signature $\{\times\}$ in the semigroup of I-algebra elementary types.

References

1. Yu.L. Ershov. Decidability of the theory of distributive lattices with relative complements and the filter theory. Algebra and Logic, vol. 3, N 3, 1964, p. 17–38.
2. A. Macintyre, J.G. Rosenstein. $\aleph_0$-categoricity for rings without nilpotent elements and for Boolean structures. J. Algebra, vol. 43, N 1, 1976, p. 129–154.
3. D.E. Pal'chunov. Countably-categorical Boolean algebras with distinguished ideals. Studia Logica, vol. XLVI, N 2, 1987, p. 121–135.
4. D.E. Pal'chunov. Finitely axiomatizable Boolean algebras with distinguished ideals. Algebra and Logic, vol. 26, N 4, 1987, p. 435–455.
5. Alain Touraille. Theories d'Algebres de Boole Munies d'Ideaux Distingues. I. Theories Elementaires. J. Symb. Log., vol. 52, N 4, 1987, p. 1027–1043.
6. Alain Touraille. Théories d'Algébres de Boole Munies d'Idéaux Distingués, II. J. Symb. Log., vol. 55, N 3, 1990, p. 1192–1212.
7. D.E. Pal'chunov. Direct summands of Boolean algebras with distinguished ideals. Algebra and Logic, vol. 31, N 5, 1992, p. 499–537.
8. D.E. Pal'chunov. Elementary type semigroup for Boolean algebras with distinguished ideals. 3-th International Conference on Algebra, Krasnoyarsk, 1993, p. 253.
9. D.E. Pal'chunov. Prime and countably-categorical Boolean algebras with distinguished ideals. SIBAM, 1994, N 3, p. 83–108.
10. D.E. Pal'chunov, Theories of Boolean algebras with distinguished ideals having no the prime model, SIBAM, 1994, N 4, p. 86–117.

11. D.E. Pal'chunov. The Lindenbaum-Tarski algebra for the class of Boolean algebras with one distinguished ideal. Algebra and Logic, vol. 33, N 2, 1994, p. 179–210.
12. D.E. Pal'chunov. The Lindenbaum-Tarski algebra for Boolean algebras with distinguished ideals. Algebra and Logic, vol. 34, N 1, 1995, p. 88–116.

INTERVAL FUZZY ALGEBRAIC SYSTEMS*

D. E. PAL'CHUNOV, G. E. YAKHYAEVA

*Institute of Mathematics,
Siberian Branch of Russian Acad. Sci.
630090, Novosibirsk, Russia;
palch@math.nsc.ru,
E-mail: gul_nara@mail.ru*

1. Introduction

The concept of fuzzy logic was introduced by Lotfi Zadeh [12] as a result of development of fuzzy set theory. A fuzzy subset A of a crisp set X is determined by a mapping (so called, membership function) which defines a membership degree of an element x in the set A, for each element $x \in X$. Similarly, if X is a set of sentences then a degree of truth for elements of X may be defined: a statement may be "absolutely true", or "absolutely false", or may have an intermediate value belonging to some partially ordered set.

There are two meanings of fuzzy logic [13]. *Fuzzy logic in a wide sense* is a tool and a methodology for fuzzy management, analysis of unprecise sentences of natural language and some other applications [5, 7, 8, 14].

Fuzzy logic in a narrow sense [4, 9, 11] is a kind of symbolic logic. It includes investigations in syntax, semantics, axiomatization, completeness and so on. Fuzzy logic in this sense may be considered as one of fields of many-valued logic.

In the paper we fix some signature σ which is a finite set of predicate symbols and constant symbols. It means that σ does not contain symbols of functions. Also we fix a set A. For the set A and the signature σ we denote $\sigma_A \rightleftharpoons \sigma \cup \{c_a \mid a \in A\}$. We assume that $c_a \notin \sigma$ for any $a \in A$.

In the present article we consider models $\mathfrak{A} = \langle A, \sigma \rangle$ of the signature σ_A with the universe A. We suppose that $c_a^{\mathfrak{A}} \rightleftharpoons a$ for any model $\mathfrak{A} = \langle A, \sigma \rangle$. By $K(A, \sigma) \rightleftharpoons \{\mathfrak{A} \mid \mathfrak{A} = \langle A, \sigma \rangle\}$ we denote the class of all such models.

We deal with sentences of the first order predicate logic of the signature σ_A without equality. Let $S(\sigma_A)$ be the set of all sentences of the signature

---

*Supported by RFBR grant N 05-01-04003-NNIO-a (DFG project COMO, GZ: 436 RUS 113/829/0-1), and by grant of the Russian Science Agency, project 2006-РИ-19.0/001/269.

σ_A and $S_a(\sigma_A) \rightleftharpoons \{ P(c_1, \ldots, c_n) \mid P, c_1, \ldots, c_n \in \sigma_A \}$ be the set of all atomic sentences of the signature σ_A.

Suppose that for a model $\mathfrak{A} \in K(A, \sigma)$ and a mapping $\mu : S_a(\sigma_A) \rightarrow \{0, 1\}$ we have $\mu(\varphi) = 1$ if and only if $\mathfrak{A} \models \varphi$ for any $\varphi \in S_a(\sigma_A)$. Note that in this case the mapping μ determines the model $\mathfrak{A}$ up to the isomorphism. It means that if we know truth values of all atomic sentences of the signature σ_A then we know truth values of all sentences of this signature (for the class $K(A, \sigma)$).

In the paper we study the following problem. Suppose that we have uncertain (i.e., fuzzy) or/and incomplete information about some model $\mathfrak{A} \in K(A, \sigma)$. What can we say about truth value of arbitrary sentence $\varphi \in S(\sigma_A)$ on the model $\mathfrak{A}$? We consider three kinds of this problem.

Problem 1 (uncertain information). Consider $\mu : S_a(\sigma_A) \rightarrow [0, 1]$ and $\varphi \in S(\sigma_A)$. What is the truth value of the sentence φ?

Problem 2 (incomplete information). Consider $U \subseteq S(\sigma_A)$, $\mu : U \rightarrow \{0, 1\}$ and $\varphi \in S(\sigma_A)$. What is the truth value of the sentence φ?

Problem 3 (uncertain and incomplete information). Consider $U \subseteq S(\sigma_A)$, $\mu : U \rightarrow [0, 1]$ and $\varphi \in S(\sigma_A)$. What is the truth value of the sentence φ?

Fuzzy logics give a decision of Problem 1. A definition of fuzzy truth values of formulas is presented by induction on construction of formulas. However, this approach has some paradoxes.

Let F be a set of formulas. For each fuzzy logic (see, for example, [1, 3, 4]) a definition of a truth function $\mu : F \rightarrow [0, 1]$ is based on functions $n : [0, 1] \rightarrow [0, 1]$ and $f, g : [0, 1]^2 \rightarrow [0, 1]$ in the following way:

$$
\begin{aligned}
\mu(\neg\varphi) &= n(\mu(\varphi)); \\
\mu(\varphi \,\&\, \psi) &= f(\mu(\varphi), \mu(\psi)); \\
\mu(\varphi \vee \psi) &= g(\mu(\varphi), \mu(\psi)).
\end{aligned}
\tag{1}
$$

Presence of these functions and such way of definition of a truth function μ lead to some logical paradoxes. Now we consider most popular fuzzy logics (from the point of view of practical applications).

I. Zadeh max-min logic:
$n(\alpha) = 1 - \alpha, f(\alpha, \beta) = \min\{\alpha, \beta\}$ and $g(\alpha, \beta) = \max\{\alpha, \beta\}$.
Paradoxes: $\mu(\varphi \,\&\, \neg\varphi) \neq 0$ and $\mu(\varphi \vee \neg\varphi) \neq 1$.

II. Lukasiewicz logic:
$n(\alpha) = 1 - \alpha, f(\alpha, \beta) = \max\{0, \alpha + \beta - 1\}$ and $g(\alpha, \beta) = \min\{1, \alpha + \beta\}$.

Paradoxes: $\mu(\varphi \,\&\, \varphi) \neq \mu(\varphi)$, $\mu(\varphi \vee \varphi) \neq \mu(\varphi)$, $\mu((\varphi \,\&\, \psi) \vee \xi) \neq \mu((\varphi \vee \xi) \,\&\, (\psi \vee \xi))$ and $\mu((\varphi \vee \psi) \,\&\, \xi) \neq \mu((\varphi \,\&\, \xi) \vee (\psi \,\&\, \xi))$.

III. Probability logic:
$n(\alpha) = 1 - \alpha, f(\alpha, \beta) = \alpha\beta$ and $g(\alpha, \beta) = \alpha + \beta - \alpha\beta$.
Probability logic has all paradoxes specified for I and II.

The main purpose of the present article is to give a definition of truth values of all sentences starting with an uncertain or/and incomplete information. In the paper we give a solution of Problems 1–3, freed of mentioned paradoxes of fuzzy logics. Our approach is based on a concept of fuzzification of Boolean-valued models, introduced in this paper.

The idea of this approach is the following. If we say that the truth value of a sentence φ is equal to $p \in [0, 1]$, we mean that there is a probability space such that a number of events for which the sentence φ is true, divided by a number of all events in this space, is equal to p. For example, if we know that $\mu(\varphi) = \frac{1}{2}$ and $\mu(\psi) = \frac{1}{2}$ then the truth value of $\mu(\varphi \,\&\, \psi)$ may be an element from the interval with the ends 0 (for $\psi \equiv \neg\varphi$) and $\frac{1}{2}$ (for $\psi \equiv \varphi$), as well as the truth value of $\mu(\varphi \vee \psi)$ is an element from the interval with the ends $\frac{1}{2}$ (for $\psi \equiv \varphi$) and 1 (for $\psi \equiv \neg\varphi$).

We have the same situation if $\varphi = P(a)$, $\psi = Q(a)$, $\mu(\varphi) = \frac{1}{2}$ and $\mu(\psi) = \frac{1}{2}$. In this case the truth sets of events for $P(a)$ and $Q(a)$ may coincide as well as these sets may be complementary.

Thus, at the first step we should defuzzificate our knowledge, i.e., transform a fuzzy model into a Boolean-valued model, and, then, fuzzificate it again, i.e., convert a Boolean-valued model into a fuzzy model.

Now we introduce our basic notations.

By $\|X\|$ we denote a cardinality of a set X.

$\mathcal{P}(X)$ is the set of all subsets of a set X.

$\wp(X)$ is the Boolean algebra $\langle \mathcal{P}(X), \cap, \cup, {}^{-}, \emptyset, X \rangle$.

By $|\mathfrak{A}|$ we denote the universe of a model $\mathfrak{A}$.

Recall that the set A and the signature σ are fixed. We denote $\sigma_A \rightleftharpoons \sigma \cup \{c_a \,|\, a \in A\}$. An atomic sentence of the signature σ is a sentence of the form $P(c_1, \ldots, c_n)$, where $P, c_1, \ldots, c_n \in \sigma_A$.

$S_a(\sigma_A)$ is the set of all atomic sentences of the signature σ_A.

$S(\sigma_A)$ is the set of all sentences of the signature σ_A.

Fuzzy algebraic systems were studied in [2, 15].

2. Fuzzification of Boolean-valued models

Definition 1. Consider a mapping $\mu : S(\sigma_A) \to [0,1]$. A triple $\widetilde{\mathfrak{A}}\langle A, \sigma, \mu \rangle$ is called **fuzzy model** and μ is called **truth function**. If $\varphi \in S(\sigma_A)$ and $\mu(\varphi) = q$ we denote $\widetilde{\mathfrak{A}} \models_q \varphi$.

Remark 1. Each fuzzy logic determines restrictions to a mapping $\mu : S(\sigma_A) \to [0,1]$ to be a fuzzy model.

Definition 2. A quadruple $\mathfrak{A}_{\mathbb{B}} = \langle A, \sigma, \mathbb{B}, \tau \rangle$ is called **Boolean-valued model** if $\mathbb{B}$ is a complete Boolean algebra and a truth function $\tau : S(\sigma_A) \to \mathbb{B}$ satisfies the following conditions:

$$\tau(\neg\varphi) = \overline{\tau(\varphi)};$$
$$\tau(\varphi \vee \psi) = \tau(\varphi) \cup \tau(\psi); \tag{2}$$
$$\tau(\varphi \,\&\, \psi) = \tau(\varphi) \cap \tau(\psi);$$
$$\tau(\varphi \to \psi) = \overline{\tau(\varphi)} \cup \tau(\psi.)$$

$$\tau(\forall x\varphi(x)) = \bigcap_{a \in A} \tau(\varphi(c_a)); \tag{3}$$
$$\tau(\exists x\varphi(x)) = \bigcup_{a \in A} \tau(\varphi(c_a)).$$

A Boolean-valued model $\mathfrak{A}_{\mathbb{B}} = \langle A, \sigma, \mathbb{B}, \tau \rangle$ is called **finitary** if the Boolean algebra $\mathbb{B}$ is finite.

Recall that a mapping $\nu : \mathbb{B} \to \mathbb{R}$ is a measure on a Boolean algebra $\mathbb{B}$ [6] if

(1) $\nu(a) \geq 0$ for any $a \in \mathbb{B}$,
(2) $\nu(0) = 0$,
(3) $\nu(a \cup b) = \nu(a) + \nu(b) - \nu(a \cap b)$ for any $a, b \in \mathbb{B}$.

Further we assume that a measure ν is not trivial, i.e., $\nu(1) \neq 0$.

Definition 3. A system $\mathfrak{A}_{\mathbb{B}}^{\nu} \rightleftharpoons \langle A, \sigma, \mathbb{B}, \tau, \nu \rangle$ is called **measured Boolean-valued model** (or **Boolean-valued model with a measure ν**) if $\langle A, \sigma, \mathbb{B}, \tau \rangle$ is a Boolean-valued model and ν is a measure on the Boolean algebra $\mathbb{B}$.

Definition 4. A fuzzy model $\widetilde{\mathfrak{A}}_{\mathbb{B}}^{\nu} = \langle A, \sigma, \mu \rangle$ is called **fuzzification of a measured Boolean-valued model** $\mathfrak{A}_{\mathbb{B}}^{\nu} = \langle A, \sigma, \mathbb{B}, \tau, \nu \rangle$ if

$$\mu(\varphi) = \frac{\nu(\tau(\varphi))}{\nu(1)} \quad \text{for any } \varphi \in S(\sigma_A).$$

As is shown in Introduction, each fuzzy logic has some of the following paradoxes: $\mu(\varphi \vee \neg\varphi) \neq 1$, $\mu(\varphi \,\&\, \neg\varphi) \neq 0$, $\mu(\varphi \vee \varphi) \neq \varphi$, $\mu(\varphi \,\&\, \varphi) \neq \varphi$. The next statement shows that the definition of fuzzification is freed of these paradoxes.

Statement 1. Let $\widetilde{\mathfrak{A}}_B^\nu$ be a fuzzification of a measured Boolean-valued model $\mathfrak{A}_{\mathbb{B}}^\nu$. Then for any $\varphi, \psi, \xi \in S(\sigma_A)$ we have:

(1) $\mu(\varphi \,\&\, \neg\varphi) = 0$;

(2) $\mu(\varphi \vee \neg\varphi) = 1$;

(3) $\mu(\neg\neg\varphi) = \mu(\varphi)$;

(4) $\mu(\varphi \,\&\, \psi) = \mu(\psi \,\&\, \varphi)$, $\mu(\varphi \vee \psi) = \mu(\psi \vee \varphi)$;

(5) $\mu((\varphi \,\&\, \psi) \,\&\, \xi) = \mu(\varphi \,\&\, (\psi \,\&\, \xi))$, $\mu((\varphi \vee \psi) \vee \xi) = \mu(\varphi \vee (\psi \vee \xi))$;

(6) $\mu((\varphi \,\&\, \psi) \vee \xi) = \mu((\varphi \vee \xi) \,\&\, (\psi \vee \xi))$, $\mu((\varphi \vee \psi) \,\&\, \xi) = \mu((\varphi \,\&\, \xi) \vee (\psi \,\&\, \xi))$;

(7) $\mu(\varphi \,\&\, \varphi) = \mu(\varphi)$, $\mu(\varphi \vee \varphi) = \mu(\varphi)$;

(8) $\mu(\neg\varphi \,\&\, \neg\psi) = \mu(\neg(\varphi \vee \psi))$, $\mu(\neg\varphi \vee \neg\psi) = \mu(\neg(\varphi \,\&\, \psi))$;

(9) $\mu(\varphi \to \varphi) = 1$.

Proof. We prove (1), other statements may be proved similarly.

$$\mu(\varphi \,\&\, \neg\varphi) = \frac{\nu(\tau(\varphi \,\&\, \neg\varphi))}{\nu(1)} \frac{\nu(\tau(\varphi) \cap \overline{\tau(\varphi)})}{\nu(1)} = \frac{\nu(0)}{\nu(1)} = 0. \qquad \square$$

Definition 5. Let $\mathbb{B}$ be a Boolean algebra. Denote $At(\mathbb{B}) \rightleftharpoons \{a \in \mathbb{B} \mid a \text{ is an atom}\}$ and $At(b) \rightleftharpoons \{a \in At(\mathbb{B}) \mid a \leq b\}$ for $b \in \mathbb{B}$.

For a finite Boolean algebra $\mathbb{B}$ a mapping $\eta : \mathbb{B} \to \mathbb{R}$ is called as **natural measure** on the Boolean algebra $\mathbb{B}$ if $\eta(b) = \|At(b)\|$ for any $b \in \mathbb{B}$.

Remark 2. The natural measure η is a measure on a finite Boolean algebra.

Definition 6. Let $\mathfrak{A}_{\mathbb{B}} = \langle A, \sigma, \mathbb{B}, \tau \rangle$ be a finitary Boolean-valued model (i.e., the Boolean algebra $\mathbb{B}$ be finite). Consider the natural measure η of the Boolean algebra $\mathbb{B}$. The fuzzification $\widetilde{\mathfrak{A}}_{\mathbb{B}}^\eta$ of the measured Boolean-valued model $\mathfrak{A}_{\mathbb{B}}^\eta = \langle A, \sigma, \mathbb{B}, \tau, \eta \rangle$ is called **natural fuzzification of the Boolean-valued model** $\mathfrak{A}_{\mathbb{B}}$ and is denoted as $\widetilde{\mathfrak{A}}_{\mathbb{B}}$.

Recall that in Zadeh logic and in Lukasiewicz logic $\mu(\neg\varphi) = 1 - \mu(\varphi)$. If $\mu(\varphi) = \alpha$ and $\mu(\psi) = \beta$ then in Zadeh logic $\mu(\varphi \vee \psi) \max\{\alpha, \beta\}$ and $\mu(\varphi \,\&\, \psi) = \min\{\alpha, \beta\}$; in Lukasiewicz logic $\mu(\varphi \vee \psi) = \min\{1, \alpha + \beta\}$ and $\mu(\varphi \,\&\, \psi) = \max\{0, \alpha + \beta - 1\}$.

Statement 2. Let $\widetilde{\mathfrak{A}}_{\mathbb{B}} = \langle A, \sigma, \mu \rangle$ be a natural fuzzification of $\mathfrak{A}_{\mathbb{B}} = \langle A, \sigma, \mathbb{B}, \tau \rangle$ and $\varphi, \psi \in S(\sigma_A)$. If $\mu(\varphi) = \alpha$ and $\mu(\psi) = \beta$ then

(1) $\mu(\neg\varphi) = 1 - \mu(\varphi)$;

(2) $\mu(\varphi \vee \psi) \in [\max\{\alpha, \beta\}, \min\{1, \alpha + \beta\}]$

(3) $\mu(\varphi \,\&\, \psi) \in [\max\{0, \alpha + \beta - 1\}, \min\{\alpha, \beta\}]$

Thus, our fuzzy values lie "between" Zadeh fuzzy values and Lukasiewicz fuzzy values.

Proof. Denote $B \rightleftharpoons At(\mathbb{B})$. So,

$$B = At(1^{\mathbb{B}}), \quad \alpha = \frac{\|At(\tau(\varphi))\|}{\|B\|} \quad \text{and} \quad \beta = \frac{\|At(\tau(\psi))\|}{\|B\|}.$$

$$(1)\ \mu(\neg\varphi) = \frac{\|At(\tau(\neg\varphi))\|}{\|B\|} = \frac{\|B\| - \|At(\tau(\varphi))\|}{\|B\|} = 1 - \frac{\|At(\tau(\varphi))\|}{\|B\|}$$
$$= 1 - \mu(\varphi);$$

$$(2)\ \mu(\varphi \vee \psi) = \frac{\|At(\tau(\varphi \vee \psi))\|}{\|B\|} = \frac{\|At(\tau(\varphi)) \cup At(\tau(\psi))\|}{\|B\|}.$$

Clearly, $\|At(\tau(\varphi))\|, \|At(\tau(\psi))\| \le \|At(\tau(\varphi)) \cup At(\tau(\psi))\| \le (\|At(\tau(\varphi))\| + \|At(\tau(\psi))\|), \|B\|$.

$$(3)\ \mu(\varphi \,\&\, \psi) = \frac{\|At(\tau(\varphi \,\&\, \psi))\|}{\|B\|} = \frac{\|At(\tau(\varphi)) \cap At(\tau(\psi))\|}{\|B\|}.$$

Clearly, $0 \le \|At(\tau(\varphi)) \cap At(\tau(\psi))\| \le \|At(\tau(\varphi))\|, \|At(\tau(\psi))\|$. Moreover,

$$\|At(\tau(\varphi)) \cap At(\tau(\psi))\|$$
$$= \|At(\tau(\varphi)) \backslash \overline{At(\tau(\psi))}\| \ge \|At(\tau(\varphi))\| - (\|B\| - \|At(\tau(\psi))\|)$$
$$= \|At(\tau(\varphi))\| + \|At(\tau(\psi))\| - \|B\|.$$

3. Generalized fuzzy models

In the present section we deal with the main problem of our consideration.

If we know fuzzy truth values for all atomic sentences $P(c_1, \ldots, c_n)$ of the signature σ_A, how to determine fuzzy truth values for all sentences of this signature?

More general, if we know fuzzy truth values of sentences $\varphi \in U$, where $U \subseteq S(\sigma_A)$, how to determine fuzzy truth values for all sentences of the signature σ_A?

Standard definitions of fuzzy truth values, given by induction on formula construction, presented by various fuzzy logics, lead to paradoxes mentioned above: $\mu(\varphi \vee \neg\varphi)\ \not\equiv 1, \mu(\varphi \,\&\, \neg\varphi) \neq 0, \mu(\varphi \vee \varphi) \neq \varphi, \mu(\varphi \,\&\, \varphi) \neq \varphi,$ $\mu(\varphi \to \varphi) \neq 1.$

We present an approach based on fuzzification of Boolean-valued models, which is freed of these paradoxes.

The main semantic idea behind our approach is the following. The fact that we know only numbers — fuzzy truth values of atomic sentences, means that we don't know real situation generating this evaluation. Namely, for an atomic sentence $\varphi \in S_a(\sigma_A)$ (or for $\varphi \in U \subseteq S(\sigma_A)$) we don't know a concrete set of cases in which the given sentence is true. So, to obtain a relevant estimation of all sentences $\varphi \in S(\sigma_A)$ we should consider all possible situations. It means that we should consider a class of all Boolean-valued models corresponding to the given evaluation of $\varphi \in S_a(\sigma_A)$ (of $\varphi \in U \subseteq S(\sigma_A)$, resp.). For example, we can say that a finitary Boolean-valued model is relevant to a fuzzy evaluation of sentences in U if the natural fuzzification of this model has exactly the same evaluation of these sentences.

Note that for practical goals, it is sufficient to consider only finitary Boolean-valued models. In fact, let $\widetilde{\mathfrak{A}}_{\mathbb{B}} = \langle A, \sigma, \mu \rangle$ be the natural fuzzification of a finitary Boolean-valued model $\mathfrak{A}_{\mathbb{B}} = \langle A, \sigma, \mathbb{B}, \tau >$. It is clear that for any sentence φ the truth value $\mu(\varphi)$ is a rational number from the interval $[0,1]$, i.e. $\mu(\varphi) = \frac{m}{n}$, where $m \leq n$ and $m, n \in \mathbb{N}$. Moreover, if $\|B\| = n$ then the truth values are elements of the set $\{0, \frac{1}{n}, \frac{2}{n} \ldots, \frac{n-1}{n}, 1\}$. So, if we take a sufficiently great number n, we can approximate the real estimation to any desired degree of precision.

Definition 7. Consider a fuzzy algebraic system $\widetilde{\mathfrak{A}}\langle A, \sigma, \mu \rangle$ and a mapping $\nu : U \to [0,1]$, where $U \subseteq S(\sigma_A)$. We say that $\widetilde{\mathfrak{A}}$ **is concordant** with ν, and denote $\widetilde{\mathfrak{A}} \uparrow \nu$, if $\mu(\varphi)\nu(\varphi)$ for any sentence $\varphi \in U$.

Definition 8. Consider a set $U \subseteq S(\sigma_A)$ and a mapping $\nu : U \to [0,1]$. Let $\mathfrak{A}_{\mathbb{B}} = \langle A, \sigma, \mathbb{B}, \tau \rangle$ be a finitary Boolean-valued model and let $\widetilde{\mathfrak{A}}_{\mathbb{B}} = \langle A, \sigma, \mu \rangle$ be the natural fuzzification of $\mathfrak{A}_{\mathbb{B}}$. We say that $\mathfrak{A}_{\mathbb{B}}$ **is concordant** with the evaluation ν, and denote $\mathfrak{A}_{\mathbb{B}} \uparrow \nu$, if $\widetilde{\mathfrak{A}}_{\mathbb{B}}$ is concordant with ν, i.e., $\widetilde{\mathfrak{A}}_{\mathbb{B}} \uparrow \nu$.

We denote $\mathbb{K}_\nu \rightleftharpoons \{ \mathfrak{A}_{\mathbb{B}} = \langle A, \sigma, \mathbb{B}, \tau \rangle \mid \mathfrak{A}_{\mathbb{B}} \uparrow \nu \}$ — the set of all finitary Boolean-valued models which are concordant with the evaluation ν.

We say that $\nu : U \to [0,1]$ is a **consistent** evaluation if $\mathbb{K}_\nu \neq \emptyset$.

Remark 3. If for $U \subseteq S(\sigma_A)$ an evaluation $\nu : U \to [0,1]$ is consistent then the set $\nu(U)$ is finite and $\nu(U) \subseteq \mathbb{Q}$.

The class K_ν completely describes the set of all possible situations for which the incomplete information, presented by the mapping ν, takes place.

Recall that for a fuzzy model $\widetilde{\mathfrak{A}} = \langle A, \sigma, \mu \rangle$ and a sentence $\varphi \in S(\sigma_A)$ we denote $\widetilde{\mathfrak{A}} \models_q \varphi$ if $\mu(\varphi) = q$.

Definition 9. Consider $U \subseteq S(\sigma_A)$, a consistent evaluation $\nu : U \to [0,1]$ and a mapping $\xi : S(\sigma_A) \to \wp([0,1] \cap \mathbb{Q})$. A system $\mathfrak{A}_\nu = \langle A, \sigma, \xi \rangle$ is called **generalized fuzzy model** (generated by the evaluation ν) if

$$\xi(\varphi) = \{\, q \mid \widetilde{\mathfrak{A}}_\mathbb{B} \models_q \varphi \text{ for some model } \mathfrak{A}_\mathbb{B} \in \mathbb{K}_\nu \,\}.$$

for any sentence $\varphi \in S(\sigma_A)$.

Now we will give exact formulations of problems 1–3 presented in Introduction, in terms of generalized fuzzy models.

Recall that the signature σ consists of predicate and constant symbols. Consider a sentence $\varphi \in S(\sigma_A)$.

Problem 1. Let $\nu : S_a(\sigma_A) \to [0,1]$ be a consistent evaluation. What is the set $\xi(\varphi)$ in the generalized fuzzy model $\mathfrak{A}_\nu = \langle A, \sigma, \xi \rangle$?

Problem 2. Consider a set $U \subseteq S(\sigma_A)$ and a consistent evaluation $\nu : U \to \{0,1\}$. What is the set $\xi(\varphi)$ in the generalized fuzzy model $\mathfrak{A}_\nu = \langle A, \sigma, \xi \rangle$?

The third problem generalizes Problems 1 and 2:

Problem 3. Consider a set $U \subseteq S(\sigma_A)$ and a consistent evaluation $\nu : U \to [0,1]$. What is the set $\xi(\varphi)$ in the generalized fuzzy model $\mathfrak{A}_\nu = \langle A, \sigma, \xi \rangle$?

The problem 1 asks how to define fuzzy truth values of all sentences if we know fuzzy truth values of all sentences of the kind $P(c_1 \ldots, c_n)$.

The problem 2 questions how to recover whole information starting from an incomplete precise information.

The following theorem gives an answer to problems 1–3. It shows that the sets of possible fuzzy truth values of sentences are intervals.

Theorem 1. *Consider a set $U \subseteq S(\sigma_A)$, a consistent evaluation $\nu : U \to [0,1]$ and a generalized fuzzy model $\mathfrak{A}_\nu = \langle A, \sigma, \xi \rangle$. Then for any sentence $\varphi \in S(\sigma_A)$ the set $\xi(\varphi)$ is an interval of rational numbers. Namely, if $q_1, q_2 \in \xi(\varphi), q \in \mathbb{Q}$ and $q_1 < q < q_2$, then $q \in \xi(\varphi)$.*

To prove this theorem we will study finitary Boolean-valued models.

4. Boolean-valued models with atomic Boolean algebras

Remark 4. Let $\mathbb{B}$ be a complete atomic Boolean algebra. Then

$$\mathbb{B} \cong \wp(At(\mathbb{B})) \rightleftharpoons \langle \mathcal{P}(At(\mathbb{B})), \cup, \cap, ^-, \emptyset, At(\mathbb{B}) \rangle,$$

furthermore, the mapping $h(b) \rightleftharpoons At(b)$ realizes this isomorphism.

Further, without loss of generality, every complete atomic Boolean algebra $\mathbb{B}$ will be considered as a power-set Boolean algebra $\mathbb{B} = \wp(X)$ for a set X.

Definition 10. Let $\mathbb{B}' = \wp(X')$ and $\mathbb{B}'' = \wp(X'')$, let $\mathfrak{A}_{\mathbb{B}'} = \langle A, \sigma, \mathbb{B}', \tau' \rangle$ and $\mathfrak{A}_{\mathbb{B}''} = \langle A, \sigma, \mathbb{B}'', \tau'' \rangle$ be Boolean-valued models. We assume that $X' \cap X'' \emptyset$ (perhaps, after renaming). A model $\mathfrak{A}_{\mathbb{B}} = \langle A, \sigma, \mathbb{B}, \tau \rangle$ is called as **union** of $\mathfrak{A}_{\mathbb{B}'}$ and $\mathfrak{A}_{\mathbb{B}''}$ if $\mathbb{B} = \wp(X' \cup X'')$ and

$$\tau(\varphi) = \tau'(\varphi) \cup \tau''(\varphi)$$

for any $\varphi \in S(\sigma_A)$. We denote $\mathfrak{A}_{\mathbb{B}} = \mathfrak{A}_{\mathbb{B}'} * \mathfrak{A}_{\mathbb{B}''}$.

Statement 3. $\mathfrak{A}_{\mathbb{B}} = \mathfrak{A}_{\mathbb{B}'} * \mathfrak{A}_{\mathbb{B}''}$ is a Boolean-valued model.

Proof. Let $\varphi, \psi \in S(\sigma_A)$. Then

(1) $\tau(\neg\varphi) = \tau'(\neg\varphi) \cup \tau''(\neg\varphi) = (X'\backslash\tau'(\varphi)) \cup (X''\backslash\tau''(\varphi)) = (X' \cup X'')\backslash(\tau'(\varphi) \cup \tau''(\varphi)) = \overline{\tau(\varphi)}$.

(2) $\tau(\varphi \vee \psi) = \tau'(\varphi \vee \psi) \cup \tau''(\varphi \vee \psi) = \tau'(\varphi) \cup \tau'(\psi) \cup \tau''(\varphi) \cup \tau''(\psi) = \tau(\varphi) \cup \tau(\psi)$.

(3) $\tau(\varphi \,\&\, \psi) = \tau'(\varphi \,\&\, \psi) \cup \tau''(\varphi \,\&\, \psi) = (\tau'(\varphi) \cap \tau'(\psi)) \cup (\tau''(\varphi) \cap \tau''(\psi)) = (\tau'(\varphi) \cup \tau''(\varphi)) \cap (\tau'(\psi) \cup \tau''(\psi))\tau(\varphi) \cap \tau(\psi)$,
because $\tau'(\varphi) \cap \tau''(\psi) = \emptyset$ and $\tau'(\psi) \cap \tau''(\varphi) = \emptyset$.

(4) $\tau(\varphi \to \psi) = \tau(\neg\varphi \vee \psi)\tau(\neg\varphi) \cup \tau(\psi) = \overline{\tau(\varphi)} \cup \tau(\psi)$.

(5) $\tau(\exists x \varphi(x)) = \tau'(\exists x \varphi(x)) \cup \tau''(\exists x \varphi(x)) = (\bigcup_{a \in A} \tau'(c_a)) \cup (\bigcup_{a \in A} \tau''(c_a)) = \bigcup_{a \in A} (\tau'(c_a) \cup \tau''(c_a)) = \bigcup_{a \in A} \tau(c_a)$.

(6) $\tau(\forall x \varphi(x)) = \tau'(\forall x \varphi(x)) \cup \tau''(\forall x \varphi(x)) = (\bigcap_{a \in A} \tau'(c_a)) \cup (\bigcap_{a \in A} \tau''(c_a)) = \bigcap_{a \in A} \left(\tau'(\varphi(c_a)) \cup \bigcap_{b \in A} \tau''(\varphi(c_a)) \right) = \bigcap_{a \in A} \bigcap_{b \in A} \left(\tau'(\varphi(c_a)) \cup \tau''(\varphi(c_b)) \right)$
$= \bigcap_{a \in A} \left(\tau'(\varphi(c_a)) \cup \tau''(\varphi(c_b)) \right)$,

by virtue of infinite distributivity for complete Boolean algebras [10].

Prove the last equality. Let $d \in \bigcap_{a \in A} \bigcap_{b \in A} \left(\tau'(\varphi(c_a)) \cup \tau''(\varphi(c_b)) \right)$, so $d \in X'$ or $d \in X''$. Suppose that $d \in X'$. Then $d \in \tau'(\varphi(c_a)) \cup \tau''(\varphi(c_b))$ for any $a, b \in A$, hence $d \in \tau'(\varphi(c_a))$ or $d \in \tau''(\varphi(c_b))$ for any $a, b \in A$. However, $d \notin X''$, so $d \in \tau'(\varphi(c_a))$, and $d \in (\tau'(\varphi(c_a)) \cup \tau''(\varphi(c_a)))$ for any $a \in A$. Therefore $d \in \bigcap_{a \in A} \left(\tau'(\varphi(c_a)) \cup \tau''(\varphi(c_b)) \right)$. The case $d \in X''$ is proved similarly.

Inversely, let $d \in \bigcap_{a \in A} \left(\tau'(\varphi(c_a)) \cup \tau''(\varphi(c_b)) \right)$. Suppose that $d \in X'$. Then $d \in \left(\tau'(\varphi(c_a)) \cup \tau''(\varphi(c_a)) \right)$ for any $a \in A$. We have $d \notin X''$, therefore $d \in \tau'(\varphi(c_a))$ for any $a \in A$. Consequently, $d \in \left(\tau'(\varphi(c_a)) \cup \tau''(\varphi(c_b)) \right)$ for any $a, b \in A$. Hence $d \in \bigcap_{a \in A} \bigcap_{b \in A} \left(\tau'(\varphi(c_a)) \cup \tau''(\varphi(c_b)) \right)$. The case $d \in X''$ is proved similarly.

Thus $\tau(\forall x \varphi(x)) = \bigcap_{a \in A} \left(\tau'(\varphi(c_a)) \cup \tau''(\varphi(c_a)) \right) \bigcap_{a \in A} \tau(\varphi(c_a))$. $\qquad\square$

Remark 5. Let $\mathfrak{A}_\mathbb{B}$ be a Boolean-valued model. Then as the union $\mathfrak{A}_\mathbb{B} * \mathfrak{A}_\mathbb{B}$ we consider $\mathfrak{A}_\mathbb{B} * \mathfrak{A}_{\mathbb{B}'}$, where $\mathfrak{A}_{\mathbb{B}'}$ is an isomorphic copy of $\mathfrak{A}_\mathbb{B}$, and $\mathbb{B}' = \wp(X')$ is obtained from $\mathbb{B} = \wp(X)$ by renaming of all elements of the set X such that $X' \cap X = \emptyset$.

Remark 6. The operation $*$ on Boolean-valued models is commutative and associative. (We consider Boolean-valued models $\mathfrak{A}_\mathbb{B} = \langle A, \sigma, \mathbb{B}, \tau \rangle$ with the common set A and the common signature σ, and with atomic Boolean algebras $\mathbb{B}$.)

Statement 4. Let $\mathfrak{A}_\mathbb{B} = \mathfrak{A}_{\mathbb{B}'} * \mathfrak{A}_{\mathbb{B}''}$, and Boolean algebras $\mathbb{B}'$ and $\mathbb{B}''$ be finite. Let $\widetilde{\mathfrak{A}} = \langle A, \sigma, \mu \rangle$, $\widetilde{\mathfrak{A}}' = \langle A, \sigma, \mu' \rangle$ and $\widetilde{\mathfrak{A}}'' = \langle A, \sigma, \mu'' \rangle$ be natural fuzzifications of Boolean-valued models $\mathfrak{A}_\mathbb{B}, \mathfrak{A}_{\mathbb{B}'}$ and $\mathfrak{A}_{\mathbb{B}''}$ respectively. Let $\mathbb{B}' = \wp(X')$ and $\mathbb{B}'' = \wp(X'')$. Then

$$\mu(\varphi) = \frac{\mu'(\varphi) \cdot \|X'\| + \mu''(\varphi) \cdot \|X''\|}{\|X'\| + \|X''\|}$$

for any sentence $\varphi \in S(\sigma_A)$.

In particular, if $\mu'(\varphi) = \mu''(\varphi)$ then $\mu(\varphi) = \mu'(\varphi)$, for any $\varphi \in S(\sigma_A)$.

Proof. Assume that $\mathbb{B}' = \wp(X')$, $\mathbb{B}'' = \wp(X'')$, $\|X'\| = m$ and $\|X''\| = n$. Then $X' \cap X'' = \emptyset$ and $\mathbb{B} = \wp(X)$, where $X = X' \cup X''$. So, $\|X\| = m+n$.

Let $\mathfrak{A}_{\mathbb{B}'} = \langle A, \sigma, \mathbb{B}', \tau' \rangle$, $\mathfrak{A}_{\mathbb{B}''} = \langle A, \sigma, \mathbb{B}'', \tau'' \rangle$ and $\mathfrak{A}_\mathbb{B} = \langle A, \sigma, \mathbb{B}, \tau \rangle$. Then for $\varphi \in S(\sigma_A)$ we have $\tau(\varphi) = \tau'(\varphi) \cup \tau''(\varphi)$, $\mu'(\varphi) \frac{\|At(\tau'(\varphi))\|}{m}$,

$\mu''(\varphi)\frac{\|At(\tau''(\varphi))\|}{n}$ and $\mu(\varphi) = \frac{\|At(\tau(\varphi))\|}{m+n}$. Therefore,

$$\mu(\varphi) = \frac{\|At(\tau(\varphi))\|}{m+n} \frac{\|At(\tau'(\varphi)) \cup At(\tau''(\varphi))\|}{m+n} \frac{\|At(\tau'(\varphi))\| + \|At(\tau''(\varphi))\|}{m+n}$$

$$= \frac{m \cdot \mu'(\varphi) + n \cdot \mu''(\varphi)}{m+n} \frac{\mu'(\varphi) \cdot \|X'\| + \mu''(\varphi) \cdot \|X''\|}{\|X'\| + \|X''\|}.$$

Obviously, if $\mu'(\varphi) = \mu''(\varphi)$ then $\mu(\varphi) = \mu'(\varphi)$.

The statement is proved. $\qquad\square$

Corollary 1. *If* $\mathfrak{A}_{\mathbb{B}'} \uparrow \nu$ *and* $\mathfrak{A}_{\mathbb{B}''} \uparrow \nu$, *then* $(\mathfrak{A}_{\mathbb{B}'} * \mathfrak{A}_{\mathbb{B}''}) \uparrow \nu$.

Lemma 1. *Consider* $U \subseteq S(\sigma_A)$ *and* $\nu : U \to [0,1]$. *Suppose that* $\widetilde{\mathfrak{A}}_{\mathbb{B}'} = \langle A, \sigma, \mu' \rangle$ *and* $\widetilde{\mathfrak{A}}_{\mathbb{B}''} = \langle A, \sigma, \mu'' \rangle$ *are natural fuzzifications of Boolean-valued models* $\mathfrak{A}_{\mathbb{B}'}$ *and* $\mathfrak{A}_{\mathbb{B}''}$ *respectively (in particular, Boolean algebras* $\mathbb{B}'$ *and* $\mathbb{B}''$ *are finite). Let* $\mathbb{B}' = \wp(X')$, $\mathbb{B}'' = \wp(X'')$, $\mathfrak{A}_{\mathbb{B}'} \uparrow \nu$, $\mathfrak{A}_{\mathbb{B}''} \uparrow \nu$, $\varphi \in S(\sigma_A)$, $\mu'(\varphi) = q_1$, $\mu''(\varphi) = q_2$, $q \in \mathbb{Q}$, $q_1 < q < q_2$, $q_1 = \frac{p_1}{l_1}$, $q_2 = \frac{p_2}{l_2}$ *and* $q = \frac{p}{l}$.

Put $i \rightleftharpoons (p_2\, l l_1 - p\, l_1 l_2)\, n$, $j \rightleftharpoons (p\, l_1 l_2 - p_1 l l_2)\, m$ *and*

$$\mathfrak{A}_{\mathbb{B}} \rightleftharpoons \underbrace{\mathfrak{A}_{\mathbb{B}'} * \ldots * \mathfrak{A}_{\mathbb{B}'}}_{i\ times} * \underbrace{\mathfrak{A}_{\mathbb{B}''} * \ldots * \mathfrak{A}_{\mathbb{B}''}}_{j\ times}.$$

Then $\mathfrak{A}_{\mathbb{B}} \uparrow \nu$ *and for the natural fuzzification* $\widetilde{\mathfrak{A}}_{\mathbb{B}} = \langle A, \sigma, \mu \rangle$ *of the model* $\mathfrak{A}_{\mathbb{B}}$ *we have* $\mu(\varphi) = q$.

Proof. Let $\mathbb{B}' = \wp(X')$ and $\mathbb{B}'' = \wp(X'')$. First, we prove that $\mathfrak{A}_{\mathbb{B}} \uparrow \nu$. Let $\psi \in U$. Then $\mu'(\psi) = \nu(\psi)$ and $\mu''(\psi) = \nu(\psi)$. Therefore, using induction on the number of occurrences of the operation $*$ (i.e., on the number $i + j + 1$), by Corollary 1 we obtain $\widetilde{\mathfrak{A}}_{\mathbb{B}} \uparrow \nu$ and $\mathfrak{A}_{\mathbb{B}} \uparrow \nu$.

Let $\|X'\| = m$ and $\|X''\| = n$. By virtue of Statement 4, we have

$$\mu(\varphi) = \frac{\mu'(\varphi) \cdot \|X'\| \cdot i + \mu''(\varphi) \cdot \|X''\| \cdot j}{i \cdot \|X'\| + j \cdot \|X''\|}$$

$$= \frac{\frac{p_1}{l_1} \cdot m \cdot (p_2 l l_1 - p l_1 l_2) \cdot n + \frac{p_2}{l_2} \cdot n \cdot (p l_1 l_2 - p_1 l l_2) \cdot m}{(p_2 l l_1 - p l_1 l_2) \cdot n \cdot m + (p l_1 l_2 - p_1 l l_2) \cdot m \cdot n} = \frac{p}{l}.$$

Hence $\mu(\varphi) = q$. The lemma is proved. $\qquad\square$

Now we prove Theorem 1.

Consider a set $U \subseteq S(\sigma_A)$, a consistent evaluation $\nu : U \to [0,1]$, a generalized fuzzy model $\mathfrak{A}_\nu = \langle A, \sigma, \xi \rangle$ and a sentence $\varphi \in S(\sigma_A)$. We show that the set $\xi(\varphi)$ is an interval of rational numbers. Let $q_1, q_2 \in \xi(\varphi)$, $q \in \mathbb{Q}$ and $q_1 < q < q_2$. Prove, that $q \in \xi(\varphi)$.

By the definition of generalized fuzzy model, there exist Boolean-valued models $\mathfrak{A}_{\mathbb{B}_1}$, $\mathfrak{A}_{\mathbb{B}_2} \in \mathbb{K}_\nu$ such that $\widetilde{\mathfrak{A}}_{\mathbb{B}_1} \models_{q_1} \varphi$ and $\widetilde{\mathfrak{A}}_{\mathbb{B}_2} \models_{q_2} \varphi$ for natural fuzzifications $\widetilde{\mathfrak{A}}_{\mathbb{B}_1}$ and $\widetilde{\mathfrak{A}}_{\mathbb{B}_2}$ of $\mathfrak{A}_{\mathbb{B}_1}$ and $\mathfrak{A}_{\mathbb{B}_2}$ respectively. Hence $\mathfrak{A}_{\mathbb{B}_1} \uparrow \nu$ and $\mathfrak{A}_{\mathbb{B}_2} \uparrow \nu$.

Then, by Lemma 1 , there exists a Boolean-valued model $\mathfrak{A}_{\mathbb{B}}$ such that $\mathfrak{A}_{\mathbb{B}} \uparrow \nu$ and $\mu(\varphi) = q$ for the natural fuzzification $\widetilde{\mathfrak{A}}_{\mathbb{B}} = \langle A, \sigma, \mu \rangle$ of $\mathfrak{A}_{\mathbb{B}}$. So, $\mathfrak{A}_{\mathbb{B}} \in \mathbb{K}_\nu$ and $\widetilde{\mathfrak{A}}_{\mathbb{B}} \models_q \varphi$. Therefore, $q \in \xi(\varphi)$.

Theorem 1 is proved. $\qquad\square$

Consider a consistent evaluation $\nu : U \to [0,1]$ and a generalized fuzzy model $\mathfrak{A}_\nu = \langle A, \sigma, \xi \rangle$.

Question 1. Is it true, that for any sentence $\varphi \in S(\sigma_A)$ the set $\xi(\varphi)$ is a closed interval of rational numbers, i.e., $\xi(\varphi) = [p, q]$ for some $p, q \in \mathbb{Q}$?

Question 2. Is it true, that $\sup \xi(\varphi)$ and $\inf \xi(\varphi)$ are rational numbers for any sentence $\varphi \in S(\sigma_A)$?

References

1. I.Z. Batirshin. *The basic operations of the fuzzy logic and their generalization.*, Kazan, Otechestvo, 2001.
2. V.P. Dobritsa , G.E. Yakhyaeva : *About possible approaches in fuzziness research.* Proceedings of yearly scientific conference of AGY, Almaty, pp. 58–59, 1999, in Russian.
3. Fuzzy sets in models of management and an artificial intellect, D.A. Pospelov, ed. Moscow, Nauka, 1986,
4. P. Hajek. *Metamathematics of fuzzy logic.*, Kluwer akademic publishers, 1998.
5. G.J. Klir ,B. Yuan , eds. *Fuzzy sets, fuzzy logic and fuzzy system: Selected papers by Lotfi A. Zadeh.* World Scientific Singapore, 1996.
6. S.C. Loo. *Measures of fuzziness.* Cybernetica, 1977, v. 3, p. 201–207.
7. H.T. Nguyen, A. Walker. *First course in fuzzy logic.* CRC Press, 1999.
8. V. Novak. *Fuzzy sets and their applications.* Adam Hilger, Bristol, 1989.
9. V. Novak, I. Perfilieva, J. Mockor. *Mathematical principles of fuzzy logic.* Kluwer 2000.
10. R. Sikorski. *Boolean algebras.* Springer-verlag, 1964.
11. E. Turunen. *Mathematics behind fuzzy logic.* Physica Verlag 1999.
12. L.A. Zadeh. *Fuzzy sets*, Inform. and control, 8 (1965), pp. 338–353.
13. L.A. Zadeh. *Fuzzy logic and approximate reasoning.* Synthese, 30, pp. 407–428, 1975.
14. H.-J. Zimmermann. *Fuzzy set theory and its applications.* Second ed. Kluwer 1991.
15. G.E. Yahkyaeva. *A fuzzy logic of finite number of experts.* Abstracts of the 9-th Asian Logic Conference, Novosibirsk, 2005, pp. 138–139.

ON ORIENTABILITY AND DEGENERATION OF BOOLEAN BINARY RELATION ON A FINITE SET

VLADISLAV POPLAVSKI

Saratov State University
83 Astrakhanskaya, 410012 Saratov, Russia
Email: poplavskivb@mail.ru

We introduce the notions of exterior, interior, positive and negative parts of Boolean square matrices with elements from an arbitrary Boolean algebra and prove the theorem on the uniqueness of Boolean matrix expansion into the union of these four parts. Exterior and interior parts form a degenerate part of a Boolean matrix. Positive and negative parts are oriented constituents of a matrix. The properties of oriented and degenerate parts of Boolean matrices are analyzed. Numerous examples of orientability and degeneration of Boolean binary relations on a finite set defined by a certain square Boolean matrix are given.

1. Introduction

In this paper, theory of matrices and square matrix determinants over Boolean algebras is considered as an instrument for Boolean binary relations analysis on a finite set. Mapping of ordered pairs of elements from some finite set onto arbitrary Boolean algebra is called *Boolean binary relation* on this finite set. Such Boolean binary relation defines certain square Boolean matrix up to simultaneous and similar permutations of rows and columns in this matrix. The opposite is also true: Boolean square matrices which differ from each other by mentioned permutations of rows and columns develop equivalence classes, which define certain Boolean binary relation.

In particular, every binary relation on a finite set can be represented by square matrix over two-element Boolean algebra to simultaneous and similar permutations of rows and columns in this matrix.

Semipermanents (or bideterminants), permanents and oriented determinants of square Boolean matrices are used in the paper. More details about it can be found in [1–6]. Permanents and determinants give examples of Boolean-valued invariants for Boolean binary relations on a finite set. Besides, they allow establishing a certain analogy with matrices over the field of real numbers for which zero, positive and negative values of determinants play a certain role in the choice of oriented bases in finite-dimensional real linear spaces.

It is necessary to mention that two-element Boolean matrices (or Boolean relations on a finite set) can be included only into one of four types (interior, exterior, positive and negative matrices) defined in this paper. Matrices over arbitrary Boolean algebra are more complicated. Nevertheless, any of them in a natural and single way can be expanded into the union of four matrices two of which give degenerate parts of the matrix (exterior and interior), and the other two (positive and negative) are its oriented constituents. It is shown that invertible linear transformations of Boolean binary relation on a finite set, if they can be represented in the form of the products (right or left) of an invertible matrix into the given Boolean matrix, preserve a corresponding type of degeneration of this Boolean binary relation. Besides, positive invertible matrices, when multiplied by the given matrix, preserve its orientation. Positive and negative parts are transferred to positive and negative accordingly. Negative invertible matrices change the orientation of the given matrix onto the opposite.

As it has been said above, degenerate Boolean matrices are composed of matrices of two types: exterior and interior. Exterior nonzero Boolean matrix is determined as a matrix the permanent of which is equal to zero. Such matrices over various semirings are well described and play an important role in combinatorial mathematics. In this paper only matrices with interior type of degeneracy and their properties are described. For such interior matrices the Boolean determinant is equal to zero and the permanent is not equal to zero. The interest to the interiority is connected, firstly, with the fact that there is a similarity with the corresponding topological concept. Secondly, binary relations on a finite set widely used in mathematics frequently are interior. For example, reflexive binary relations (of equivalence, tolerance, partial order, etc.) on a finite set are either positive or, as it is shown further, more often interior.

The examples used in the paper are given to illustrate degeneracy or orientability applicable to various problems connected with the inverse problem, asymptotic forms of Boolean matrices, and transitivity and reflexivity of Boolean binary relations.

2. Permanents and determinants of Boolean matrices

Let $\langle B, \cup, \cap, ', \emptyset, I \rangle$ be an arbitrary Boolean algebra, where $\cup$, $\cap$, $'$ denote operations of union, intersection, complementation and $\emptyset$, I are zero and identity elements accordingly.

Definition 2.1. *Oriented semipermanents* $\overset{+}{\nabla} A$ *and* $\overset{-}{\nabla} A$ *of a Boolean* $n \times n-$ *matrix* A $(n \geq 2)$ *are calculated by* $\overset{\pm}{\nabla} A = \bigcup_{(\alpha_1,\ldots,\alpha_n)\in\overset{\pm}{P}} (a_1^{\alpha_1} \cap a_2^{\alpha_2} \cap \ldots \cap a_n^{\alpha_n})$, *where* a_j^i *are elements of* A, *both even and odd n-permutations of upper indices are denoted as* $\overset{+}{P}$, $\overset{-}{P}$ *correspondingly. Oriented semipermanents allow to introduce the* permanent $Per A = \overset{+}{\nabla} A \cup \overset{-}{\nabla} A$ *and the* common part of oriented semipermanents $\triangle A = \overset{+}{\nabla} A \cap \overset{-}{\nabla} A$. Right *and* left determinants *are defined as* $RDet A = \overset{+}{\nabla} A \setminus \overset{-}{\nabla} A \overset{+}{\nabla} A \cap (\overset{-}{\nabla} A)'$ *and* $LDet A = \overset{-}{\nabla} A \setminus \overset{+}{\nabla} A = \overset{-}{\nabla} A \cap (\overset{+}{\nabla} A)'$ *accordingly. Boolean sum* $Det A = RDet A \cup LDet A$ *of right and left determinants is called* the determinant of Boolean matrix.

Let us point out some properties of the introduced matrix functions.

It is obvious that even permutations of rows and columns of square Boolean matrix do not change oriented semipermanents. Odd permutations of rows and columns of the given matrix transfer the semipermanents $\overset{+}{\nabla} A$, $\overset{-}{\nabla} A$ into each other.

Let intersection and union of elements of a Boolean algebra and a Boolean matrix be defined elementwise. Then

$$\overset{\pm}{\nabla} (\lambda \cap A) \quad \bigcup_{(\alpha_1,\ldots,\alpha_n)\in\overset{\pm}{P}} ((\lambda \cap a_1^{\alpha_1}) \cap (\lambda \cap a_2^{\alpha_2}) \cap \ldots \cap (\lambda \cap a_n^{\alpha_n})) = \lambda \cap \overset{\pm}{\nabla} A$$

and

$$\overset{\pm}{\nabla} (\lambda \cup A) \quad \bigcup_{(\alpha_1,\ldots,\alpha_n)\in\overset{\pm}{P}} ((\lambda \cup a_1^{\alpha_1}) \cap (\lambda \cup a_2^{\alpha_2}) \cap \ldots \cap (\lambda \cup a_n^{\alpha_n})) = \lambda \cup \overset{\pm}{\nabla} A$$

hold for oriented semipermanents of intersection and union of matrix A and a certain element λ of Boolean algebra. It allows to show the truth of the following identities by simple verification.

Proposition 2.1. *For any element* $\lambda \in B$ *and square Boolean matrix* A, *we have*

$$Per(\lambda \cap A) = \lambda \cap Per A, \triangle(\lambda \cap A) = \lambda \cap \triangle A,$$
$$RDet(\lambda \cap A) = \lambda \cap RDet A, LDet(\lambda \cap A) = \lambda \cap LDet A,$$
$$Per(\lambda \cup A) = \lambda \cup Per A, \triangle(\lambda \cup A) = \lambda \cup \triangle A,$$
$$RDet(\lambda \cup A) = (RDet A) \cap \lambda' = (RDet A) \setminus \lambda = RDet(\lambda' \cap A),$$
$$LDet(\lambda \cup A) = (LDet A) \cap \lambda' = (LDet A) \setminus \lambda = LDet(\lambda' \cap A).$$

Definition 2.2. Determine a *conjunctive product* of $n \times n$-matrices A and B as the matrix $C = A \sqcap B$ of the same size whose elements c_j^i are given by the formula $c_j^i = \bigcup_{t=1}^{n}(a_t^i \cap b_j^t)$.

It is obvious that a *disjunctive* product $C = A \sqcup B$ of matrices A and B whose elements are defined as $c_j^i = \bigcap_{t=1}^{n}(a_t^i \cup b_j^t)$ could be obtained in a dual way. However, as we are going to deal with one conjunctive product, it is that operation which is referred to as a "product". Besides, we are going to consider conjunctive degrees $A^k \prod_{t=1}^{k} A$ only, and it is that conjunctive degree which is supposed to be called *a degree of Boolean matrices*. A set of square Boolean matrices under a product forms a semigroup with an identity element $E = (\delta_j^i)$, where δ_j^i receives the value I, if $i = j$ and the value $\emptyset$, if $i \neq j$ in Boolean algebra $\langle B, \cup, \cap, ', \emptyset, I \rangle$. It is assumed that $A^0 = E$.

The deriving of the following formulas which are the analogues of the Cauchy-Binet formulas can be found in [3] and [4]. Reasoning given in the article [2] and the monograph [6] for oriented semipermanents of matrices over commutative semiring can be used.

Proposition 2.2. *If A, B are any square Boolean matrices of the same size, then*

$$Per(A \sqcap B) \supseteq Per A \cap Per B, \tag{1}$$

$$\Delta(A \sqcap B) \supseteq \Delta A \cap \Delta B, \tag{2}$$

$$RDet(A \sqcap B) \subseteq (RDet A \cap RDet B) \cup (LDet A \cap LDet B), \tag{3}$$

$$LDet(A \sqcap B) \subseteq (LDet A \cap RDet B) \cup (RDet A \cap LDet B), \tag{4}$$

$$Det(A \sqcap B) \subseteq Det A \cap Det B. \tag{5}$$

3. Permanent expansion of Boolean matrices and its uniqueness

Definition 3.1. Let us denote $\check{A} = (Per A)' \cap A$, $\hat{A} = \Delta A \cap A$, $\overset{+}{A} = RDet A \cap A$, $\bar{A} = LDet A \cap A$. Call the union $A = \check{A} \cup \hat{A} \cup \overset{+}{A} \cup \bar{A}$ a *permanent expansion* of a Boolean matrix A.

It is clear that the oriented semipermanents of a linear combination of matrices $A = (\alpha^1 \cap A) \cup (\alpha^2 \cap A) \cup (\alpha^3 \cap A) \cup (\alpha^4 \cap A)$ with pairwise non-intersecting (disjoint) coefficients $\alpha^1 = (Per A)'$, $\alpha^2 = \Delta A$, $\alpha^3 = RDet A$, $\alpha^3 = LDet A$ and condition $\bigcup_{i=1}^{k} \alpha^i = I$ are expanded into pairwise disjoint semipermanents of Boolean summands $\check{A}$, $\hat{A}$, $\overset{+}{A}$, $\bar{A}$.

Then we get the equalities $Per A = Per \check{A} \cup Per \hat{A} \cup Per \overset{+}{A} \cup Per \overline{A}$, $\triangle A = \triangle \check{A} \cup \triangle \hat{A} \cup \triangle \overset{+}{A} \cup \triangle \overline{A}$, $RDet A = RDet \check{A} \cup RDet \hat{A} \cup RDet \overset{+}{A} \cup RDet \overline{A}$, $LDet A = LDet \check{A} \cup LDet \hat{A} \cup LDet \overset{+}{A} \cup LDet \overline{A}$ which give disjunctive expansion of permanent, common part of semipermanents, and determinants of matrix A, according to matrices $\check{A}$, $\hat{A}$, $\overset{+}{A}$, $\overline{A}$.

Besides, the following statement the proof of which immediately follows from the definition of matrices $\check{A}$, $\hat{A}$, $\overset{+}{A}$, $\overline{A}$ and the identities of Proposition 2.1 is valid.

Theorem 3.1. *Permanent, common part of semipermanents, and determinants of matrices* $\check{A}$, $\hat{A}$, $\overset{+}{A}$, $\overline{A}$ *possess the following properties:*

$$Per \check{A} = \emptyset, \quad Per \hat{A} = \triangle \hat{A} = \triangle A,$$

$$Per \overset{+}{A} = RDet \overset{+}{A} = RDet A, \quad Per \overline{A} = LDet \overline{A} = LDet A,$$

$$\triangle \check{A} = \triangle \overset{+}{A} \triangle \overline{A} = RDet \hat{A} = LDet \hat{A} = RDet \check{A}$$

$$= LDet \check{A} = RDet \overline{A} = LDet \overset{+}{A} = \emptyset.$$

Definition 3.2. Call the matrix $\check{A} \cup \hat{A}$ a *degenerate* part of a Boolean matrix A, which consists of the *exterior* part $\check{A}$ and the *interior* part $\hat{A}$ of A. Call the matrix $\overset{+}{A} \cup \overline{A}$ a *nondegenerate* part, consisting of the *positive* $\overset{+}{A}$ and *negative* $\overline{A}$ (*oriented*) parts correspondingly, and call the matrices $\check{A} \cup \hat{A} \cup \overset{+}{A}$ and $\check{A} \cup \hat{A} \cup \overline{A}$ *nonnegative* and *nonpositive* parts of matrix A accordingly. Call the matrix A *degenerate, nondegenerate, exterior, interior, positive, negative, nonpositive, nonnegative* if it coincides with its corresponding degenerate, nondegenerate, exterior, interior, positive, negative, nonpositive, nonnegative part.

The following theorem shows the uniqueness of permanent expansion of a Boolean matrix.

Theorem 3.2. Let $A = (\alpha^1 \cap A) \cup (\alpha^2 \cap A) \cup (\alpha^3 \cap A) \cup (\alpha^4 \cap A) = A_1 \cup A_2 \cup A_3 \cup A_4$ be a linear combination of matrices with pairwise disjoint coefficients α^i $(i = 1, \ldots, 4)$, where $A_1 = \alpha^1 \cap A = \check{A}_1$ is an exterior Boolean matrix, $A_2 = \alpha^2 \cap A = \hat{A}_2$ is an interior Boolean matrix, $A_3 = \alpha^3 \cap A = \overset{+}{A}_3$ is a positive and $A_4 = \alpha^4 \cap A = \overline{A}_4$ is a negative matrix. Then A_1, A_2, A_3, A_4 are the exterior, interior, positive and negative parts of matrix A correspondingly. The converse implication also holds.

208

Proof. From the fact that α^i ($i = 1, \ldots, 4$) are pairwise disjoint elements and that any element of A is included into the union $\bigcup_{i=1}^4 \alpha^i$, it follows that

$$\triangle A = \triangle A_1 \cup \triangle A_2 \cup \triangle A_3 \cup \triangle A_4 = \emptyset \cup \triangle A_2 \cup \emptyset \cup \emptyset = \triangle A_2,$$
$$RDet A = RDet A_1 \cup RDet A_2 \cup RDet A_3 \cup RDet A_4$$
$$= \emptyset \cup \emptyset \cup RDet A_3 \cup \emptyset = RDet A_3,$$
$$LDet A = LDet A_1 \cup LDet A_2 \cup LDet A_3 \cup LDet A_4$$
$$= \emptyset \cup \emptyset \cup \emptyset \cup LDet A_4 = LDet A_4.$$

That $A_2 = \hat{A}_2 = \alpha_2 \cap A_2 \subset \alpha_2 \cap J$ permits to write down $\triangle A = \triangle A_2 \subset \alpha_2$. Here the matrix whose all elements are equal to the unity element I, is designated by J . Similarly, from $A_3 \alpha_3 \cap A_3 \subset \alpha_3 \cap J$ we get $RDet A = RDet A_3 \subset \alpha_3$ and from $A_4 = \alpha_4 \cap A_4 \subset \alpha_4 \cap J$ we get $LDet A = LDet A_4 \subset \alpha_4$.

Then, $\alpha_1 \cap Per A = \alpha_1 \cap (\triangle A \cup RDet A \cup LDet A)\emptyset$ as α^i ($i = 1, \ldots, 4$) are pairwise disjoint. It means that $(Per A)' \cap A_1 = (Per A)' \cap \alpha_1 \cap A = \alpha_1 \cap A = A_1$. Then we obtain

$$\check{A} = (Per A)' \cap A = (Per A)' \cap (A_1 \cap A_2 \cap A_3 \cap A_4)(Per A)' \cap A_1 = A_1.$$

The verification of equalities $\hat{A} = A_2$, $\overset{+}{A} = A_3$ and $\overline{A} = A_4$ is done similarly. It is obvious that the inverse implication of theorem 3.2 is valid. $\square$

4. The group of invertible Boolean matrices and permanent expansion

Let G be a subgroup of square invertible matrices in the semigroup $B_{n \times n}$ of all square Boolean $n \times n-$ matrices relative to multiplication, that is a subgroup of such matrices $S \in G$ for which matrices $S^{-1} \in G$ exist and the equalities $S \sqcap S^{-1} = S^{-1} \sqcap S = E$ hold. A rather full characteristic of properties and signs of invertible matrices can be found, for example, in [5] and [7].

Definition 4.1. The *left* $L : G \times B_{n \times n} \longrightarrow B_{n \times n}$ and *right* $R : B_{n \times n} \times G \longrightarrow B_{n \times n}$ *actions* of the group G of invertible matrices on the semigroup $B_{n \times n}$ are defined on elements $S \in G$ and $A \in B_{n \times n}$ as $L(S, A) = S \sqcap A$ and $R(A, S) = A \sqcap S$.

The mappings L, R are linear in every argument as it follows from the corresponding properties of operations. It means that for some $S \in G$ the mappings $L_S : B_{n \times n} \longrightarrow B_{n \times n}$ and $R_S : B_{n \times n} \longrightarrow B_{n \times n}$ defined on elements $A \in B_{n \times n}$ by the equalities $L_S(A) = L(S, A) = S \sqcap A$ and $R_S(A) = R(A, S) = A \sqcap S$ are linear and invertible as $L_S^{-1} = L_{S^{-1}}$ and $R_S^{-1} = R_{S^{-1}}$.

The following statement shows that L_S and R_S preserve degeneracy and determinacy of the given matrix. Besides, they preserve a corresponding type of degeneracy of the matrix mapping exterior into exterior and interior into interior. If an invertible matrix is positive, that is $S = \overset{+}{S}$ then L_S and R_S preserve its orientation, that is positive and negative parts are transferred into positive and negative accordingly. Negative invertible matrices $S = \overline{S}$ when multiplied change the orientation of the given matrix into the opposite transferring positive and negative parts into negative and positive accordingly.

Theorem 4.1. Let $A = \check{A} \cup \hat{A} \cup \overset{+}{A} \cup \overline{A}$ be a permanent expansion of a Boolean square matrix A and S is an invertible matrix of the same size. Then $L_S(\overset{+}{A} \cup \overline{A}) = L_S^{+}A \cup L_S^{-}A$, $L_S\check{A} = L_S^{\vee}A$, $L_S\hat{A} = L_S^{\wedge}A$. If S is positive, that is $S = \overset{+}{S}$ then $L_S \overset{+}{A} = L_S^{+}A$ and $L_S \overline{A} = L_S^{-}A$. If S is negative, that is $S = \overline{S}$ then $L_S \overline{A} = L_S^{+}A$ and $L_S \overset{+}{A} = L_S^{-}A$. The same holds for R_S.

Proof. First of all, from the equalities $S \sqcap S^{-1} = S^{-1} \sqcap S = E$, formulas (3), (5) (Proposition 2.2) and from $DetE = RDetE = I$ we receive $DetS = DetS^{-1} = I$. Thus, invertible Boolean relations are nondegenerate matrices. Besides, $RDetS = RDetS^{-1}$, $LDetS = LDetS^{-1}$.

Now, let $L_S(A) = S \sqcap A = B$. Then $L_{S^{-1}}(B) = S^{-1} \sqcap B = A$. It follwos from Proposition 2.2 that $\triangle A = \triangle B$, $PerA = PerB$ and $DetA = DetB$. Besides, if $S = \overset{+}{S}$ then $RDetS = RDetS^{-1} = I$, $LDetS = LDetS^{-1} = \emptyset$. That is why $RDetA = RDetB$ and $LDetA = LDetB$. If $S = \overline{S}$ then $LDetS = LDetS^{-1} = I$, $RDetS = RDetS^{-1} = \emptyset$, but in this case $RDetA = LDetB$ and $LDetA = RDetB$. Finally,

$$L_S\check{A} = L_S((PerA)' \cap A) = (PerA)' \cap L_SA = (PerB)' \cap B = \check{B}L_S^{\vee}A.$$

The other equalities of Theorem 3.3 are verified in the same way. $\square$

Remark. Permanents, determinants and common parts of semipermanents are invariant for all matrices from one Green's D-class (or from the coinciding with it J-class) of the semigroup of Boolean $n \times n$−matrices.

It should be noted that for generalized invertible Boolean matrices A and $\tilde{A}$, that is, such that $A \sqcap \tilde{A} \sqcap A = A$ and $\tilde{A} \sqcap A \sqcap \tilde{A} = \tilde{A}$, the equalities of oriented determinants also hold: $RDetA = RDet\tilde{A}$, $LDetA = LDet\tilde{A}$. It is not difficult to show this fact using formulas of Proposition 2.2.

5. Examples of nonnegative Boolean binary relations

Example 5.1. The set of nonnegative matrices defined by the requirement $LDetA = \emptyset$ forms a subsemigroup of Boolean matrices with the identity element E. It is evident from the formula for the left determinant in Proposition 2.2.

Even degrees of an arbitrary Boolean relation on a finite set [4] are examples of nonnegative matrices, that is left determinants are zero: $LDetA^{2k} = \emptyset$, $k = 0, 1, 2, \dots$. It follows immediately from the formula (2) for the left determinant. Thus, all idempotent Boolean matrices are nonnegative.

Example 5.2. Boolean matrices with the condition of transitivity $A^2 \subseteq A$, defining any transitive Boolean binary relations can be considered as examples of nonnegative matrices. Moreover, the requirements of transitivity are tough for constancy of semipermanents of such matrices and their powers [8].

The next matrix over Boolean algebra of intervals of real numbers is a degenerate transitive binary relation of three elements:

$$T = \begin{pmatrix} [0;1] & [0;3] & [0;2] \\ [0;1] & [0;2] & [0;2] \\ [0;1] & [0;2] & [-1;2] \end{pmatrix}.$$

In fact,

$$T \supset T^2 = \begin{pmatrix} [0;1] & [0;2] & [0;2] \\ [0;1] & [0;2] & [0;2] \\ [0;1] & [0;2] & [-1;2] \end{pmatrix} = T^3 = T^4 = \dots.$$

For all degrees of matrix T, semipermanents are equal $\overset{\pm}{\nabla} T^k[0;1]$ and $RDetT^k = LDetT^k = \emptyset$ $(k = 1, 2, \dots)$.

Example 5.3. The sequence of inclusions $E \subseteq A \subseteq A^2 \subseteq \dots \subseteq A^{n-1} = A^n = \dots$ is well known for any reflexive Boolean binary relation on a finite set. The proof of this fact can be found, for example, in [9], [10]. For semipermanents of powers of any reflexive relation, we have $\overset{+}{\nabla} A^k = I$ $(k = 0, 1, 2, \dots)$ and $\overline{\nabla} A \subseteq \overline{\nabla} A^2 \subseteq \dots \subseteq \overline{\nabla} A^{n-1} = \overline{\nabla} A^n \dots$ [8]. It follows from $\overset{+}{\nabla} E = I$ and that oriented semipermanents are isotonic. Hence, left determinants of any reflexive relation and its degrees are equal to zero, that is, all reflexive relations are nonnegative Boolean relations. Moreover, if $Per A^k = I$ $(k = 0, 1, 2, \ldots)$ then reflexive relations are relations with zero exterior.

Remark. It is obvious that nonempty binary relations on a finite set represented by square matrices over Boolean algebra $\{\emptyset, I\}$ can be of four types: exterior, interior, positive and negative. Thus, from example 5.3 we deduce that widely used in mathematics reflexive relations on a finite set (equivalence, tolerance, partial order etc.) might be either positive or interior.

6. Interiorities and their properties

The set of degenerate matrices defined by the requirement $Det A = \emptyset$ forms an ideal in the semigroup of square matrices relative to Boolean matrix multiplication. It follows from the formula (5) of Proposition 2.2.

Let us concentrate only at the following properties of interior parts of matrices. We will show below interior parts which form a semiring relative to additive operation $\cup$ and multiplicative operation $\sqcap$.

Theorem 6.1. *For interior of Boolean matrices, the following holds:*
1) $\hat{\Theta} = \Theta, \hat{J} = J$;
2) $\hat{A} \subset A$;
3) $\hat{\hat{A}} = \hat{A}$;
4) $(A \subset B) \to (\hat{A} \subset \hat{B})$;
5) $\widehat{\hat{A} \cap \hat{B}} = \widehat{A \cap B} \subset \hat{A} \cap \hat{B}$;
6) $\widehat{\hat{A} \cup \hat{B}} = \hat{A} \cup \hat{B} \subset \widehat{A \cup B}$;
7) $\widehat{\hat{A} \sqcap \hat{B}} = \hat{A} \sqcap \hat{B}$.

Here the matrices Θ and J are zero, and the universal element of secondary Boolean algebra is formed by square Boolean matrices of given size. Properties $2 - 7$ hold for any matrices A and B.

Proof. The first property holds in view of $\hat{\Theta} = \triangle\Theta \cap \Theta\emptyset \cap \Theta = \Theta$ and $\hat{J} = \triangle J \cap J = I \cap J = J$. The second property follows from the definition of interior. The third property is verified immediately:

$$\hat{\hat{A}} = \triangle\hat{A} \cap \hat{A}\triangle\hat{A} \cap (\triangle A \cap A) = \triangle A \cap (\triangle A \cap A) = \triangle A \cap A = \hat{A}.$$

The fourth property follows from the fact that inclusion $A \subseteq B$ implies $\triangle A \subseteq \triangle B$. Therefore, $\triangle A \cap A \subseteq \triangle B \cap B$, that is $\hat{A} \subseteq \hat{B}$.

To verify the fifth property note that from properties 2 and 4 we obtain $\widehat{\hat{A} \cap \hat{B}} \subset \widehat{A \cap B}$. On the other hand, it could be easily verified that $\triangle(A \cap$

212

$B) \subset \triangle A \cap \triangle B$ holds. Then

$$\widehat{A \cap B} = \triangle(A \cap B) \cap (A \cap B) \subseteq (\triangle A \cap \triangle B \cap A \cap B) = \hat{A} \cap \hat{B}.$$

Thus, there is a chain of inclusions $\hat{A} \cap \hat{B} \subseteq \widehat{A \cap B} \subseteq \hat{A} \cap \hat{B}$ from which and from property 4 the inclusions

$$\widehat{\hat{A} \cap \hat{B}} \subseteq \widehat{\widehat{A \cap B}} \subseteq \widehat{\hat{A} \cap \hat{B}}$$

follow. Applying property 3 to the last expression, we get the equality $\widehat{A \cap B} = \widehat{\hat{A} \cap \hat{B}}$ stated in 5.

Property 6 is deduced from

$$\hat{A} \cup \hat{B} \supseteq \widehat{\hat{A} \cup \hat{B}} \triangle (\hat{A} \cup \hat{B}) \cap (\hat{A} \cup \hat{B}) \supseteq (\triangle \hat{A} \cup \triangle \hat{B}) \cap (\hat{A} \cup \hat{B})$$
$$\supseteq (\triangle \hat{A} \cap \hat{A}) \cup (\triangle \hat{B} \cap \hat{B}) \hat{A} \cup \hat{B}.$$

Equality $\widehat{\hat{A} \sqcap \hat{B}} = \hat{A} \sqcap \hat{B}$ is proved with the help of Proposition 2.2 and the following properties of matrix multiplication

$$\lambda \cap (A \sqcap B) = (\lambda \cap A) \sqcap B = A \sqcap (\lambda \cap B) = (\lambda \cap A) \sqcap (\lambda \cap B)$$

that hold for any square Boolean matrices A, B of given size and for an arbitrary element λ of Boolean algebra. Then we obtain

$$\widehat{A \sqcap B} = \triangle(A \sqcap B) \cap (A \sqcap B) \supseteq (\triangle A \cap \triangle B) \cap (A \sqcap B)$$
$$= (\triangle A \cap A) \sqcap (\triangle B \cap B) = \hat{A} \sqcap \hat{B}.$$

The last expression holds, in particular, for interior matrices A, B. We obtain $\widehat{\hat{A} \sqcap \hat{B}} \supseteq \hat{A} \sqcap \hat{B} = \hat{A} \sqcap \hat{B}$. On the other hand, the inverse inclusion is true. Hence, the equality $\widehat{\hat{A} \sqcap \hat{B}} = \hat{A} \sqcap \hat{B}$ holds for any square Boolean matrices A and B. $\square$

Properties 1–5 of Theorem 6.1 show how close are the concept of "interior" of a Boolean binary relation on a finite set introduced in this paper and the concept of "interior of a subset of a topological space". They are different from the axiom of Kuratowski, where topology is defined by the notion of "interior of a set", because of inclusion 5, given instead of equality. Nevertheless, it explains the choice of such term as "interior" for Boolean matrix.

Example 6.1. It is not difficult to show that interior matrices of size 2×2 are matrices of the form $A = \left(\begin{smallmatrix} a & a \\ a & a \end{smallmatrix}\right)$. Then property 5 of Theorem 6.1 is turned into the equality $\hat{A} \cap \hat{B} = \widehat{A \cap B} = \hat{A} \cap \hat{B}$. This is the reason why a set of interior 2×2-matrices forms a certain open-closed topology, obviously isomorphic to the given Boolean algebra. However, in case of a large size of a matrix one can find out an example of interior matrices, where an intersection of interiors is not an interior.

Example 6.2. Notice that if for a matrix B and a nonzero interior matrix $\hat{A}$ over Boolean algebra $\{\emptyset, I\}$ the inclusion $\hat{A} \subseteq B$ holds, than B is a nonzero interior matrix. It follows that $\hat{A} \subseteq B$ implies $\triangle\hat{A} \subseteq \triangle B$ and therefore $\triangle\hat{A} = I$ implies $\triangle B = I$. That is, $B = \triangle B \cap B = \hat{B}$.

Obviously, the next types of matrices

$$A_{2\times 2} = \begin{pmatrix} I & I \\ I & I \end{pmatrix}, A_{3\times 3} \begin{pmatrix} I & I & \emptyset \\ I & I & \emptyset \\ \emptyset & \emptyset & I \end{pmatrix}, \ldots, A_{n\times n} \begin{pmatrix} I & I & \emptyset & \ldots & \emptyset \\ I & I & \emptyset & \ldots & \emptyset \\ \emptyset & \emptyset & I & \ldots & \emptyset \\ \vdots & \vdots & \vdots & \ddots & \vdots \\ \emptyset & \emptyset & \emptyset & \ldots & I \end{pmatrix}, \quad n > 3,$$

are the minimal nonzero interior matrices. Consequently, any reflexive binary relation on a finite set with at least one pair of equivalent elements is an interior binary relation.

References

1. Chesley, D. S., Bevis, J. H., Determinants for matrices over lattices. Proc. Roy. Soc. Edinburgh, A 68, 2, p. 138–144, 1969,
2. Golan, J. S. Semirings and their Applications. Dordrecht: Kluwer Academic Publishers. xi, 381 p., 1999.
3. Poplavski, V. B. Oriented determinants and composition of Boolean matrices. Mathematics. Mechanics. Saratov: Publishing house of Saratov state university, 6, p. 111–114, 2004.
4. Poplavski, V. B. Determinants of degrees of Boolean matrices. Chebishev's collection. Vol. 5, 3 (11), p. 98–111, 2004.
5. Poplavski, V. B. Invertible and joined Boolean matrices. Chebyshev's collection. Vol. 6 (1), p. 174–181, 2005.
6. Poplin, P. L.; Hartwig, R. E. Determinantal identities over commutative semirings. Linear Algebra Appl. 387, 99–132, 2004.
7. Rudeanu, S. Boolean functions and equations. Amsterdam–London: North-Holland Publishing Company; New York: American Elsevier Publishing Company, Inc. XIX, 442 p., 1974.
8. Poplavski, V. B. Volumes and determinants of the powers of transitive and reflexive Boolean relations on a finite set. Izvestiya of the Tula State University. Ser. Mathematics. Mechanics. Informatics. Tula: TSU, V. 10 (1), p. 134–141, 2004.

9. Give'on, Y. Lattice matrices. Inform. And Control. 7 (4), p. 477–484, 1964.

10. Loontz, A. G. Application of matrix Boolean Algebra in the analysis and synthesis of contact-relay circuits. Reports by Academy of Sciences USSR, 3, p. 70, 1950.

HIERARCHIES OF RANDOMNESS TESTS

JAN REIMANN

Institut für Informatik
Ruprecht-Karls-Universität Heidelberg, Germany
Email: reimann@math.uni-heidelberg.de

FRANK STEPHAN*

Departments of Mathematics and Department of Computer Science
National University of Singapore, Republic of Singapore
Email: fstephan@comp.nus.edu.sg

It is well known that Martin-Löf randomness can be characterized by a number of equivalent test concepts, based either on effective nullsets (Martin-Löf and Solovay tests) or on prefix-free Kolmogorov complexity (lower and upper entropy). These equivalences are not preserved as regards the partial randomness notions induced by effective Hausdorff measures or partial incompressibility. Tadaki [20] and Calude, Staiger and Terwijn [2] studied several concepts of partial randomness, but for some of them the exact relations remained unclear. In this paper we will show that they form a proper hierarchy of randomness notions, namely for any ρ of the form $\rho(x) = 2^{-|x|s}$ with s being a rational number satisfying $0 < s < 1$, the Martin-Löf ρ-tests are strictly weaker than Solovay ρ-tests which in turn are strictly weaker than strong Martin-Löf ρ-tests. These results also hold for a more general class of ρ introduced as unbounded premeasures.

1. Introduction

The correspondence between effective nullsets in the sense of measure theoretic tests and compressible initial segments in terms of (prefix-free) Kolmogorov complexity is one of the cornerstones of algorithmic information theory. Furthermore, the concept of randomness itself appears thereby very robust, as several variants of measure theoretic tests (Martin-Löf and Solovay tests) and complexity theoretic properties all yield the same notion of randomness; that is, a sequence A is Martin-Löf random iff one of the following equivalent conditions hold:

(1) A is not covered by any Martin-Löf test;
(2) A is not covered by any Solovay test;

---

*the second author is supported in part by NUS research grant R252-000-212-112.

(3) for some constant c and for all n, $K(A\restriction_n) \geq n - c$;

(4) $\lim_n K(A\restriction_n) - n = \infty$.

On the other hand, the complexity theoretic formulations suggests not only a qualitative, but also a quantitative classification of randomness. For instance, a sequence A for which $K(A\restriction_n) \geq n/2 + c$ for all n and some constant c might be classified as being 1/2-random.

This idea is, in particular, reflected in the study of *relative randomness* initiated by Solovay [17] and later leading to a variety of *reducibility notions* for random sequences. The forthcoming book by Downey and Hirschfeldt [5] will provide for a detailed account.

One may ask whether it is possible to catch "partial" randomness not only within a rather fine-grained hierarchy of relative randomness, but as an absolute notion in terms of measure theoretic tests and their complexity theoretic counterparts.

The problem here is that the property of being an (effective) nullset is rather qualitative. What is therefore needed is a further ramification of the effective Lebesgue nullsets (which constitute the non-random sequences).

Such a ramification can be given in terms of *Hausdorff measures*, which are an essential tool in fractal geometry. In particular, they allow for a definition of a non-integral notion of dimension, *Hausdorff dimension*. Works by Ryabko [14, 15], Staiger [19] and Cai and Hartmanis [1] established a close connection between Hausdorff dimension and the (lower) asymptotic complexity of sequences given as $\liminf_n K(A\restriction_n)/n$.

Later, Lutz [7] used the martingale characterization of nullsets to define an effective variant of Hausdorff measure and dimension. This yields a quantitative classification of randomness for individual sequences in terms of measure. Indeed, Mayordomo [9] showed that Lutz's approach leads to the same effective dimension as the number $\liminf_n K(A\restriction_n)/n$ mentioned above.

It is straightforward to transform the martingale approach to Hausdorff measures into a Martin-Löf-style test concept (as done by Reimann and Stephan [11], Tadaki [20] or Calude, Staiger and Terwijn [2]). Besides, one may also define partial randomness notions based on the other randomness criteria (2)-(4). It could be shown that in the generalized framework Martin-Löf tests still coincide with the complexity criterion derived from (3), referred to as *weak Chaitin randomness* by Tadaki [20] and Calude, Staiger and Terwijn [2]. Furthermore, Solovay tests and *strong Chaitin randomness* (4) still remain equivalent [2, 20].

However, it remained unclear whether the full robustness of (Lebesgue) randomness (in the sense of complete equivalence of all test notions) prevailed.

In this paper we show that this is indeed not the case. Not only with respect to the usual Hausdorff measures, but also a very wide family of measures given by *unbounded premeasures*, Martin-Löf tests and Solovay tests are not equivalent. Another test notion proposed by Calude, Staiger and Terwijn [2] called *strong Martin-Löf randomness* yields an even stronger notion of randomness.

The paper is structured as follows. In Section 2 we give a detailed introduction of effective tests derived from a general class of outer measures on Cantor space. Section 3 will treat the connection between tests and Kolmogorov complexity. Finally, in Section 4 we will show that the test notions introduced in Section 2 form a proper hierarchy of randomness notions for unbounded premeasures. The latter include the non-integral Hausdorff measures.

Notation: Most notation is standard. $\{0,1\}^\star$ denotes the set of finite binary strings, $\{0,1\}^\omega$ the set of all infinite binary strings. $\sqsubset$ is the partial prefix order on strings, which extends to $\{0,1\}^\star \cup \{0,1\}^\omega$ in a natural way. $x \sqsubseteq y$ holds if either $x \sqsubset y$ or $x = y$. Given a set $V \subseteq \{0,1\}^\star$ and a string x, we write V_x for the set $\{w \in V : x \sqsubseteq w\}$ and V_x^+ for the set $\{w \in V : x \sqsubset w\}$.

We assume the reader is acquainted with the basic definitions and results of Recursion Theory and the theory of Kolmogorov complexity. We refer to the textbooks of Li and Vitányi [6], Odifreddi [12] and Soare [18] for any background on this.

2. Effective Randomness Tests for Outer Measures

There are mainly two ways to devise measures on Cantor space (as on any other suitable topological/metric space). One can start with an additive set function on a (semi)algebra of sets (usually comprising a basis of the topology) and then use Caratheodory's extension theorem, which ensures that there is a unique extension of this set function to a σ-algebra (which includes the Borel sets if the starting (semi)algebra included the basic open sets).

Alternatively, measures can be obtained by restricting outer measures to a suitable family of sets in $\{0,1\}^\omega$. Outer measures are often defined via *premeasures* and *coverings*. A premeasure is a non-negative (possibly infinite) set function ρ on a family $\mathfrak{C}$. In most cases, $\mathfrak{C}$ will consist of the family of *basic open cylinders* which are defined as

$$[\![x]\!] = \{X \in \{0,1\}^\omega : x \sqsubset X\}$$

Therefore, it is convenient to regard premeasures as functions

$$\rho : \{0,1\}^\star \to \mathbb{R}_0^+$$

from which one can obtain an outer measure μ_ρ by letting

$$\mu_\rho(\mathcal{X}) = \inf \left\{ \sum_i \rho(x_i) : \bigcup_i [\![x_i]\!] \supseteq \mathcal{X} \right\}.$$

It is not hard to show that $\mu = \mu_\rho$ is a countably subadditive, monotone set function. If one restricts μ to those sets $\mathcal{A}$ which satisfy

$$(\forall \mathcal{Y}) \, [\mu(\mathcal{Y}) = \mu(\mathcal{Y} \cap \mathcal{A}) + \mu(\mathcal{Y} \setminus \mathcal{A})],$$

called the *measurable sets*, the measurable sets form a σ-algebra and μ is an additive set function on this σ-algebra.

If the underlying space is a metric space, the method of passing from a premeasure to an outer measure can be refined in a geometrical way. The standard metric d on $\{0,1\}^\omega$ (which yields a topology compatible with the one generated by the cylinder sets defined above) is defined as

$$d(X,Y) = \inf\{2^{-n} : (\forall m < n)[X(m) = Y(m)]\}.$$

That is, if $X \neq Y$ then $d(X,Y) = 2^{-n}$ for the least n with $X(n) \neq Y(n)$. The diameter of a set $\mathcal{X} \subseteq \{0,1\}^\omega$ is $d(\mathcal{X}) = \sup\{d(X,Y) : X,Y \in \mathcal{X}\}$. If ρ is a premeasure, we can define

$$\mu_\delta(\mathcal{X}) = \inf \left\{ \sum_i \rho(x_i) : \bigcup_i [\![x_i]\!] \supseteq \mathcal{X} \wedge (\forall i)[d([\![x_i]\!]) \leq \delta] \right\}$$

and

$$\mu^\rho(\mathcal{X}) = \sup\{\mu_\delta(\mathcal{X}) : \delta > 0\}.$$

Here it is the "fine covers" that determine the value of μ^ρ. It can be shown that μ^ρ is also an outer measure and that it behaves, in geometric sense, more stable than measures constructed via the first method. Note that $d([\![x_i]\!]) \leq \delta$ if and only if $|x_i| \geq -\log \delta$.

An extensive treatment of constructing measures via premeasures is found in the book by Rogers [13].

It can be shown that every *nullset*, a set for which μ^ρ takes the value zero, is measurable. It was Martin-Löf's groundbreaking idea to use the concept of an *effective nullset* to define a notion of randomness for individual sequences. Basically, a sequence is random with respect to a measure if it is not contained in an effectively presented nullset with respect to the measure. As the nullsets are precisely the sets which have outer measure zero, it suffices to study effective nullsets with respect to premeasures. It is not hard to see that Martin-Löf's approach works for arbitrary (outer) measures which are derived from computable premeasures.

A well-known group of outer measures is obtained from the premeasures $\rho(x) = 2^{-|x|s}$ where $0 \leq s \leq 1$, the s-dimensional Hausdorff measures. For $s = 1$, we obtain the uniform distribution $\rho(x) = 2^{-|x|}$, which generates a measure isomorphic to Lebesgue measure on the unit interval.

We will study a certain class of premeasures. These premeasures can be thought of as "geometrically well behaved". Among the measures they induce are the usual probability measures on $\{0,1\}^\omega$ as well as the family of s-dimensional Hausdorff measures. To be able to effectivize, we will always assume premeasures to be computable.

Definition 2.1. A (*geometrical*) *premeasure* is a computable function $\rho : \{0,1\}^\star \to \mathbb{R}_0^+$ such that $\rho(\epsilon) = 1$ and there are (computable) real numbers p, q with

- $1/2 \leq p < 1$ and $1 \leq q < 2$;
- $(\forall x \in \{0,1\}^\star)\,(\forall i \in \{0,1\})\,[\rho(xi) \leq p\rho(x)]$;
- $(\forall x \in \{0,1\}^\star)\,[q\rho(x) \leq \rho(x0) + \rho(x1)]$.

We will call such ρ a (p,q)-premeasure. ρ is called an *unbounded premeasure* if it is (p,q)-premeasure for some $q > 1$. A premeasure is called *length-invariant* if

$$(\forall x, y)\,[|x| = |y| \;\Rightarrow\; \rho(x) = \rho(y)]$$

and is called *additive* if

$$(\forall x \in \{0,1\}^\star)\,[\rho(x) = \rho(x0) + \rho(x1)].$$

Remarks and Examples: (a) Additive premeasures induce non-atomic *probability measure* on $\{0,1\}^\omega$. Among the probability measures induced by additive premeasures are standard uniform distribution $\rho(x) = 2^{-|x|}$, which we will denote by λ, as well as all non-degenerate *Bernoulli measures* μ_p with $0 < p < 1$. μ_p is the measure obtained by setting

$$\rho(x) = p^{|\{i : x(i)=0\}|}(1 - p)^{|\{i : x(i)=1\}|}$$

220

for all x.

(b) It is not hard to see that, if ρ is unbounded, the corresponding measure μ constructed by the second method mentioned above will be infinite $(\mu(\{0,1\}^\omega) = \infty)$. This motivates the term "unbounded" (see Proposition 2.4).

(c) The most common unbounded and length invariant premeasures are the functions of the form $\rho(x) = 2^{-|x|s}$ with $0 < s \le 1$. The corresponding constants are $p = 2^{-s}$ and $q = 2^{1-s}$. These premeasures give rise to the *s-dimensional Hausdorff measures*. In general, length-invariant premeasures are often called *dimension functions*, as the induce a generalized type of Hausdorff measure.

(d) Other less orthodox examples of premeasures are $\rho(x)p(|x|)2^{-s|x|}$, where p is a suitable polynomial, or $\rho(x)2^{-p|\{i:x(i)=0\}|-(1-p)|\{i:x(i)=1\}|}$, where p is a real number satisfying $0 < p < 1$.

In the following, if we consider measures derived from premeasures, we will always assume they are constructed via the second method. We will be particularly interested in *nullsets*, that is, sets for which μ^ρ takes the value zero. The following proposition states some equivalent characterization of nullsets for a measure μ^ρ. These will be important for the definition of effective nullsets later on. It will be convenient to introduce some further notation: Given $W \subseteq \{0,1\}^*$, let $\rho(W)$ stand for $\sum_{x \in W} \rho(x)$.

Proposition 2.2. *Given a premeasure ρ and a set $\mathcal{X} \subseteq \{0,1\}^\omega$, the following are equivalent:*

(a) $\mu^\rho(\mathcal{X}) = 0$;

(b) *For every $n \in \mathbb{N}$ there exists a set $U_n \subseteq \{0,1\}^*$ such that*

$$\mathcal{X} \subseteq [\![U_n]\!] \quad \text{and} \quad \rho(U_n) \le 2^{-n};$$

(c) *There exists a set $W \subseteq \{0,1\}^*$ such that $\rho(W) < \infty$ and for any $X \in \mathcal{X}$ there are infinitely many $w \in W$ such that $w \sqsubset X$.*

Proof. By definition, $\mu^\rho(\mathcal{X})$ is the infimum of all numbers $\rho(W)$ with $\mathcal{X} \subseteq [\![W]\!]$. This gives the equivalence of (a) and (b). Furthermore, taking W to be the union of all U_n gives the direction from (b) to (c).

For the missing direction from (c) to (b), let $w_0, w_1, \ldots$ be an one-one enumeration of W. Now one defines $U_n = \{w_m, w_{m+1}, \ldots\}$ for the first m such that $\sum_{k \ge m} \rho(w_k) < 2^{-n}$. This m exists as $\sum_{k \ge 0} \rho(w_k)$ is finite. As for every $X \in \mathcal{X}$ there are infinitely many k with $w_k \sqsubset X$, it holds that $X \in [\![U_n]\!]$ for all n. This completes the direction from (c) to (b). $\qquad\square$

We now introduce the effective variants of measure zero sets and the corresponding randomness concepts. This generalizes earlier work. The underlying tests are defined as below but were mostly restricted to $\rho(x) = 2^{-|x|}$ (or some additive premeasure, i.e. probability measures on $\{0,1\}^\omega$). Martin-Löf [8] introduced the notion of effective tests now named after him. Solovay [17] showed that $X \in \{0,1\}^\omega$ is Martin-Löf random iff there is no r.e. set W of strings such that $\rho(W) < \infty$ and $w \sqsubseteq X$ for infinitely many $w \in W$. Schnorr [16] contrasted Martin-Löfs general condition to more restrictive randomness-tests where X is random iff there is no uniformly enumerable sequence $(U_n)_{n \in \mathbb{N}}$ of sets with $X \in [\![U_n]\!]$ and $\rho(U_n) = 2^{-n}$ for all n. This condition had also many natural characterizations and is now known as Schnorr randomness. Based on this work, Lutz [7] initiated the study effectivizations of Hausdorff measures via modified martingales. Building on these notions, further investigations were carried out by Tadaki [20], Reimann [10], Calude, Staiger and Terwijn [2].

Definition 2.3. Let $\rho : \{0,1\}^\star \to \mathbb{R}_0^+$ be a geometrical premeasure.

(a) A *Martin-Löf ρ-test* is a uniformly enumerable sequence $(U_n)_{n \in \mathbb{N}}$ of sets of strings such that

$$(\forall n)[\rho(U_n) \leq 2^{-n}].$$

The test $(U_n)_{n \in \mathbb{N}}$ *covers* a set $\mathcal{X} \subseteq \{0,1\}^\omega$ if

$$\mathcal{X} \subseteq \bigcap_{n \in \mathbb{N}} [\![U_n]\!].$$

In this case $\mathcal{X}$ is called *Martin-Löf ρ-null*. A sequence $A \in \{0,1\}^\omega$ is called *Martin-Löf ρ-random* if $\{A\}$ is not covered by any Martin-Löf ρ-test.

(b) A *strong Martin-Löf ρ-test* is a uniformly enumerable sequence $(U_n)_{n \in \mathbb{N}}$ such that

$$(\forall n)(\forall V \subseteq U_n)[V \text{ prefix-free } \Rightarrow \rho(V) \leq 2^{-n}].$$

Again, the test $(U_n)_{n \in \mathbb{N}}$ *covers* $\mathcal{X} \subseteq \{0,1\}^\omega$ if

$$\mathcal{X} \subseteq \bigcap_{n \in \mathbb{N}} [\![U_n]\!].$$

Accordingly, a $\mathcal{X}$ is called *strongly Martin-Löf ρ-null* and A is called *strongly Martin-Löf ρ-random* if $\{A\}$ is not covered by any strong Martin-Löf ρ-test.

222

(c) A *Solovay ρ-test* is a recursively enumerable set W such that

$$\rho(W) < \infty.$$

The test W *covers* $\mathcal{X} \subseteq \{0,1\}^\omega$ if, for any $X \in \mathcal{X}$, W contains infinitely many prefixes of X. $A \in \{0,1\}^\omega$ is called *Solovay ρ-random* if it is not covered by any Solovay ρ-test.

Note that the name "strong Martin-Löf test" might be misleading at first, since every Martin-Löf test is also a strong Martin-Löf test. In fact, the use of "strong" makes more sense from the viewpoint of random sequences since every strongly Martin-Löf ρ-random sequence is also Martin-Löf ρ-random.

Also note that for additive premeasures, the notions of Martin-Löf ρ-randomness, Solovay ρ-randomness and strong Martin-Löf ρ-randomness coincide, yielding an effective analog of Proposition 2.2. If ρ is additive, any Solovay ρ-test W can be converted effectively into a Martin-Löf ρ-test (V_n) by letting x enter V_n if and only if 2^n proper prefixes of x have been enumerated into W prior to x. Furthermore, we can pass from a strong Martin-Löf ρ-test (U_n) to an ordinary Martin-Löf ρ-test (V_n) covering the same set of sequences by the following effective procedure:

If x is enumerated into U_n at some stage, check whether a prefix of x has already been enumerated into V_n. If so, discard x. Otherwise check whether some extensions of x have already been enumerated into V_n. If not, enumerate x into V_n. Otherwise assume W is the finite set of strings extending x and already in V_n. Enumerate into V_n a finite, prefix-free set $\widetilde{W}$ such that $[\![\widetilde{W}]\!] \cap [\![W]\!] = \emptyset$ and $[\![\widetilde{W}]\!] \cup [\![W]\!] = [\![x]\!]$.

The so constructed V_n covers the same set of sequences as a (maximal, with respect to covering) prefix-free subset V of U_n. If ρ is additive, $\rho(V_n) = \rho(V) \leq 2^{-n}$.

It is not clear how these procedures may be transferred to unbounded premeasures, since a string may not be substituted by a cover-equivalent set of longer strings of the same ρ-measure. In fact, we will show later that this is in general not possible.

We conclude this section by showing that unbounded, length-invariant premeasures induce measures that are incompatible with the uniform distribution λ.

Proposition 2.4. *For every unbounded premeasure ρ there exists a set $\mathcal{X}$ such that some Martin-Löf λ-test covers all sequences $X \in \mathcal{X}$ but $\mathcal{X}$ does not have μ^ρ-measure zero.*

Proof. Assume ρ is an unbounded (p,q)-premeasure. Let $R_n(x) = \{xy : y \in \{0,1\}^n\}$. It follows from an easy induction that,

$$\rho(R_n(x)) \geq q^n \rho(x) \quad \text{and} \quad (\forall w \in R_n(x)) \, [\rho(w) \leq p^n \rho(x)]. \tag{1}$$

Let $U_0 = \{\epsilon\}$. Given U_n, we can use (1) and the fact that ρ is computable to (effectively) find prefix-free sets V_n^0, V_n^1 such that $[\![V_n^i]\!] \subset [\![U_n]\!]$, $[\![V_n^1]\!] \cap [\![V_n^1]\!] = \emptyset$ and $\rho(V_n^i) \geq 1$. But obviously, for some i we must have $\lambda([\![V_n^i]\!]) \leq \frac{1}{2}\lambda([\![U_n]\!])$. Pick such i and let $U_{n+1} = V_n^i$. Then, for $\mathcal{U} = \bigcap[\![U_n]\!]$, by choice of the sets $U_0, U_1, \ldots$, the test $(U_n)_{n \in \mathbb{N}}$ is a Martin-Löf λ-test and covers $\mathcal{U}$ but $\mu^\rho(\mathcal{U}) \neq 0$. $\qquad\square$

As a corollary we get that the (weak) randomness notions with respect to λ on the one hand and with respect to unbounded measure functions on the other hand differ.

Corollary 2.5. *For every unbounded premeasure ρ there exists a sequence X such that X is Martin-Löf ρ-random but not Martin-Löf λ-random.*

Proof. Any set of non-zero μ^ρ-measure has to contain a Martin-Löf ρ-random sequence, so the set $\mathcal{U}$ constructed above has to contain one, but cannot contain a Martin-Löf λ-random sequence, for it is covered by some Martin-Löf λ-test. $\qquad\square$

3. Nullsets and Kolmogorov complexity

In this section we will study to what extent the correspondence between effective nullsets and compressibility extends from the well-known characterization of randomness with respect to the uniform distribution to nullsets with respect to (length-invariant) premeasures.

For this purpose, we generalize a definition by Chaitin [3] to arbitrary premeasures. For the case $\rho(x) = 2^{-|x|s}$, this was first done by Tadaki [20].

Definition 3.1. Let ρ be a premeasure.

(a) A sequence $A \in \{0,1\}^\omega$ is *weakly Chaitin ρ-random* if there exists a constant c such that

$$(\forall n) \, [K(A \upharpoonright_n) \geq -\log \rho(A \upharpoonright_n) - c].$$

(b) A set $A \in \{0,1\}^\omega$ is *strongly Chaitin ρ-random*[a] if

$$(\forall c)(\overset{\infty}{\forall} n)[K(A \upharpoonright_n) \geq -\log \rho(A \upharpoonright_n) + c].$$

---

[a]The notion of strong Chaitin ρ-randomness should not be confused with strong Chaitin randomness as defined in [5], meaning $(\exists^\infty n)[K(A \upharpoonright_n) \geq n + K(n) - \mathcal{O}(1)]$.

The well-known Kolmogorov characterization for random sequences using prefix-free complexity K generalizes to a characterization for Martin-Löf ρ-nullsets. For one direction, though, it seems one has to presuppose length-invariance.

In the case $\rho(x) = 2^{-|x|s}$, the following two propositions were shown by Tadaki [20]. Reimann [10] obtained related results for the more general case of computable *dimension functions*.

Proposition 3.2. *Let ρ be a premeasure. If $A \in \{0,1\}^\omega$ is weakly Chaitin ρ-random then it is also Martin-Löf ρ-random.*

Proof. Assume A is not Martin-Löf ρ-random. Thus there exists a computable sequence $C_1, C_2, C_3, \ldots$ of enumerable sets of strings such that for all n

$$(\exists w \in C_n)\,[w \sqsubset A] \quad \text{and} \quad \rho(C_n) \leq 2^{-n}. \tag{2}$$

Define functions $m_n : \{0,1\}^\star \to \mathbb{Q}$ by

$$m_n(w) = \begin{cases} n\rho(w) & \text{if } w \in C_n, \\ 0 & \text{otherwise,} \end{cases}$$

and let

$$m(w) = \sum_{n=1}^{\infty} m_n(w).$$

Obviously, all m_n and thus m are enumerable from below. Furthermore,

$$\sum_{w \in \{0,1\}^\star} m(w) = \sum_{w \in \{0,1\}^\star} \sum_{n=1}^{\infty} m_n(w) = \sum_{n=1}^{\infty} n \sum_{w \in C_n} \rho(w) \leq \sum_{n=1}^{\infty} \frac{n}{2^n} < \infty.$$

It follows from the fundamental *Coding Theorem*, due to Levin (see the book by Li and Vitányi [6]), that there exists a constant c_m such that $m(w) \leq c_m 2^{-K(w)}$ for almost every w. Now let $c > 0$ be any constant. If we set $k = \lceil c \rceil + 1$, then, by (2), there is some $w \in C_k$ with $w \sqsubset A$, say $w = A \restriction_n$. This implies $m(A \restriction_n) \geq k\rho(A \restriction_n) > c\rho(A \restriction_n)$ and therefore $\limsup m(A \restriction_n)/\rho(A \restriction_n) = \infty$. Hence A is not weakly Chaitin ρ-random. $\square$

For length-invariant premeasures, we can prove the converse of Proposition 3.2.

Proposition 3.3. *Let ρ be a length-invariant premeasure. If $A \in \{0,1\}^\omega$ is Martin-Löf ρ-random, then it is weakly Chaitin ρ-random.*

Proof. The proof is an adaptation of a standard proof that every Martin-Löf random sequence is incompressible with respect to prefix-free Kolmogorov complexity (see for instance the book by Downey and Hirschfeldt [5]). It is based on a fundamental result by Chaitin [4] which establishes that for any l,

$$|\{x \in \{0,1\}^n : K(x) \le n + K(n) - l\}| \le 2^{n+c-l}, \tag{3}$$

where c is a constant independent of n, l. Remember that the natural numbers are identified with their binary representation.

Let ρ be a length-invariant premeasure and assume A is not weakly Chaitin ρ-random. As ρ is length-invariant, there exists a computable function $h : \mathbb{N} \to \mathbb{R}_0^+$ such that

$$h(n) = \alpha \iff (\forall x)[|x| = n \Rightarrow \rho(x) = \alpha].$$

Now choose a c for which (3) holds. Define

$$V_n = \{x \in \{0,1\}^\star : K(x) \le -\log h(x) - n - c\}.$$

Then each V_n covers A, since for every l there is some prefix x of A such that $K(x) \le -\log \rho(x) - l$. Furthermore, each V_n is r.e., since K is enumerable from above. Finally, using (3), we have for each n,

$$\sum_{w \in V_n} \rho(w) = \sum_{k=0}^{\infty} \sum_{\substack{w \in V_n \\ |w|=k}} \rho(w) = \sum_{k=0}^{\infty} h(k)|\{0,1\}^k \cap V_n|$$

$$\le 2^{-n} \sum_{k=0}^{\infty} 2^{-K(k)} \le 2^{-n}$$

and this bound completes the proof. $\square$

It is also possible to characterize Solovay randomness in terms of complexity. Namely, Solovay and strong Chaitin randomness coincide. The following proposition generalizes a result by Tadaki [20] and Calude, Staiger and Terwijn [2].

Proposition 3.4. *Given a premeasure ρ, a sequence $A \in \{0,1\}^\omega$ is Solovay ρ-random if and only if it is strongly Chaitin ρ-random.*

Proof. Assume A is not strongly Chaitin ρ-random. Then there is a constant c such that for infinitely many n, $\rho(A \restriction n) < 2^{c-K(A\restriction n)}$; without loss of generality c is a natural number. Then the set $W = \{x : \rho(x) < 2^{c-K(x)}\}$ is recursively enumerable since ρ is computable, K is enumerable from above and thus $2^{c-K(x)}$ is enumerable from below. Furthermore, $\rho(W) <$

$\sum_{x \in W} 2^{c-K(x)} < 2^c$. So W is a Solovay ρ-test. By choice of c, W covers every A. So A is not Solovay ρ-random.

The converse direction can easily be seen by using the Kraft-Chaitin Theorem (see for instance Li and Vitányi's book [6]). Assume A is covered by a Solovay ρ-test W. Then $\sum_{x \in W} \rho(x)$ is finite and one can consider $\{\langle x, -\log \rho(x)\rangle : x \in W\}$ as a Kraft-Chaitin axiom set. Thus there is a constant c with $K(x) < c - \log \rho(x)$. Since A is covered by W, there are infinitely many n with $K(A \restriction_n) < c - \log \rho(A \restriction_n)$. $\qquad\square$

4. A Hierarchy of Randomness Tests

Since a Martin-Löf ρ-test (V_n) can be transformed into a Solovay ρ-test W covering all the sets covered by (V_n) by letting

$$W = \bigcup_{n \in \mathbb{N}} V_n,$$

one obtains that every Martin-Löf ρ-nullset is contained in a Solovay ρ-nullset. Thus every Solovay ρ-random set is also Martin-Löf ρ-random.

Proposition 4.1. *For every premeasure ρ, every Martin-Löf ρ-nullset is covered by a Solovay ρ-test.*

In the case that $\rho = \lambda$, the converse direction of above proposition is also true. The next result shows that this is not the case when ρ is an unbounded premeasure.

Theorem 4.2. *For every unbounded premeasure ρ there exists a sequence A which is Solovay ρ-null but not Martin-Löf ρ-null.*

Proof. Assume ρ is an unbounded (p, q)-premeasure. Let $F(x) = K(x) + \log \rho(x)$. Note that for all $x \in \{0,1\}^*$ and $i \in \{0,1\}$, $(q-p)\rho(x) \le \rho(xi) \le p\rho(x)$ and $q - p > 0$. Thus there is a constant bounding the absolute value of the difference $\rho(xi) - \rho(x)$ for all $x \in \{0,1\}^*$ and $i \in \{0,1\}$. The same applies to K and F. Using Lemma 4.3 below, there is a constant c and a sequence A which satisfies for all n the following three properties:

- if $F(A \restriction_{cn}) \ge 0$ then $F(A \restriction_{cn+c}) \le F(A \restriction_{cn}) - 1$;
- if $F(A \restriction_{cn}) < 0$ then $F(A \restriction_{cn+c}) \ge F(A \restriction_{c_n}) + 1$.

It follows that the there is a constant c' with $-c' \le F(A \restriction_m) \le c'$ for all m. So A is weakly but not strongly Chaitin ρ-random. It follows from the results in the previous section that A is Solovay ρ-null and Martin-Löf ρ-random. $\qquad\square$

Lemma 4.3. *Let ρ be an unbounded premeasure. There exists a constant c with the following property: For all strings x there exist strings y, z of length c such that*

$$K(xy) + \log \rho(xy) \geq K(x) + \log \rho(x) + 1;$$
$$K(xz) + \log \rho(xz) \leq K(x) + \log \rho(x) - 1.$$

Proof. Again assume ρ is an unbounded (p, q)-premeasure. We first outline an algorithm which treats c as an input. The result is obtained by fixing c to a sufficiently large value.

Given x, c, the strings $y, z \in \{0, 1\}^c$ are determined as follows. Order the 2^c strings in $\{0, 1\}^c$ according to the size of $\rho(xu)$, $u \in \{0, 1\}^c$, in descending order and let, for $b \in \{1, 2, \ldots, 2^c\}$, $g(x, c, b)$ be the b-th string in this ordering prefixed by x, so $\rho(g(x, c, 1)) \geq \rho(g(x, c, 2)) \geq \ldots \geq \rho(g(x, c, 2^c))$. Note that

$$q^c \rho(x) \leq \sum_{u \in \{0,1\}^c} \rho(xu)$$
$$\leq \rho(g(x, c, 1)) + \sum_{a \in \{0,1,\ldots,c-1\}} 2^a \cdot \rho(g(x, c, 2^a + 1))$$

and $\rho(g(x, c, 1)) \leq p^c \rho(x)$. Now choose $a \in \{0, 1, \ldots, c-1\}$ such that the $2^a \cdot \rho(g(x, c, 2^a + 1))$ is within a sufficiently small constant of the maximal value of all $2^a \cdot \rho(g(x, c, 2^a + 1))$, $a \in \{0, 1, \ldots, c-1\}$. This can be done effectively, since ρ is recursive. Furthermore, choose y such that $xy = g(x, c, b)$ for the $b \in \{1, 2, \ldots, 2^a\}$ for which $K(g(x, c, b))$ is maximal. Furthermore, let $z = 0^c$, that is, z consists of c zeroes. Now, the following statements hold for all sufficiently large c, all x and the a, y, z chosen for them as above.

- $K(xy) \geq K(x) + a - K(c) - K(a) - \log(c) \geq K(x) + a - 4\log(c)$ where c has to be sufficiently large so that $\log(c)$ absorbs the constants involved;
- $\rho(xy) \geq q^c \rho(x) 2^{-a} \cdot c^{-2}$ where c is sufficiently large that $q^c - p^c \geq q^c/c$;
- $K(xy) + \log \rho(xy) \geq K(x) + \log \rho(x) + \log(q)c - 6\log(c)$ where $\log(q) > 0$ as ρ is unbounded;
- $K(xz) \leq K(x) + K(c) + \log(c) \leq K(x) + 3\log(c)$ as z can be computed from x and c and $\log(c)$ absorbs the involved constants if c is sufficiently large;
- $\log \rho(xz) \leq \log \rho(x) + c\log(p)$ where $\log(p) < 0$;
- $K(xz) + \log \rho(xz) \leq K(x) + \log \rho(x) + c\log(p) + 3\log(c)$.

Now one can choose a constant c sufficiently large so that all of the above hold, that $\log(q)c - 6\log(c) > 1$, and that $c\log(p) + 3\log(c) < -1$. With c

thus chosen, one can find for every x some $y, z \in \{0,1\}^c$ with the desired properties. $\square$

Though not equivalent, Solovay ρ-tests and Martin-Löf ρ-tests induce the same notion of effective dimension. They distinguish between sequences on a rather "fine" level of complexity oscillations. For details on effective dimension see, for example, Reimann [10].

Proposition 4.4. *Let ρ be a premeasure. If $X \in \{0,1\}^\omega$ is covered by an effective Solovay ρ-test, then X is Martin-Löf ρ'-null for any premeasure ρ' such that $\lim_n \rho'(X \restriction_n)/\rho(X \restriction_n) = 0$.*

Proof. Assume that C is an effective Solovay ρ-cover for X and let ρ' be a premeasure with $\lim_n \rho'(X \restriction_n)/\rho(X \restriction_n)0$. Deleting a finite number of strings from C does not change the covering properties of a Solovay test, so we may assume that $\sum_{w \in C} \rho(w) \leq 1$. Given $n \geq 0$, we define a r.e. set C_n by enumerating only those elements of C for which

$$\frac{\rho'(w)}{\rho(w)} \leq 2^{-n}.$$

Then X is covered by C_n and it holds that

$$\sum_{w \in C_n} \rho(w) = \sum_{w \in C_n} \frac{\rho'(w)}{\rho(w)}\rho(w) \leq 2^{-n} \sum_{w \in C_n} \rho(w) \leq 2^{-n}.$$

Hence, (C_n) is a Martin-Löf ρ'-test for X. $\square$

On the other hand, Solovay tests can always be covered by strong Martin-Löf tests.

Theorem 4.5. *For every premeasure ρ, every Solovay ρ-nullset is covered by a strong Martin-Löf ρ-test.*

Proof. Let W be a Solovay ρ-test. Again, we may assume that $\rho(W) < 1$. Assume further that for $A \in \{0,1\}^\omega$ there are infinitely many $w \in W$ such that $w \sqsubset A$. We distinguish two cases:

Case 1: It holds that

$$(\forall n)(\exists x \in W)[x \sqsubset A \ \wedge \ \rho(W_x^+) \geq \rho(x)2^n]. \tag{4}$$

Define

$$V_n = \{x \in W : \rho(W_x^+) \geq \rho(x)2^n\}.$$

We claim that $(V_n)_{n\in\mathbb{N}}$ is a strong Martin-Löf ρ-test that covers A. Obviously, A is covered by each V_n, due to the assumption above. Furthermore, if $V \subseteq V_n$ is prefix-free, then

$$2^n \rho(V) = 2^n \sum_{x\in V} \rho(x) \le \sum_{x\in V} \rho(W_x^+) \le \sum_{w\in W} \rho(w) < 1,$$

so $\rho(V) < 2^{-n}$.

Case 2: We have

$$(\exists n)(\forall x \in W)[x \sqsubset A \;\Rightarrow\; \rho(W_x^+) < \rho(x)2^n]. \tag{5}$$

By taking r to be a rational number a tiny bit below the value $\limsup_{x\sqsubset A} \rho(W_x^+)/\rho(x)$, we can strengthen this to

$$(\exists^\infty x \in W)[x \sqsubset A \;\wedge\; \rho(W_x^+) > r\rho(x)]$$
$$\text{and } (\forall^\infty x)[x \sqsubset A \;\Rightarrow\; \rho(W_x^+) < (r+\tfrac{1}{2})\rho(x)].$$

By removing finitely many elements from W we can even assume

$$(\forall x \sqsubset A)[\rho(W_x^+) < (r+\tfrac{1}{2})\rho(x)].$$

Now let $x_0, x_1, \ldots$ be an enumeration of W and construct inductively sets $T_n \subseteq \{0,1\}^\star$ starting with $T_0 = \{x_0\}$. If $Q = T_n \cup \{x_{n+1}\}$ satisfies

$$(\forall y \in Q)[\rho(Q_y^+) < (r+\tfrac{1}{2})\rho(y)]$$

then let $T_{n+1} = T_n \cup \{x_n\}$ else let $T_{n+1} = T_n$. It is easy to see that the resulting union $T = \bigcup T_n$ satisfies

$$(\forall y \in T)[\rho(T_y^+) < (r+\tfrac{1}{2})\rho(y)].$$

Furthermore, every prefix of A in W is also in T. To see this, assume that x_n is a prefix of A. If $n = 0$ then $x_n \in T$ anyway. If $n > 0$ then consider $Q = T_{n-1} \cup \{x_n\}$ and any $y \in Q$. If $y \not\sqsubseteq x_n$ then $\rho(Q_y^+) = \rho((Q - \{x_n\})_y^+)$ and y does prevent x_n from being added to T. If $y \sqsubseteq x_n$ then $y \sqsubset A$ and $\rho(Q_y^+) \le \rho(W_y^+) \le (r+\tfrac{1}{2})\rho(y)$ and again y does prevent x_n from being enumerated into T. Thus $x_n \in T_n$ and $x_n \in T$. So all prefixes of A in W are also in T and T covers A. The set T is obviously enumerable.

From T one enumerates $S = \{x \in T : \rho(W_x^+) > r\rho(x)\}$. The set S contains infinitely many prefixes of A. Furthermore, for every $x \in S$, every prefix-free subset Q of S_x^+ and every $y \in Q$, the following inequalities hold.

$$\rho(W_y^+) > r\rho(y);$$
$$\rho(W_x^+) \geq \rho(Q) + \sum_{y \in Q} \rho(W_y^+) > (1+r)\rho(Q);$$
$$\rho(W_x^+) \leq (r + \tfrac{1}{2})\rho(x);$$
$$\rho(Q) \leq \tfrac{1+2r}{2+2r}\rho(x).$$

Now define inductively

$$Q_m = \{x \in S : \forall y \in S\,(y \sqsubset x \Rightarrow y \in \cup_{k<m} Q_k)\}.$$

So Q_m is the set of all $x \in S$ such that the cardinality of $\{y \in S : y \sqsubset x\}$ is exactly m. Now $\rho(Q_{m+1}) \leq \tfrac{2r+1}{2r+2}\rho(Q_m)$ for all m. Now one can compute numbers $m_0, m_1, m_2, \ldots$ such that for all n, $(\tfrac{2r+1}{2r+2})^{m_n} < 2^{-n}$ and thus $\rho(Q_{m_n}) < 2^{-n}$. Now let

$$V_n = \{x \in S : \exists y \in Q_{m_n}\,(y \sqsubset x)\}\, S - Q_0 - Q_1 - Q_2 - \ldots - Q_{m_n}.$$

The sets V_n are uniformly enumerable. Furthermore, if Q is a prefix-free subset of V_n, then there is for every $x \in Q$ some $y \in Q_{m_n}$ with $y \sqsubset x$. Thus Q is the union of prefix free sets Q_y^+ with $y \in Q_{m_n}$. By choice of S and V_n, $\rho(Q_y^+) < \rho(y)$ and $\rho(Q) < \rho(Q_{m_n}) < 2^{-n}$. This completes the proof. $\qquad\square$

Corollary 4.6. *For $s \geq 0$ and ρ_s given as $\rho_s(x) = 2^{-|x|s}$, every Solovay ρ_s-nullset is covered by a strong Martin-Löf ρ_s-test.*

For unbounded, length-invariant premeasures, strong Martin-Löf tests are strictly more powerful then Solovay tests.

Theorem 4.7. *For any unbounded, length-invariant premeasure ρ there is a set A which is covered by a strong Martin-Löf ρ-test but not by a Solovay ρ-test.*

Proof. Let $I_0, I_1, \ldots$ be a recursive sequence of disjoint intervals such that

$$(\forall i)\,(\exists j \in I_i)\,(\forall k)\,[|\{x \in \{0,1\}^j : K(x) \leq k\}| < 2^{k-2i}].$$

Using Lemma 4.3, we can construct a sequence A such that, up to a constant c,

$$(\forall i)\,(\forall j \in I_i)\,[K(A \restriction_j) = -\log \rho(A \restriction_j) + i].$$

By construction A is Solovay ρ-random. Let

$$W_i = \{x \in \{0,1\}^{\max I_i + c} : \forall y \sqsubseteq x\,(K(y) \leq -\log \rho(y) + i + 2c)\}.$$

Obviously, every W_i is finite and covers A. Furthermore, the W_i are uniformly enumerable. We show how to modify (W_i) to obtain a strong Martin-Löf ρ-test that covers A.

We initialize with $V_i = \emptyset$. Every time some w is enumerated into W_i, we check whether there exists a $v \in V_i$ such that $v \sqsubseteq w$. If so, we let V_i unchanged. Otherwise we pick the longest $v \sqsubseteq w$ such that for all prefix-free subsets $Q \subseteq V_i \cup \{v\}$ and every $u \sqsubseteq w$ it holds that

$$\rho(Q_u^+) < \rho(u).$$

Enumerate v into V_i. It is clear that the V_i still cover A. It remains to show that for every prefix-free subset Q of V_i, $\rho(Q) \leq 2^{-i}$.

Let $j \in I_{i+c}$ be such that for all k,

$$|\{x \in \{0,1\}^j : K(x) \leq k\}| < 2^{k-2i}.$$

Consider the cover $U = \{u \in \{0,1\}^j : \exists w \in W_i(u \sqsubseteq w)\}$. ρ is length-invariant, so let r be the unique value of all $\rho(u)$, $u \in \{0,1\}^j$. It follows by the choice of j that

$$\rho(U) \leq r 2^{-\log(r)+i+2c-2(i+c)} = 2^{-i}.$$

We claim that for any prefix-free set $Q \subseteq V_i$, $\rho(Q) \leq \rho(U)$. Assume this is not the case for some prefix-free $Q \subseteq V_i$. Let $g : Q \to \{1,\ldots,n\}$ and $h : U \to \{1,\ldots,n\}$ such that $g(x) = h(u)$ if and only if x and u are compatible, i.e. $x \sqsubseteq u$ or $u \sqsubseteq x$. Since $\rho(Q) > \rho(U)$, there must be an $m \leq n$ such that, for $Q_m g^{-1}(\{m\})$ and $U_m = h^{-1}(\{m\})$

$$\rho(Q_m) > \rho(U_m).$$

Note that at least one of Q_m, U_m must contain at most one element. If $U_m = \{u\}$ and for all $x \in Q_m$, $x \sqsupseteq u$, then $\rho(u) < \rho(Q_m) \leq \rho(Q_u^+)$, so one of the elements of Q_m could not have been enumerated in the construction of V_i. If, on the other hand, $Q_m = \{x\}$ and for all $u \in U_m$, $u \sqsupseteq x$, then this contradicts the choice of x as the longest string possible. $\qquad\square$

References

1. Jin-Yi Cai and Juris Hartmanis. On Hausdorff and topological dimensions of the Kolmogorov complexity of the real line. *Journal of Computer and System Sciences*, 49(3):605–619, 1994.
2. Cristian Calude, Ludwig Staiger and Sebastiaan A. Terwijn. On partial randomness. *Annals of Pure and Applied Logic*, to appear.
3. Gregory J. Chaitin. A theory of program size formally identical to information theory. *Journal of the ACM*, 22:329–340, 1975.

4. Gregory J. Chaitin. Information-theoretic characterizations of recursive infinite strings. *Theoretical Computer Science*, 2(1):45–48, 1976.
5. Rodney G. Downey and Denis R. Hirschfeldt. *Algorithmic randomness and complexity*. Springer, to appear.
6. Ming Li and Paul Vitányi. *An Introduction to Kolmogorov Complexity and its Applications*. Second Edition. Springer, 1997.
7. Jack H. Lutz. Gales and the constructive dimension of individual sequences. In *Automata, languages and programming* (Geneva, 2000), volume 1853 of Lecture Notes in Computer Science, pages 902–913. Springer, Berlin, 2000.
8. Per Martin-Löf. The definition of random sequences. *Information and Control*, 9:602–619, 1966.
9. Elvira Mayordomo. A Kolmogorov complexity characterization of constructive Hausdorff dimension. *Information Processing Letters*, 84:1–3, 2002.
10. Jan Reimann. *Computability and fractal dimension*. Doctoral Dissertation, Universität Heidelberg, 2004.
11. Jan Reimann and Frank Stephan. Effective Hausdorff dimension. In *Logic Colloquium '01*, volume 20 of *Lecture Notes in Logic*, pages 369–385. Association of Symbolic Logic, Urbana, IL, 2005.
12. Piergiorgio Odifreddi. *Classical Recursion Theory*. North-Holland, 1989.
13. Claude Ambrose Rogers. *Hausdorff Measures*. Cambridge University Press, 1970.
14. Boris Y. Ryabko. Coding of combinatorial sources and Hausdorff dimension. *Doklady Akademii Nauk SSSR*, 277(5):1066–1070, 1984.
15. Boris Y. Ryabko. Noise-free coding of combinatorial sources, Hausdorff dimension and Kolmogorov complexity. *Problemy Peredachi Informatsii*, 22(3):16–26, 1986.
16. Claus-Peter Schnorr. *Zufälligkeit und Wahrscheinlichkeit*. Lecture Notes in Mathematics 218, Springer, 1971.
17. Robert M. Solovay. *Draft of a paper on Chaitin's work*. Manuscript, IBM Thomas J. Watson Research Center, New York, May 1975, 215 pages.
18. Robert I. Soare. *Recursively Enumerable Sets and Degrees*. Springer, 1987.
19. Ludwig Staiger. Kolmogorov complexity and Hausdorff dimension. *Information and Computation*, 103(2):159–194, 1993.
20. Kohtaro Tadaki. A generalization of Chaitin's halting probability Ω and halting self-similar sets. *Hokkaido Mathematical Journal*, 31(1):219–253, 2002.
21. Alexander K. Zvonkin and Leonid A. Levin. The complexity of finite objects and the basing of the concepts of information and randomness on the theory of algorithms. *Uspekhi Matematicheskikh Nauk*, 25(6):85–127, 1970.

INTRANSITIVE LINEAR TEMPORAL LOGIC BASED ON INTEGER NUMBERS, DECIDABILITY, ADMISSIBLE LOGICAL CONSECUTIONS

V. V. RYBAKOV

Manchester Metropolitan University,
John Dalton Building, Chester Street,
Manchester M1 5GD, U.K.
E-mail: V.Rybakov@mmu.ac.uk

We investigate the intransitive linear temporal logic $\mathcal{L}_{t,y}(\mathbf{Z})$ generated by the frame of all integer numbers $\mathbf{Z}$ with alternative accessibility relations *Previos* and *Next* (intended to model *tomorrow* and *yesterday*). The prime problem we are dealing with is description of logical consequence in $\mathcal{L}_{t,y}(\mathbf{Z})$ via admissible logical consecutions (inference rules). We use a reduction of logical consecutions and formulas to an equivalent simple consecutions consisting of formulas of temporal degree 1. Using it, as an initial technical instrument, we find special necessary and sufficient conditions for a consecution to be not admissible in $\mathcal{L}_{t,y}(\mathbf{Z})$. These conditions lead to an algorithm which recognizes admissible in $\mathcal{L}_{t,y}(\mathbf{Z})$ consecutions through model checking at Kripke structures of size linear in the reduced normal forms of the starting consecutions.

Keywords: temporal logic, linear temporal logic, logical consequence, inference, inference rules, consecutions, admissible rules

1. Introduction

In present time, temporal logic is an active issue, it has found numerous applications in Artificial Intelligence (AI) and Computer Science (CS). Tense Logic was introduced by Arthur Prior (1957, 1967, cf. [31, 32]) as a result of an interest in the relationship between tense and modality attributed to the Megarian philosopher Diodorus Cronus (ca. 340–280 B.C.). Tense Logic is obtained by joining tense operators to an existing logic; above this was tacitly assumed to be the classical Propositional Calculus. Initial stage in the research concerned with mathematical symbolic modeling of time (cf. [9]) and finding approach to model temporal logic by using varieties of temporal algebras and Kripke/Hintikka models (cf. S.K. Thomason [45]).

Nowadays there are various temporal logics circumscribing multifarious properties of the flow of possible events. They have numerous applications in AI and CS (cf. Manna and Pnueli [23, 24], Pnueli [29, 30], Clark E. et al., [5], Goldblatt [13]) and philosophy (cf. van Benthem [46]).

Among applications to CS and AI, linear temporal logic LTL has been shown to be an effective tool. Linear temporal logic can be viewed as modal logic with linear alternative relations (cf. K. Segerberg [44]). Any temporal logic is a multi-modal logic (cf. Gabbay et. al., [11]) with two modal operations for representation of future and past. Since natural modeling of sequences of steps in computational processes, linear temporal logic LTL has been quite successful in dealing with applications to systems specifications and verification (cf. [29, 20]), with model checking (cf. [2, 5]). Pnueli [29] introduced special temporal logics for reasoning about concurrent programs. Those logics provide powerful methods for specifying and verifying properties of reactive systems. Temporal logic is a natural logic for hardware verification (cf. Cyrluk, Natendran [6]), it has various applications in formal specifications and verifications of concurrent and distributed systems (cf. Pnueli [29, 30]). Temporal logic has numerous applications to safety, liveness and fairness (cf. Manna, Pnueli [23], Emerson [7]). A good reference point is the book "Advances in Temporal Logic", edited by Barringer, Fisher, Gabbay and Gough [1], which contains many techniques focused on applications of temporal logic to AI and CS. Model checking (cf. [22]) is another active area which uses temporal logic. In general terms, model checking is an approach to the automatic verification of finite state systems, often it involves system specifications expressed in temporal logic, e.g. linear temporal logic or branching temporal logic [22, 3]. Temporal operations themselves can also be revised from distinct viewpoints, for instance, two-dimensional temporal logics are described in [27].

It would be not an exaggeration to say that one of prime questions concerning temporal logics is the question about decidability (cf. [19]). To cite earliest results, axiomatic systems and decision procedures for some particular linear temporal logics were constructed in (Segerberg, [44]).

In this paper we will study the question about decidability of linear intransitive temporal logics, but in more general form, — decidability w.r.t. admissible logical consecutions. Any formula A can be viewed as an unconditional statement; if A is provable, then A must be true in all situations. But we can extend the language of temporal logic by considering conditional statements of the form $A_1, \ldots, A_n/B$, which have the meaning *if all* $A_1, \ldots, A_n$ *hold, then* B *also holds.* We often want to know *what will follow* from given statements (assumptions) $A_1, \ldots, A_n$ *rather then what is always true.* For this we need to know *what are correct logical consequences* from given assumptions. Nowadays it is rather widely accepted that neither implication nor any kind of deduction theorems express adequately the notion of logical consequence. The problem: what *logical consequence* means, is crucial; this question does not have an evident and definite answer.

We will study logical consequence in terms of admissible and valid consecutions (or, synonymously, inference rules). Usage of inference rules gives opportunity to describe subtle properties of models which are problematic to be expressed by formulas. A good example is the Gabbay's *irreflexivity rule* (cf. [10])

$$(ir) := \frac{\neg(p \to \Diamond p) \to \varphi}{\varphi}$$

(where p does not occur in the formula φ). This rule is actually saying that any element of a model, where φ is not valid, should be irreflexive; it was implemented in [10] for the proof of the completeness theorem. Axiomatizations of various logics generated by classes of multi-modal frames by Sahlqvist formulas and special derivation rules were suggested by Venema [47].

To construct axiomatic systems we have to know which inference rules can be applied. The admissible rules form a greatest class of rules which can be implemented (which are compatible with the set of valid formulas), therefore it seems important to identify admissible rules. They, potentially, can be used in CS for construction of new provers and descriptions of structure of modal/temporal frames.

Admissible consecutions have been investigated reasonably deep for numerous modal and superintuitionistic logics. The history could be dated since Harvey Friedman's question (1975, [8]) about existence of algorithms which could distinguish rules admissible in the intuitionistic propositional logic IPC, and since Harrop's examples [15] of rules admissible but not derivable in standard Hilbert-style axiomatic systems for IPC. In the middle of 70-th, G. Mints [28] found strong sufficient conditions for derivability in IPC admissible rules in special form. H. Friedman's question was answered affirmatively by V. Rybakov (1984, [33]) and later S. Ghilardi [12] found another algorithm recognizing admissibility.

A. Kuznetsov (1973) raised the question, whether IPC has a finite basis for admissible rules. First positive results towards this question were obtained by A. Citkin [4] who found a basis for all admissible in IPC quasi-characteristic rules, but the question, in total, was solved by Rybakov [35] in negative: it was shown that IPC even does not have bases in finitely many variables. Later on, Iemhoff [17] found an explicit basis for rules admissible in IPC and constructed a characterization of IPC by means of these basis-rules [16]. Earlier, an implicit basis — a merely recursive infinite one— was proposed in Rybakov, Terziler, Rimazki [40]. An explicit basis for rules admissible in the modal logic $S4$ was found in Rybakov [41].

It was discovered that inference rules in a special generalized form, — inference rules with meta-variables, — allow to describe unifiable formulas. Rybakov in [36] found an algorithm which, for any inference rule with meta-variables determines whether this rule is admissible in IPC. This algorithm can check unification in IPC and verify solvability of logical equations. Later S. Ghilardi [12] discovered a new approach to unification in IPC via projective formulas. Similar results for inference rules with meta-variables admissible in modal transitive logics are presented, in particular, in the book (Rybakov, [38]). A logic $\mathcal{L}$ is structurally complete if the classes of all rules admissible in $\mathcal{L}$ and all rules derivable in $\mathcal{L}$ coincide. Using ideas of W. Dziobiak, a complete characterization of modal hereditarily structurally complete logics extending $K4$ was found in Rybakov [37]. It was established that there is a big difference between refuting of formulas and refuting of inference rules. For instance, in Rybakov, Kiyatkin, Oner [39] it was shown that the great majority of the logics with the finite model property (fmp) extending the logic IPC or the modal logic S4 do not have the fmp w.r.t. inference rules. The technique developed for description of rules admissible in modal logics works also, for instance, for specific rules of common knowledge logics as well (cf. [42]).

For the case of temporal logics, relatively few results concerning admissible inference rules are known — in these logics it is more difficult to construct decision procedures for admissibility. The same could be related to intransitive modal logics. All techniques used before have as a prerequisite the transitivity of the accessibility relations. Recently in [43] the intransitive logic generated by finite intervals of integer numbers was investigated, and an algorithm checking admissibility of inference rules was found.

In this paper we will extend results from [43] to intransitive linear temporal logic $\mathcal{L}_{t,y}(\mathbf{Z})$ generated by whole frame of integer numbers. The main problem we focused is to prove that $\mathcal{L}_{t,y}(\mathbf{Z})$ is decidable w.r.t. admissible consecutions. We start by Small Models Theorem for $\mathcal{L}_{t,y}(\mathbf{Z})$, which gives us a linear bound on the size of models refuting non-theorems, and decidability of $\mathcal{L}_{t,y}(\mathbf{Z})$ itself, it is about immediate observation. Then we turn to the main question of our research: study of logical consequence in $\mathcal{L}_{t,y}(\mathbf{Z})$. We consider valid and admissible consecutions, demonstrate their distinction and concentrate on admissible consecutions since all valid consecutions are admissible but not vise versa.

Initially we describe admissible consecutions in a semantic manner via consecutions valid in special temporal, linear and infinite Kripke/Hinttikka models. Then we prepare some technique, show that any temporal consecution has a normal reduced form, which is given in terms of temporal formulas of temporal/modal degree at most 1. Finally, we construct an algorithm recognizing consecutions admissible in $\mathcal{L}_{t,y}(\mathbf{Z})$. An estimation of the complexity of obtained explicit algorithms is given. In the sequel we prefer the term *consecution* rather than *inference rule* in order to emphasize that we are interested in the direct logical consequence, — to know immediately what follows from assumptions.

2. General Definitions, Notation, Preliminary Facts

The *language of temporal propositional logic* can be viewed as the language of bi-modal logic. It consists of the language of the standard propositional Boolean logic and two additional temporal operations: $\Box_F$ and $\Box_P$. The formation rules for temporal formulas are as usual for propositional logic and, for any formula $\mathcal{A}$, $\Box_F\mathcal{A}$ and $\Box_P\mathcal{A}$ are wffs. $\Box_P\mathcal{A}$ has meaning: $\mathcal{A}$ *has always been true*, and $\Box_F\mathcal{A}$ can be read $\mathcal{A}$ *will always be true*.

A temporal Kripke frame is a tuple $\mathcal{F} := \langle F, R_F, R_F^{-1} \rangle$, where $F \neq \emptyset$ is a set of possible worlds, $R_F, \subseteq F \times F$ and $R_F^{-1} := \{(a,b) \mid (b,a) \in R_F\}$ are accessibility relations in F imitating flow of time (R_F — towards future, R_F^{-1} — converse relation). Writing $a \in \mathcal{F}$ means $a \in |\mathcal{F}|$, i.e. a is a world from the base set of $\mathcal{F}$.

Suppose we have a valuation V of all letters from a formula $\mathcal{A}$ in a temporal frame $\mathcal{F}$, i.e., for any propositional letter p occurring in $\mathcal{A}$, $V(p) \subseteq F$. The valuation V can be extended to all subformulas of $\mathcal{A}$ and $\mathcal{A}$ itself in standard way: (i) for any $a \in \mathcal{F}$, a propositional letter p is true at a (abbreviation $(\mathcal{F},a) \Vdash_V p$, or just $a \Vdash_V p$, if $\mathcal{F}$ is known from a context) if $a \in V(p)$;

(ii) steps for Boolean operations are standard and (iii):

$$a \Vdash_V \Box_F \mathcal{B} \Leftrightarrow \forall c \in \mathcal{F}((aR_Fc) \Rightarrow (c \Vdash_V \mathcal{B}));$$

$$a \Vdash_V \Box_P \mathcal{B} \Leftrightarrow \forall c \in \mathcal{F}((aR_F^{-1}c) \Rightarrow (c \Vdash_V \mathcal{B})).$$

In the sequel instead of $aR_F^{-1}c$ we will just write cR_Fa. If $a \Vdash_V \mathcal{B}$ holds we say $\mathcal{B}$ is *true at a w.r.t. V*. A temporal model $\mathcal{M} := \langle \mathcal{F}, V \rangle$ is a frame with a given valuation V. A formula $\mathcal{A}$ is *true in $\mathcal{M}$* (abbreviation $\mathcal{M} \Vdash \mathcal{A}$), if $\mathcal{A}$ is true at each world of $\mathcal{M}$ w.r.t. the valuation of $\mathcal{M}$. For a frame $\mathcal{F}$, a formula $\mathcal{A}$ is *valid* in $\mathcal{F}$ (abbreviation $\mathcal{F} \Vdash \mathcal{A}$) if $\mathcal{A}$ is true in any model based upon $\mathcal{F}$.

For a class of frames $\mathcal{K}$, the logic $L(\mathcal{K})$ generated by $\mathcal{K}$ is the set of all formulas which are valid in any frame from $\mathcal{K}$. This case, we say $L(\mathcal{K})$ is generated by $\mathcal{K}$. Very often temporal propositional logics are generated by certain classes of transitive frames $\mathcal{F}$ (i.e. where R_F is transitive). We will investigate a logic based on heavily intransitive frames.

To represent the flow of time, we use special Kripke/Hintikka models based at the frame $\mathbf{Z} := \langle \mathbf{Z}, Next, Prev \rangle$ of integer numbers. The relation $Next$ is the binary relation *next natural number*, i.e. $Next(k, m)$ is true iff $m = k + 1$, and $Prev$ is the binary relation *previous natural number*, i.e. $Prev(k, m)$ is true iff $m = k - 1$. We can also understand $Next$ as the one-to-one function where $Next(n) := n + 1$, the same regarding $Prev$, with $Prev(n) := n - 1$. In the sequel, for $n, m \in \mathbf{Z}, n \leq m$, $[n, m]$ is the interval of all integer numbers which are situated between n and m.

The frame $\mathbf{Z}$ is a natural model for flow of time. Any integer number n can be viewed as a time-point, $n + 1$ is the next time-point, *tomorrow*, for n, and $n - 1$ is the previous time-point, *yesterday*, w.r.t. n. From computational viewpoint, $\mathbf{Z}$ can be viewed as a particular unbounded computation (run) with possibility to analyse intermediate results of computation and backtraking. In the sequel we will name temporal formulas, temporal frames and models just formulas, frames and models for short.

Definition 1. The intransitive temporal logic $\mathcal{L}_{t,y}(\mathbf{Z})$ is the set of all formulas which are valid in the frame $\mathbf{Z}$, i.e. $\mathcal{L}_{t,y}(\mathbf{Z}) : \mathcal{L}(\mathbf{Z})$.

We call the formulas contained in $\mathcal{L}_{t,y}(\mathbf{Z})$ by theorems of $\mathcal{L}_{t,y}(\mathbf{Z})$, so, we identify the logic $\mathcal{L}_{t,y}(\mathbf{Z})$ with the set of its theorems, in particular, the notation $\mathcal{A} \in \mathcal{L}_{t,y}(\mathbf{Z})$ means $\mathcal{A}$ *is a theorem of* $\mathcal{L}_{t,y}(\mathbf{Z})$. As always for any new logic, the one of most important questions is decidabilty: *how to efficiently determine whether a formula is a theorem of* $\mathcal{L}_{t,y}(\mathbf{Z})$. This question is easy to answer using standard reasoning based on temporal degrees of formulas.

Recall that the *temporal degree* — $tdeg(\mathcal{A})$ — of a formula $\mathcal{A}$ is the maximal number of nested occurrences in $\mathcal{A}$ of the temporal operations: $\Box_F$ and $\Box_p$. To formulate the result we need some auxiliary finite frames. For any $n, m \in \mathbf{Z}$, where $n \leq m$, $\mathcal{F}[n, m]$ is the frame based on the interval $[n, m]$ with the accessibility relations $Next$ and $Prev$ taken from $\mathbf{Z}$.

For a Kripke frame F, a formula $\mathcal{A}$, $a \in \mathcal{F}$ and a valuation V, $(F, a) \Vdash_V \mathcal{A}$ stands to say that $\mathcal{A}$ is valid at a in F w.r.t. valuation V; $(F, a) \Vdash \mathcal{A}$ denotes that $\mathcal{A}$ is valid at a in F w.r.t. any valuation.

Theorem 2. *Small Models Theorem. For any formula $\mathcal{A}$,*

$$\mathcal{A} \notin \mathcal{L}_{t,y}(\mathbf{Z}) \Leftrightarrow (\mathcal{F}[-m-1, m+1], 0) \not\Vdash \mathcal{A},$$

where m is the temporal degree of $\mathcal{A}$, i.e. m is linear in the length of $\mathcal{A}$.

Proof is a standard inductive routine reasoning based on temporal degrees of subformulas of the formula $\mathcal{A}$. $\qquad\square$

This statement allows us to efficiently recognize theorems of $\mathcal{L}_{t,y}(\mathbf{Z})$ by models of size linear in the length of the testifying formulas. In particular we get

Corollary 3. *The temporal logic $\mathcal{L}_{t,y}(\mathbf{Z})$ is decidable.*

3. Admissibility of Logical Consecutions, Preliminary Discussion

The main question of our investigation is the study of logical consequence relation in $\mathcal{L}_{t,y}(\mathbf{Z})$. Assume we are given a collection of certain formulas $\varphi_1(x_1,\ldots,x_m)$, $\ldots$, $\varphi_n(x_1,\ldots,x_m)$ and a formula $\psi(x_1,\ldots,x_m)$ constructed from letters x_1, $\ldots$, x_m. We would like to analyze what means that ψ *is a logical consequence of formulas* $\varphi_1,\ldots,\varphi_n$. A *consecution*, (or, synonymously, — a rule) **c** is an expression

$$\mathbf{c} := \frac{\varphi_1(x_1,\ldots,x_n),\ldots,\varphi_m(x_1,\ldots,x_n)}{\psi(x_1,\ldots,x_n)},$$

where $\varphi_1(x_1,\ldots,x_n),\ldots,\varphi_m(x_1,\ldots,x_n)$ and $\psi(x_1,\ldots,x_n)$ are some formulas. Consecutions are supposed to describe the logical consequences, an informal meaning of a consecution is that the conclusion logically follows from the premises. The questions what *logically follows* means is crucial and has no evident and unique answer.

Given a frame $\mathcal{F}$ and a valuation V in $\mathcal{F}$ of all variables from a consecution $\mathbf{c} := \varphi_1,\ldots\varphi_n/\psi$, we say $\mathbf{c}$ is valid in the model $\langle \mathcal{F}, V \rangle$ (notation $\langle \mathcal{F}, V \rangle \Vdash \mathbf{c}$, or $\mathcal{F} \Vdash_V \mathbf{c}$) if

$$\left(\langle \mathcal{F}, V \rangle \Vdash \bigwedge_{1 \leq i \leq m} \varphi_i \right) \Rightarrow \left(\langle \mathcal{F}, V \rangle \Vdash \psi \right).$$

Otherwise we say $\mathbf{c}$ is *refuted, or refuted by V*, and write $\mathcal{F} \not\Vdash_V \mathbf{c}$. A consecution $\mathbf{c}$ is *valid* in a frame $\mathcal{F}$ (notation $\mathcal{F} \Vdash \mathbf{c}$) if, for any valuation V, $\mathcal{F} \Vdash_V \mathbf{c}$. A consecution $\mathbf{c} := \varphi_1,\ldots\varphi_n/\psi$ is *valid* for a logic $\mathcal{L}(\mathcal{K})$ generated by a class of frames $\mathcal{K}$ if $\forall \mathcal{F} \in \mathcal{K}(\mathcal{F} \Vdash \mathbf{c})$.

It is relevant to note here that our definition of valid consecutions equivalent to the notion of valid modal sequent from [18], where a theory of sequent-axiomatic classes is developed. Also the notion of valid consecutions can be reduced to validity of formulas in the extension of the language with universal modality (cf. Goranko and Passy [14]). Based on these, some relevant results concerning validity of consecutions can be derived.

It is easy to agree that valid consecutions correctly describe logical consequence. In particular, $\mathcal{L}(\mathcal{K})$ is closed w.r.t. valid consecutions. But a reasonable question is whether we could restrict ourselves by only such consecutions studying a given logic $\mathcal{L}(\mathcal{K})$. In particular, Lorenzen (1955, [21]) proposed to consider admissible consecutions. They can be defined as follows (for a given logic $\mathcal{L}$, $Form_L$ is the set of all formulas in the language of $\mathcal{L}$).

Definition 4. A consecution $\mathbf{c} := \varphi(x_1, \ldots, x_n), \ldots, \varphi_m(x_1, \ldots, x_n)$ $/\psi(x_1, \ldots, x_n)$, is said to be *admissible* for a logic $\mathcal{L}$ if, $\forall \eta_1 \in Form_L, \ldots, \forall \eta_n \in Form_L$

$$\bigwedge_{1 \leq i \leq m} [\varphi_i(\eta_1, \ldots, \eta_n) \in \mathcal{L}] \Longrightarrow [\psi(\eta_1, \ldots, \eta_n) \in \mathcal{L}].$$

Thus, for any admissible consecution, any instance into the premises making all of them theorems of $\mathcal{L}$ also makes the conclusion to be a theorem also. It is *most strong type of structural logical consecutions:* a consecution $\mathbf{c}$ is admissible in $\mathcal{L}$ iff $\mathcal{L}$, as the set of its own theorems, is closed with respect to $\mathbf{c}$. It is evident that any valid consecution is admissible. The converse is not always true. Before to discuss it, we would like to describe another sort of consecutions: derivable consecutions.

For a logic $\mathcal{L}$ with a fixed axiomatic system Ax_L and a certain consecution $\mathbf{cs} := \varphi_1, \ldots, \varphi_n/\psi$, $\mathbf{cs}$ is said to be *derivable* if $\varphi_1, \ldots, \varphi_n \vdash_{Ax_L} \psi$.

The derivable consecutions are safely correct. But it could happen, that, for a logic $\mathcal{L}$, with a given axiomatic system, a formula ψ is not derivable from the premises $\varphi_1, \ldots, \varphi_m$, but still the rule $\mathbf{cs} := \varphi_1, \ldots, \varphi_m/\psi$ is admissible: $\mathbf{cs}$ derives $\mathcal{L}$-provable conclusions from $\mathcal{L}$-provable premises. The earliest example of a consecution which is admissible in the intuitionistic propositional logic (IPC, in sequel) but not derivable in the Heyting axiomatic system for IPC is the Harrop's rule (1960, [15]):

$$r := \frac{\neg x \to y \vee z}{(\neg x \to y) \vee (\neg x \to z)}.$$

That is, $\neg x \to y \lor z \not\vdash_{IPC} (\neg x \to y) \lor (\neg x \to z)$, were $\vdash_{IPC}$ is the notation for derivability in the Heyting axiomatic system for IPC. But, for any formulas α, β and γ, as soon as $\vdash_{IPC} \neg\alpha \to \beta \lor \gamma$, it follows that $\vdash_{IPC} (\neg\alpha \to \beta) \lor (\neg\alpha \to \gamma)$.

G. Mints (1976, [28]) proposed the consecution

$$\frac{(x \to y) \to x \lor y}{((x \to y) \to x) \lor ((x \to y) \to z)}$$

which is also not derivable but admissible in IPC.

The Lemmon-Scott rule (cf. [38])

$$\frac{\Box(\Box(\Box\Diamond\Box p \to \Box p) \to (\Box p \lor \Box\neg\Box p))}{\Box\Diamond\Box p \lor \Box\neg\Box p}$$

is admissible but not derivable in the modal logics $S4$, $S4.1$, Grz.

Turning back to our logic $\mathcal{L}_{t,y}(\mathbf{Z})$, firstly, we consider the question whether admissible consecutions say more then just valid ones, valid consecutions must be admissible, if the converse true? Recall that $\bot := x \land \neg x$.

Lemma 5. *The consecution* $c_1 := \dfrac{(x \to \Box_F \neg x \land \Box_P \neg x) \land (\neg x \to \Box_F x \land \Box_P x)}{\bot}$ *is admissible but invalid in* $\mathcal{L}_{t,y}(\mathbf{Z})$.

Proof. Indeed, the frame $\mathbf{Z}$ refutes c_1 by the valuation $V(x) := \{2n \mid n \in \mathbf{Z}\}$. Assume that for a formula φ,

$$(\varphi \to \Box_F\neg\varphi \land \Box_P\neg\varphi) \land (\neg\varphi \to \Box_F\varphi \land \Box_P\varphi) \in \mathcal{L}_{t,y}(\mathbf{Z}).$$

Consider the frame $\mathbf{Z}$ with the valuation $V(p) = \emptyset$ for all letters p from φ. Then by induction on the length of formulas ψ constructed from the same letters as φ, it is easy to see that for any $a \in \mathbf{Z}$,

$$(\mathbf{Z}, a) \Vdash_V \psi \Leftrightarrow \forall b \in \mathbf{Z}[(\mathbf{Z}, b) \Vdash_V \psi].$$

Therefore we get

$$(\mathbf{Z}, 0) \not\Vdash_V (\varphi \to \Box_F\neg\varphi \land \Box_P\neg\varphi) \land (\neg\varphi \to \Box_F\varphi \land \Box_P\varphi),$$

which contradicts the assumption above. $\qquad\Box$

Thus, as we see, admissible consecutions are stronger than just valid ones for the logic $\mathcal{L}_{t,y}(\mathbf{Z})$ also. And we turn to the question whether $\mathcal{L}_{t,y}(\mathbf{Z})$ is decidable w.r.t. admissible consecutions. We need to recall and prepare some technique.

4. Decidability of $\mathcal{L}_{t,y}(\mathbf{Z})$ w.r.t. Admissible Consecutions

Definition 6. A model $\mathcal{M}$ is said to be definable if any world $c \in \mathcal{M}$ is definable in $\mathcal{M}$, i.e. there is a formula ϕ_a which in true in $\mathcal{M}$ only at element a.

Definition 7. Given a model $\mathcal{M} := \langle \mathcal{F}, V \rangle$ based upon the frame $\mathcal{F}$ and a new valuation V_1 in $\mathcal{F}$ of a set of propositional letters q_i, V_1 is definable in the model $\mathcal{M}$ if, for any q_i, $V_1(q_i) = V(\phi_i)$ for some formula ϕ_i.

Definition 8. Given a logic $\mathcal{L}$ and a model $\mathcal{M}$ with a valuation defined for a set of letters $p_1, \ldots, p_k$, $\mathcal{M}$ is said to be k-*characterizing* for $\mathcal{L}$ if the following holds. For any formula $\varphi(p_1, \ldots, p_k)$ built using letters $p_1, \ldots, p_k$, $\varphi(p_1, \ldots, p_k) \in \mathcal{L}$ iff $\mathcal{M} \Vdash \varphi(p_1, \ldots, p_k)$.

Lemma 9. (cf., for instance, [38]) *A consecution* **cs** *is not admissible for a logic $\mathcal{L}$ iff, for any sequence $Ch_{\mathcal{L}}(k), k \in N$ of k-characterizing for $\mathcal{L}$ models, there are a number n and an n-characterizing model $Ch_{\mathcal{L}}(n)$ from this sequence, such that the frame of $Ch_{\mathcal{L}}(n)$ refutes* **cs** *by a certain definable in $Ch_{\mathcal{L}}(n)$ valuation.*

The construction of n-characterizing models for $\mathcal{L}_{t,y}(\mathbf{Z})$, comparing with similar ones for modal logics, is surprisingly simple (though we will need to pay a cost for this simplicity). Indeed, as models $\mathcal{M}_i$, consider the frame $\mathbf{Z}$ with all possible valuations V of letters $p_1, \ldots, p_k$. Take the disjoint union $\bigsqcup_{i \in I} \mathcal{M}_i, \|I\| = 2^\omega$ of all such non-isomorphic models $\mathcal{M}_i$. It is a model which we denote by $Ch_k(\mathcal{L}_{t,y}(\mathbf{Z}))$.

Lemma 10. *The model $Ch_k(\mathcal{L}_{t,y}(\mathbf{Z}))$ is k-characterizing for $\mathcal{L}_{t,y}(\mathbf{Z})$.*

Proof immediately follows from the definitions. $\qquad\qquad\square$

Now note that our simple definition of $Ch_k(\mathcal{L}_{t,y}(\mathbf{Z}))$ has a negative consequence:

Lemma 11. *The model $Ch_k(\mathcal{L}_{t,y}(\mathbf{Z}))$ is not definable.*

Proof. Indeed, consider the component $\mathbf{Z}$ of the disjoint union $\bigsqcup_{i \in I} \mathcal{M}_i$, where the valuation V for any $p_j, 1 \leq j \leq k$ is $V(p_i) = \emptyset$. Then, again, by induction on the length of formulas ϕ constructed from the letters p_i it is easy to see that for any $a \in \mathbf{Z}$,

$$(\mathbf{Z}, a) \Vdash_V \phi \Leftrightarrow \forall b \in \mathbf{Z}[(\mathbf{Z}, b) \Vdash_V \phi].$$

Therefore we cannot distinguish elements of this component $\mathbf{Z}$ of $Ch_k(\mathcal{L}_{t,y}(\mathbf{Z}))$ by formulas. $\qquad\square$

Thus we cannot apply directly the technique worked out for algorithmic description of rules admissible for transitive modal logics to the logic $\mathcal{L}_{t,y}(\mathbf{Z})$. This is the price to be paid for intransitivity. The new technique we will apply will use definable sets instead of definable elements.

A technical instrument which will be helpful is a reduction of consecutions to equivalent ones but with temporal degree 1 and with a homogeneous form. A consecution $\mathbf{c}$ is said to have the *reduced normal form* if

$$\mathbf{c} = \frac{\bigvee_{1 \le j \le m}(\bigwedge_{1 \le i \le n}[x_i^{k(j,i,0)} \wedge (\square_F x_i)^{k(j,i,1)} \wedge (\square_P x_i)^{k(j,i,2)}])}{x_1},$$

where x_s are certain variables, $k(i,j,z) \in \{0,1\}$ and, for any term f above, $f^0 := f, f^1 := \neg f$. Given a consecution $\mathbf{c_{nf}}$ in the reduced normal form, $\mathbf{c_{nf}}$ is said to be a *normal reduced form for a consecution* $\mathbf{c}$ iff, for any temporal logic $\mathcal{L}$, $\mathbf{c}$ is admissible in $\mathcal{L}$ iff $\mathbf{c_{nf}}$ is so. Using exactly same ideas and the structure of proof as for Lemma 3.1.3 and Theorem 3.1.11 in [38] we can derive

Theorem 12. *There exists a double exponential algorithm which, for any given consecution* $\mathbf{c}$, *constructs its normal reduced form* $\mathbf{c_{nf}}$.

Proof. Actually an analog of this theorem for inference rules of modal logics was proved in (Rybakov, [34]). In spite of the construction is very simple, the paper mentioned is old, therefore we give merely a sketch of the construction for our case, in order to show the complexity of the transformation.

Given a consecution $\mathbf{c} = \varphi_1(x_1,..,x_n), \ldots, \varphi_m(x_1,..,x_n)/\psi(x_1,..,x_n)$. $\mathbf{c}$ is equivalent to $\mathbf{c_0} = \varphi_1(x_1,\ldots,x_n) \wedge \ldots \wedge \varphi_m(x_1,\ldots,x_n) \wedge x_c \equiv \psi(x_1,\ldots,x_n)/x_c$, where x_c is a new variable. Therefore we can restrict the case by consideration only consecutions which have the form $\mathbf{c} = \varphi(x_1,\ldots,x_n)/x_c$.

If $\varphi = \alpha \circ \beta$, where $\circ$ is a binary logical operation and both formulas α, β are not simply variables or a unary logical operations applied to variables (which both we call atomic formulas), take two new variables x_α and x_β and the rule $\mathbf{c_1} := x_\alpha \circ x_\beta \wedge x_\alpha \equiv \alpha \wedge x_\beta \equiv \beta/x_c$. If one from formulas α, β is final and another one not, we apply this transformation to only non-final formula. It is clear that $\mathbf{c}$ and $\mathbf{c_1}$ are equivalent w.r.t. admissibility in logics and w.r.t. validity in frames.

If $\varphi = *\alpha$, where $*$ is a unary logical operation and α is not a variable, take a new variable x_α and the rule $\mathbf{c_1} := *x_\alpha \wedge x_\alpha \equiv \alpha/x_c$. And again $\mathbf{c}$ and $\mathbf{c_1}$ are equivalent in both mentioned above senses. We continue this (similar) transformation over the resulting consecutions

$$\frac{\bigwedge_{j \in J_1} \gamma_j \wedge \bigwedge_{i \in I_1} x_{\alpha_i} \equiv \alpha_i}{x_c}$$

until all formulas α_i and γ_j in the premise of the resulting consecutions will be either atomic formulas - logical operation applied to variables, or variables themselves. Evidently this transformation is *polynomial*. Next, we transform the premise of the resulting consecution in the disjunctive normal form and obtain the equivalent consecution:

$$\mathbf{c_2} = \frac{\bigvee_{1 \le j \le m}(\bigwedge_{1 \le i \le n}[\bullet x_i^{k(j,i,0)} \wedge \bullet(\square_F x_i)^{k(j,i,1)} \wedge (\bullet \square_P x_i)^{k(j,i,2)}])}{x_c},$$

were any formula with $\bullet$ could be empty formula. This transformation, unfortunately, as well as all known ones for reduction Boolean formulas to disjunctive normal forms, is *exponential*. To reduce $\mathbf{c_2}$ to the required form, it remains only *to delete* $\bullet$ everywhere by (again exponentially costly) inserting of all combinations of missed formulas. $\square$

To discuss this transformation more, in (Rybakov [38]), as an immediate consequence of the reduction of modal consecutions to normal forms (as above), it was shown (Corollary 3.1.27) that any normal modal logic extending $K4$ can be axiomatized by formulas of modal degree at most 2 (known result, Zakharyaschev, 1992). It can be done with impressive ease. Take a modal formula α, its validity is equivalent to the validity of the consecution $c := \top/\alpha$. Take its reduced normal form $rf(c) = \beta/x_1$. For any consecution $c_1 = \gamma/\delta$, its transformation to the semi-universal formula $f(c_1)$ is the formula $\square\gamma \wedge \gamma \to \delta$. And, for any modal logic $\mathcal{L}$ over $K4$, $\mathcal{L} \oplus \alpha = \mathcal{L} \oplus f(rf(c))$, so all works.

In the case of temporal logics, using the transformation of consecutions to the reduced normal forms described above, we can obtain a similar result: any temporal logic (it is unimportant transitive or not) enriched with the universal modality can be axiomatized by formulas of modal/temporal degree at most 2.

Turning to further study of consecution for our logic, fix some more notation. For any consecution $\mathbf{c_{nf}}$ in normal reduced form, $Pr(c_{nf}) = \{\varphi_i \mid i \in I\}$ is the set of all disjunctive members of the premise of $\mathbf{c_{nf}}$. $Sub(c_{nf})$ is the set of all subformulas of $\mathbf{c_{nf}}$. In the sequel we will use also the following notation. For any frame $\mathcal{F}$ and any valuation V of the set of propositional letters of a formula φ, the expression $(\mathcal{F} \Vdash_V \varphi)$ is the abbreviation for $\forall a \in \mathcal{F}((\mathcal{F}, a) \Vdash_V \varphi)$. The following lemma is evident.

Lemma 13. *If, for a frame $\mathcal{F}$ and a valuation V of all letters from formulas of $Pr(c_{nf})$ in $\mathcal{F}$, $\mathcal{F} \Vdash_V \bigvee Pr(c_{nf})$, then for any $a \in \mathcal{F}$ there is a unique $\varphi \in Pr(c_{nf})$ such that $(\mathcal{F}, a) \Vdash_V \varphi_i$.*

We will denote this φ_i by $D_{c_{nf}}(a)_V$ assuming that the frame $\mathcal{F}$ is known from a context.

Recall that for a model $\mathcal{M}$ with a valuation $\mathcal{G}$ of a set of letters $Dom(\mathcal{G})$, for any $a \in \mathcal{M}$,

$$Val_{\mathcal{G}}(a) := \{p \mid p \in Dom(\mathcal{G}), (\mathcal{M}, a) \Vdash_{\mathcal{G}} p\}.$$

For any integer number m, the model $\mathcal{Z}^- \oplus [-m, m] \oplus \mathcal{Z}^+$ has the following structure: $[-m, m]$ is the interval of integer numbers with the standard relations $Next$ and $Prev$. $\mathcal{Z}^-$ and $\mathcal{Z}^+$ are the sets of strictly negative and strictly positive integer numbers respectively, which we consider to be disjoint with $[-m, m]$. They also have the standard relations $Next$ and $Prev$. And else, $-1\ Prev\ -m$; $-m\ Next\ -1$; $m\ Prev\ -1$; $m\ Next\ 1$. So, it is simply consecutive concatenation of the initial models.

Lemma 14. *If a consecution $\mathbf{c_{nf}}$ is not admissible in $\mathcal{L}_{t,y}(\mathbf{Z})$, then there is a linear intransitive temporal model $\mathcal{M}_Z := \langle \mathbf{Z}, \mathcal{G} \rangle$ with the following properties.*
   (i) $\mathcal{M}_Z := \mathcal{Z}^- \oplus [-m, m] \oplus \mathcal{Z}^+$, m is linearly computable from the size of $\mathbf{c_{nf}}$;
   (ii) The model $\mathcal{M}_Z$ refutes $\mathbf{c_{nf}}$, that is

$$\forall a \in \mathcal{M}_Z(\mathcal{M}_Z, a) \Vdash_{\mathcal{G}} \bigvee Pr(\mathbf{c_{nf}}); \quad (\mathcal{M}_Z, 0) \not\Vdash_{\mathcal{G}} x_1;$$

   (iii) $\forall x, y \in \mathbf{Z}^-\ Val_{\mathcal{G}}(x) = Val_{\mathcal{G}}(y)$;
   (iv) $\forall x, y \in \mathbf{Z}^+\ Val_{\mathcal{G}}(x) = Val_{\mathcal{G}}(y))$;
   (v) $\forall x, y \in \mathbf{Z}^-, \forall a, b \in \mathbf{Z}^+\quad D_{c_{nf}}(x)_{\mathcal{G}} = D_{c_{nf}}(y)_{\mathcal{G}} D_{c_{nf}}(a)_{\mathcal{G}} = D_{c_{nf}}(b)_{\mathcal{G}} = D_{c_{nf}}(-m)_{\mathcal{G}} D_{c_{nf}}(m)_{\mathcal{G}}.$

Proof. Since $\mathbf{c_{nf}}$ is not admissible in $\mathcal{L}_{t,y}(\mathbf{Z})$, for some substitution s for variables x_j from $\mathbf{c_{nf}}$ by certain formulas γ_j ($s : x_j \rightarrow \gamma_j$), $\bigvee Pr(c_{nf})(x_j/\gamma_j) \in \mathcal{L}_{t,y}(\mathbf{Z})$, but $\gamma_1 \notin \mathcal{L}_{t,y}(\mathbf{Z})$.

Using Theorem 10 $Ch_k(\mathcal{L}_{t,y}(\mathbf{Z})) \Vdash_V \bigvee Pr(c_{nf})(x_j/\gamma_j)$ and $Ch_k(\mathcal{L}_{t,y}(\mathbf{Z})) \not\Vdash_V \gamma_1$, where V is the valuation of $Ch_k(\mathcal{L}_{t,y}(\mathbf{Z}))$ and k is the number of propositional letters $p_1, \ldots p_k$ occurring in formulas γ_j. Consequently, there exists a component $\mathbf{Z}$ in the decomposition of $Ch_k(\mathcal{L}_{t,y}(\mathbf{Z}))$ in the disjoint union, where

$$\mathbf{Z} \Vdash_V \bigvee Pr(c_{nf})(x_j/\gamma_j) \text{ and } \mathbf{Z} \not\Vdash_V \gamma_1.$$

Thus, for some $m_1 \in \mathbf{Z}$, $(\mathbf{Z}, m_1) \not\Vdash_V \gamma_1$, but $\forall m \in \mathbf{Z}$ $(\mathbf{Z}, m) \Vdash_V \bigvee Pr(c_{nf})(x_j/\gamma_j)$. Evidently we may assume that $m = 0$. Let n be the maximal temporal degree of the formula γ_1.

Consider any component $\mathbf{Z}_1$ of $Ch_k(\mathcal{L}_{t,y}(\mathbf{Z}))$, where the interval $[-n-1, n+1]$ has the same valuation V as it had in $\mathbf{Z}$. Basing on $Ch_k(\mathcal{L}_{t,y}(\mathbf{Z})) \Vdash_V \bigvee Pr(c_{nf})(x_j/\gamma_j)$ and using standard reasoning on temporal degrees of formulas it is not difficult to show that

$$\forall m \in \mathbf{Z}_1 \ (\mathbf{Z}_1, m) \Vdash_V \bigvee Pr(c_{nf})(x_j/\gamma_j), \text{ and } (\mathbf{Z}_1, 0) \not\Vdash_V \gamma_1. \tag{1}$$

Fix the $\mathbf{Z}_1$ where $\forall a, b > n+1$, $Val_V(a) = Val_V(b)\emptyset$ and $\forall a, b < -n-1$, $Val_V(a) = Val_V(b)\emptyset$. Let n_1 be be the maximal temporal degree of all formulas γ_j. Consider the world $m_{max} := n + n_1 + 2$ from $\mathbf{Z}_1$. Basing on temporal degrees of formulas γ_j it is not hard to prove that

$$\forall a, b \in \mathbf{Z}_1, \forall \gamma_j \ (a, b > m_{max}) \Rightarrow (\mathbf{Z}_1, a) \Vdash_V \gamma_j \Leftrightarrow (\mathbf{Z}_1, b) \Vdash_V \gamma_j \tag{2}$$

For the element $m_{min} := -n - n_1 - 2 \in \mathbf{Z}_1$, we similarly obtain

$$\forall a, b \in \mathbf{Z}_1, \forall \gamma_j \ (a, b < m_{min}) \Rightarrow (\mathbf{Z}_1, a) \Vdash_V \gamma_j \Leftrightarrow (\mathbf{Z}_1, b) \Vdash_V \gamma_j \tag{3}$$

Besides, reasoning similar to above we conclude

$$\forall a, b \in \mathbf{Z}_1, \forall \gamma_j \ (a < m_{min} \ \& \ b > m_{max})$$

$$\Rightarrow (\mathbf{Z}_1, a) \Vdash_V \gamma_j \Leftrightarrow (\mathbf{Z}_1, b) \Vdash_V \gamma_j. \tag{4}$$

Using (1)–(4) we obtain that the frame of $\mathbf{Z}_1$ with the new valuation $\mathcal{G}$, where

$$(\mathbf{Z}_1, a) \Vdash_{\mathcal{G}} x_j \Leftrightarrow (\mathbf{Z}_1, a) \Vdash_V \gamma_j,$$

satisfy all conditions of our lemma except the size of the interval $[m_{min}, m_{max}]$ is not linearly bounded. To complete our lemma it is sufficient to thin out $[m_{min}, m_{max}]$ in the standard manner using that the frame $[m_{min}, m_{max}]$ is finite, intransitive and linear. $\square$

Lemma 15. *If a consecution $\mathbf{c}_{nf}$ in the normal reduced form satisfies all conditions of Lemma 14 then $\mathbf{c}_{nf}$ is not admissible in $\mathcal{L}_{t,y}(\mathbf{Z})$.*

Proof. Take the model $\mathcal{M}_Z := \mathcal{Z}^- \oplus [-m, m] \oplus \mathcal{Z}^+$ from Lemma 14. Take the n-characterizing model $Ch_n(\mathcal{L}_{t,y}(\mathbf{Z}))$, where $n := ||Dom(\mathcal{G})|| + 2m + 4$. The model $\mathcal{M}_Z : \mathcal{Z}^- \oplus [-m, m] \oplus \mathcal{Z}^+$ is a component of $Ch_n(\mathcal{L}_{t,y}(\mathbf{Z}))$ in its decomposition in the disjoint union (with restriction of the valuation V of the model $Ch_n(\mathcal{L}_{t,y}(\mathbf{Z}))$ to $Dom(\mathcal{G})$). We denote this component by $\mathcal{M}_1$. Thus we have that the valuation $\mathcal{G}$ and the original valuation V of $Ch_n(\mathcal{L}_{t,y}(\mathbf{Z}))$ refute $\mathbf{c_{nf}}$ in $\mathcal{M}_1$, so:

$$\mathcal{M}_1 \Vdash_V \bigvee Pr(c_{nf}), \quad (\mathcal{M}_1, 0) \nVdash_V x_1,$$

$$\text{where} \quad \mathcal{M}_1 := \mathcal{Z}^- \oplus [-m, m] \oplus \mathcal{Z}^+. \quad (5)$$

The problem is to find a definable in $Ch_n(\mathcal{L}_{t,y}(\mathbf{Z}))$ valuation V_1 which would coincide with $\mathcal{G}$ in $\mathcal{M}_1$ and would make the premise of $\mathbf{c_{nf}}$ to be true in whole the model $Ch_n(\mathcal{L}_{t,y}(\mathbf{Z}))$.

Consider the formulas written down below. Let p_k be letters from the domain of the valuation V of the model $Ch_n(\mathcal{L}_{t,y}(\mathbf{Z}))$ which do not belong to $Dom(\mathcal{G})$, where $k \in [-m, m]$, and, for any $k \in [-m, m]$,

$$\varphi(k) := [p_k \bigwedge_{j \in [-m,m], j \neq k} \neg p_j \wedge \neg q_1 \wedge \neg q_2]$$

$$\wedge \, [\bigwedge_{i \in [-m,m], i < k} \Box_P^{k-i}[p_i \wedge \bigwedge_{j \in [-m,m], j \neq i} \neg p_j \wedge \neg q_1 \wedge \neg q_2]]$$

$$\wedge \, [\bigwedge_{i \in [-m,m], k < i} \Box_F^{i-k}[p_i \wedge \bigwedge_{j \in [-m,m], j \neq i} \neg p_j \wedge \neg q_1 \wedge \neg q_2]]$$

$$\wedge \Box_F^{m-k+1}[q_1 \wedge \bigwedge_{i \in [-m,m]} \neg p_i \wedge \neg q_2] \wedge \Box_F^{m-k+2}[q_2 \wedge \bigwedge_{i \in [-m,m]} \neg p_i \wedge \neg q_1]$$

$$\wedge \Box_P^{k+m+1}[q_1 \wedge \bigwedge_{i \in [-m,m]} \neg p_i \wedge \neg q_2] \wedge \Box_P^{k+m+2}[q_2 \wedge \bigwedge_{i \in [-m,m]} \neg p_i \wedge \neg q_1].$$

We set also

$$\varphi(Next(m)) := q_1 \wedge \neg q_2 \wedge \bigwedge_{i \in [-m,m]} \neg p_i \wedge \Box_F[q_2 \wedge \neg q_1 \wedge \bigwedge_{i \in [-m,m]} \neg p_i]$$

$$\wedge \bigwedge_{i \in [-m,m]} \Box_P^{m-i+1}\varphi(i) \wedge \Box_P^{2m+2} q_1 \wedge \Box_P^{2m+3} q_2;$$

$$\varphi(Next^2(m)) := q_2 \wedge \neg q_1 \bigwedge_{i \in [-m,m]} \neg p_i \wedge \Box_P \varphi(Next(m));$$

$$\varphi(Prev(-m)) := q_2 \wedge \neg q_1 \wedge \bigwedge_{i\in[-m,m]} \neg p_i \wedge \Box_P[q_1 \wedge \neg q_2 \wedge \bigwedge_{i\in[-m,m]} \neg p_i]$$

$$\wedge \bigwedge_{i\in[-m,m]} \Box_F^{m+i+1}\varphi(i) \wedge \Box_P^{2m+2}q_1 \wedge \Box_P^{2m+3}q_2;$$

$$\varphi(Prev^2(-m)) := q_1 \wedge \neg q_2 \wedge \bigwedge_{i\in[-m,m]} \neg p_i \wedge \Box_F\varphi(Prev(-m)).$$

Consider the following definable in $Ch_n(\mathcal{L}_{t,y}(\mathbf{Z}))$ valuations for all variables x_i occurring in $\mathbf{c_{nf}}$:

$$V_{1,1}(x_i) := V(\bigwedge_{i\in[-m,m]} \neg\varphi(i) \wedge \neg\varphi(Next(m)) \wedge \neg\varphi(Next^2(m))$$

$$\wedge \neg\varphi(Prev(-m) \wedge \neg\varphi(Prev^2(-m))$$

if $(\mathcal{M}_1, Next(m)) \Vdash_{\mathcal{G}} x_i$, otherwise $V_{1,1}(x_i) : \emptyset$.

$$V_{1,2}(x_i) := V(\bigvee\{\varphi(j) \mid j \in [-m,m], (\mathcal{M}_Z, j) \Vdash_{\mathcal{G}} x_i\} \vee \eta(i)),$$

where

$$\eta(i) := \phi(i, Next(m)) \vee \phi(i, Next^2(m)) \vee$$
$$\phi(i, Prev(-m)) \vee \phi(i, Prev^2(-m)),$$

and $\phi(i, X(k)) := \varphi(X(k))$ if $(\mathcal{M}_Z, X(k)) \Vdash_{\mathcal{G}} x_i$ otherwise $\phi(i, X(k)) := \bot$.
Take the following definable in $Ch_n(\mathcal{L}_{t,y}(\mathbf{Z}))$ valuation for variables x_i:

$$V_1(x_i) := V_{1,1}(x_i) \cup V_{1,2}(x_i).$$

We use below the sign $\cong_f$ to say that there is an isomorphism f of one model onto another one.

Lemma 16. *The following holds:*

(i) *If $(Ch_n(\mathcal{L}_{t,y}(\mathbf{Z})), c) \Vdash_V \varphi(k)$ then*
$\langle[c-m-2+k, c+m+2-k], V_1\rangle \cong_f \langle[-m-2, m+2], \mathcal{G}\rangle$, *where $f(c) = k$.*

(ii) *If $(Ch_n(\mathcal{L}_{t,y}(\mathbf{Z})), c) \Vdash_V \varphi(Next(m))$ then*
$\langle[c-2m-3, c+1], V_1\rangle \cong_f \langle[-m-2, m+2], \mathcal{G}\rangle$, *where $f(c) = Next(m)$.*

(iii) *If $(Ch_n(\mathcal{L}_{t,y}(\mathbf{Z})), c) \Vdash_V \varphi(Next^2(m))$ then*
$\langle[c-2m-4, c], V_1\rangle \cong_f \langle[-m-2, m+2], \mathcal{G}\rangle$, *where $f(c) = Next^2(m)$.*

(iv) *If $(Ch_n(\mathcal{L}_{t,y}(\mathbf{Z})), c) \Vdash_V \varphi(Prev(m))$ then*
$\langle[c-1, c+2m+3], V_1\rangle \cong_f \langle[-m-2, m+2], \mathcal{G}\rangle$, *where $f(c) = Prev(m)$.*

(v) *If $(Ch_n(\mathcal{L}_{t,y}(\mathbf{Z})), c) \Vdash_V \varphi(Prev^2(m))$ then*

$$\langle [c-2, c+2m+2], V_1 \rangle \cong_f \langle [-m-2, m+2], \mathcal{G} \rangle, \text{ where } f(c) = Prev^2(m).$$

Proof of this lemma is a standard verification based on our definitions of the formulas $\varphi(k), \varphi(Next(m)), \varphi(Next^2(m)), \varphi(Prev(-m))$ and $\varphi(Prev^2(-m))$, and the definition of the new valuation V_1 for variables x_i given above. Due limitations on the volume of the papers, and because the verification is a routine standard computation, we leave details to the reader. $\square$

To complete our proof of Lemma 15, using Lemma 16, property (v) from Lemma 14, (5) and the components $V_{1,1}$ and $V_{1,2}$ in the definition of the valuation V_1 is is not hard to show that the definable valuation V_1 refutes our consecution $\mathbf{c_{nf}}$ in $Ch_n(\mathcal{L}_{t,y}(\mathbf{Z}))$. Thus by Lemma 9 $\mathbf{c_{nf}}$ is not admissible in $\mathcal{L}_{t,y}(\mathbf{Z})$. $\square$

Based on Lemmas 14 and 15 we derive

Corollary 17. *A consecution $\mathbf{c_{nf}}$ is not admissible in $\mathcal{L}_{t,y}(\mathbf{Z})$, if and only if there is is a linear intransitive temporal model $\mathcal{M}_Z : \mathcal{Z}^- \oplus [-m, m] \oplus \mathcal{Z}^+$ satisfying properties of Lemma 14.*

This corollary does not refer to finite models, so we cannot use it immediately to construct an algorithm verifying admissibility, but it is easy to reduce the infinite models from this corollary to finite ones.

Lemma 18. *A model $\mathcal{M}_Z := \mathcal{Z}^- \oplus [-m, m] \oplus \mathcal{Z}^+$ refutes a consecution $\mathbf{c_{nf}}$ by a valuation $\mathcal{G}$:*

(a) $\forall a \in \mathcal{M}_Z (\mathcal{M}_Z, a) \Vdash_{\mathcal{G}} \bigvee Pr(c_{nf}), \quad (\mathcal{M}_Z, 0) \nVdash_{\mathcal{G}} x_1;$
and has the properties:
(b) $\forall x, y \in \mathbf{Z}^- \quad Val_{\mathcal{G}}(x) = Val_{\mathcal{G}}(y);$
(c) $\forall x, y \in \mathbf{Z}^+ \quad Val_{\mathcal{G}}(x) = Val_{\mathcal{G}}(y);$
(d) $\forall x, y \in \mathbf{Z}^-, \forall a, b \in \mathbf{Z}^+ \quad D_{c_{nf}}(x)_{\mathcal{G}} = D_{c_{nf}}(y)_{\mathcal{G}} D_{c_{nf}}(a)_{\mathcal{G}} = D_{c_{nf}}(b)_{\mathcal{G}} = D_{c_{nf}}(-m)_{\mathcal{G}} D_{c_{nf}}(m)_{\mathcal{G}}$

if and only if the model $\mathcal{M}_1 := [-m-2, -m-1] \oplus [-m, m] \oplus [m+1, m+2]$, where $m+2$ Next $m+2$; $m+2$ Prev $m+1$; $-m-2$ Prev $-m-2$; $-m-2$ Next $-m-1$,
with a valuation S refutes $\mathbf{c_{nf}}$, i.e.

(i) $\forall a \in \mathcal{M}_1 (\mathcal{M}_1, a) \Vdash_S \bigvee Pr(c_{nf}), \quad (\mathcal{M}_1, 0) \nVdash_S x_1;$

and has the properties:

(ii) $Val_S(-m-2) = Val_S(-m-1) = Val_S(-m)$

$$= Val_S(m) = Val_S(m+1) = Val_S(m+2);$$

$$(iii) \quad D_{c_{nf}}(-m-1)_S = D_{c_{nf}}(-m)_S D_{c_{nf}}(m)_S = D_{c_{nf}}(m+1)_S$$

Proof. If $\mathcal{M}_Z := \mathcal{Z}^- \oplus [-m, m] \oplus \mathcal{Z}^+$ refutes a consecution $\mathbf{cs_{nf}}$ by a valuation $\mathcal{G}$ and (a)-(d) hold, it is sufficient to take the valuation S in $\mathcal{M}_1 := [-m-2, -m-1] \oplus [-m, m] \oplus [m+1, m+2]$ as the restriction of $\mathcal{G}$ on $\mathcal{M}_1$. The verification of (i)-(iii) is evident.

If $\mathcal{M}_1 := [-m-2, -m-1] \oplus [-m, m] \oplus [m+1, m+2]$ with a valuation S refutes $\mathbf{c_{nf}}$ and (i)-(iii) hold, then to expand the valuation S from $\mathcal{M}_1$ onto $\mathcal{M}_Z := \mathcal{Z}^- \oplus [-m, m] \oplus \mathcal{Z}^+$, take

$$\forall x \in [-m-2, m+2] \cap \mathcal{M}_Z, \ Val_{\mathcal{G}}(x) := Val_S(x);$$

$$\forall x \in \mathcal{M}_Z, x < m - 2 \Rightarrow Val_{\mathcal{G}}(x) := Val_S(m-2);$$

$$\forall x \in \mathcal{M}_Z, x > m + 2 \Rightarrow Val_{\mathcal{G}}(x) := Val_S(m+2);$$

To show that $\mathcal{M}_Z := \mathcal{Z}^- \oplus [-m, m] \oplus \mathcal{Z}^+$ refutes $\mathbf{c_{nf}}$ by $\mathcal{G}$ and that properties (a)-(d) hold use (i)-(iii) and standard truth value computation. Due to limitations on the size of papers and primary because this computation is quite standard we leave it to the reader. $\square$

Combining Theorem 12, Corollary 17, and Lemma 18 we derive

Theorem 19. *The temporal linear logic $\mathcal{L}_{t,y}(\mathbf{Z})$ is decidable w.r.t. admissible consecutions.*

To comment the complexity of the deciding algorithm, for any consecution $\mathbf{c}$ we first transform $\mathbf{c}$ into the reduced normal form $\mathbf{c_{nf}}$ (complexity is double exponential, cf. Theorem 12). Then we verify conditions (i)-(iii) given in Lemma 18 in the frames of sort $\mathcal{M}_1 := [-m-2, -m-1] \oplus [-m, m] \oplus [m+1, m+2]$ where m is linear in the size of $\mathbf{c_{nf}}$. So, we have to perform the model checking on models of size linear in $\mathbf{c_{nf}}$.

Future work: The technique developed in this paper can be applied for a number of other similar temporal logics. We studied only one natural, maybe most intuitive, intransitive linear temporal logic, and there is a variety of temporal logics (such as branching time logics, etc.) which are widely recognized in literature and successfully applied in AI or CS. All basic technique from our paper can be applied to approach these logics. An important open question is finding of bases for admissible consecutions. It is also interesting to investigate possible links with logical unification and logical equations in temporal logics. As one more important open question, we would mention the extension of the results of this paper to the temporal logic LTL (which means to study the additional binary temporal operation

until in this context). The logic LTL had numerous fruitful applications in CS and AI. The problem to extend our results immediately is the fact that the operation **until** is binary and that the operation **until** is stronger than modal operations and cannot be expressed by them. These generate difficulties to analyze the structure of definable valuations, construction of which is a base of our approach in this paper. But we hope that there are other ways to approach this problem.

References

1. Barringer H, Fisher M, Gabbay D., Gough G. *Advances in Temporal Logic*, Vol. 16 of Applied logic series, Kluwer Academic Publishers, Dordrecht, December 1999. (ISBN 0-7923-6149-0).
2. Bloem R., Ravi K, Somenzi F, *Efficient Decision Procedures for Model Checking of Linear Time Logic Properties*. In: Conference on Compurer Aided Verification (CAV), LNCS 1633, Terento, Italy, Springer-Verlag, 1999.
3. Bruns G. and Godefroid P. *Temporal Logic Query-Checking*. In Proceedings of 16th Annual IEEE Symposium on Logic in Computer Science (LICS'01), pages 409–417, Boston, MA, USA, June 2001. IEEE Computer Society.
4. Citkin A.I. *On Admissible Rules of Intuitionistic Propositional Logic*. Math. USSR Sbornik, Vol. 31, 1977, No. 2, p. 279–288.
5. Clarke E., Grumberg O., Hamaguchi K. P. *Another look at LTL Model Checking*. In: Conference on Computer Aided Verification (CAV), LNCS 818, Stanford, California, Springer-Verlag, 1994.
6. Cyrluk David, Narendran Paliath. *Ground Temporal Logic: A Logic for Hardware Verification*, Lecture Notes in computer Science, V. 818. From Computer-aided Verification (CaV'94), Ed. David Dill, Springer-Verlag, Stanford, CA, 1994, p. 247–259.
7. Emerson E.A. *Temporal and Modal Logics*. in: Handbook of Theoretical Computer Science. J. van Leenwen, Ed., Elsevier Science Publishers, the Netherlands, 1990, p. 996–1072.
8. Friedman H., *One Hundred and Two Problems in Mathematical Logic*.- Journal of Symbolic Logic, Vol. 40, 1975, No. 3, p. 113–130.
9. Gabbay D. *Model Theory for Tense Logics*. U.S. Air Forse Office of Science Research, contract no. F61052-68-C-0036, report no. 1, April 1969.
10. Gabbay D. *An Irreflevivity Lemma with Applications to Axiomatizations of Conditions of Linear Frames*. Aspects of Phoilosophical Logic (Ed. V.Monnich), Reidel, Dordrecht, 1981, p. 67–89.
11. Gabbay D.M., Kurucz A., Wolter F., Zakharyaschev M. *Many-Dimensional Modal Logics: Theory and Applications*. Elsevier Science Pub Co, 2003, ISBN: 0444508260.
12. Ghilardi S. *Unification in Intuitionistic logic*. Journal of Symbolic Logic, Vol. 64, No. 2 (1999), pp. 859–880.
13. Goldblat R. *Logics of Time and Computation*. CSLI Lecture Notes, No. 7, Second Edition, 1992.

14. Goranko V., Passy S. *Using the Universal Modality: Gains and Questions.* J. Log. Comput. 2 (1) (1992), 5–30.

15. Harrop R. *Concerning Formulas of the Types $A \to B \vee C$, $A \to \exists x B(x)$ in Intuitionistic Formal System.* J. of Symbolic Logic, Vol. 25, 1960, pp. 27–32.

16. Iemhoff R. *A(nother) Characterization of Intuitionistic Propositional Logic.* Annals of Pure and Applied Logic, Vol. 113 (No. 1-3), 2001, pp. 161–173.

17. Iemhoff R. *On the admissible rules of Intuitionistic Propositional Logic.* Journal of Symbolic Logic Vol. 66, 2001, pp. 281–294.

18. Kapron B.M. *Modal Sequents and Definability,* J. of Symbolic Logic, Vol. 52, No.3, (1987), pp. 756–765.

19. Lichtenstein O., Pnueli A. *Propositional temporal logics: Decidability and completeness.* Logic Journal of the IGPL, 8 (1), 2000, pp. 55–85.

20. Francois Laroussinie, Nicolas Markey, Philippe Schnoebelen. *Temporal Logic with Forgettable Past* . IEEE Symp. Logic in Computer Science (LICS'2002).

21. Lorenzen P. *Einführung in die operative Logik und Mathematik.* Berlin-Göttingen, Heidelberg, Springer-Verlag, 1955.

22. McMillan K.L. *Symbolic Model Checking.* Kluwer Academic Publishers: Boston, MA, 1993.

23. Manna Z, Pnueli A. *Temporal Verification of Reactive Systems: Safety,* Springer-Verlag, 1995.

24. Manna Z., Pnueli A. *The Temporal Logic of Reactive and Concurrent Systems: Specification.* Springer-Verlag, 1992.

25. Nikolaj Bjorner, Anca Browne, Michael Colon, Bernd Finkbeiner, Zohar Manna, Henny Sipma, Tomas Uribe. *Verifying Temporal Properties of Reactive Systems: A Step Tutorial.* In Formal Methods in System Design, vol. 16, 2000, pp. 227–270.

26. Manna Z., Sipma H. *Alternating the Temporal Picture for Safety.* In Proc. 27th Intl. Colloq. Aut. Lang. Prog.(ICALP 2000). LNCS 1853, Springer-Verlag, pp. 429–450.

27. Marx M, Venema Y. *Multi-Dimensional Modal Logics.* Kluwer Academic Publishers, Dordrecht/Boston/London, 1997.

28. Mints G.E. *Derivability of Admissible Rules.* J. of Soviet Mathematics, V. 6, 1976, No. 4, pp. 417–421.

29. Pnueli A. *The Temporal Logic of Programs.* In Proc. of the 18th Annual Symp. on Foundations of Computer Science, IEEE, 1977, pp. 46–57.

30. Pnueli A., Kesten Y. *A deductive proof system for CTL^*.* In Proc. 13th Conference on Concurrency Theory, volume 2421 of Lecture Notes in Computer Science, Brno, Czech Republic, August 2002, pp. 24–40.

31. Prior A. *Time and Modaliy,* Oxford, 1957.

32. Prior A. *Past, Present and Future,* Oxford, 1967.

33. Rybakov V.V. *A Criterion for Admissibility of Rules in the Modal System S4 and the Intuitionistic Logic.* Algebra and Logic, V. 23, No 5 (1984), pp. 369–384 (Engl. Translation).

34. Rybakov V.V. *Criterin for Admissibility of Rules in the modal system S4 and the Intuitionistic Logic.* Algebra and Logic, V. 23, No. 5 (1984), pp. 369–384 (English translation).

35. Rybakov V.V. *The Bases for Admissible Rules of Logics S4 and Int.* Algebra and Logic, V. 24 (1985), pp. 55–68 (English translation).

36. Rybakov V.V. *Rules of Inference with Parameters for Intuitionistic logic.-* Journal of Symbolic Logic, Vol. 57, No. 3, 1992, pp. 912–923.

37. Rybakov V.V. *Heriditarily Structurally Complete Modal Logics.* Journal of Symbolic Logic, Vol. 60, No. 1 (1995), pp. 266–288.

38. Rybakov V.V. *Admissible Logical Inference Rules. Studies in Logic and the Foundations of Mathematics, Vol. 136, Elsevier Sci. Publ., North-Holland, New-York-Amsterdam,* 1997.

39. Rybakov V.V., Kiyatkin V.R., Oner T., *On Finite Model Property For Admissible Rules.* Mathematical Logic Quarterly, Vol. 45, No. 4 (1999), pp. 505–520.

40. Rybakov V.V. Terziler M., Remazki V. *Basis in Semi-Reduced Form for the Admissible Rules of the Intuitionistic Logic IPC.* Mathematical Logic Quarterly, Vol. 46, No. 2 (2000), pp. 207–218.

41. Rybakov V.V. *Construction of an Explicit Basis for Rules Admissible in Modal System S4.* Mathematical Logic Quarterly, Vol. 47, No. 4 (2001), pp. 441–451.

42. Rybakov V.V. *Refined Common Knowledge Logics or Logics of Common Information.* Archive for Mathematical Logic, Vol. 42 2003, pp. 179–200.

43. Rybakov V.V. *Logical Consecutions in Intransitive Temporal Linear Logic of Finite Intervals.* Journal of Logic Computation, (Oxford Press, London), Vol. 15 No. 5 (2005) pp. 633–657.

44. Segerberg K. *Modal Logics with Linear Alternative Relations.-* Theoria, Vol. 36 (1970), pp. 301–322.

45. Thomason S. *Semantic Analysis of Tense Logic.* J. of Symbolic Logic, Vol. 37, No. 1 (1972).

46. van Benthem J. *The Logic of Time* Second Revised Edition, A Model-Theoretic Investigation into the Varieties of Temporal Ontology and Temporal Discourse. Kluwer, 1991, ISBN 0-7923-1081-0.

47. Venema Y. *Derivation Rules as Anti-Axioms in Modal Logic.* J. of Symbolic Logic, Vol. 58, No. 3 (1993), pp. 1003–1034.

ISOMORPHISMS AND DEFINABLE RELATIONS ON RINGS AND LATTICES*

J. A. TUSSUPOV

*ul. Stroitelnaya 31, Taraz,
484047, Republic of Kazakhstan
E-mail: tusupov@gorodok.net*

S. S. Goncharov, V. S. Harizanov J. F. Knight, C. McCoy, R. Miller and R. Solomon showed that for any computable ordinal successor α and any finite n there exists a structure with Δ_α^0 dimension n. S. S. Goncharov and J. A. Tussupov established that for any computable ordinal successor α and any finite n there are a partial ordering and directed graph with Δ_α^0 dimension n. In the present paper we show that for each computable successor ordinal α: 1) there is a computable lattice (a ring) $\mathcal{A}_0$ that is Δ_α^0 categorical but not relatively Δ_α^0 categorical (and without formally Σ_α^0 Scott family); 2) there is a computable lattice (ring) $\mathcal{A}_0$ with a relation that is intrinsically Σ_α^0 but not relatively intrinsically Σ_α^0; 3) for any finite n there is a lattice (a ring) of Δ_α^0 dimension n; 4) there is a lattice (a ring) $\mathcal{A}_0$ with copies in just the degrees of sets X such that $\Delta_\alpha^0(X)$ is not Δ_α^0.

We are interested in computable structures and some different computable representations of these models. We will consider the problems on algorithmic complexity of isomorphic and definable properties on models and connections with Scott families.

Let $\mathcal{A}$ be a computable structure. We say that $\mathcal{A}$ is Δ_α^0 *categorical* if for all computable $\mathcal{B} \cong \mathcal{A}$, there is a Δ_α^0 isomorphism from $\mathcal{A}$ to $\mathcal{B}$. We say that $\mathcal{A}$ is *relatively* Δ_α^0 categorical if for all computable $\mathcal{B} \cong \mathcal{A}$, there is a $\Delta_\alpha^0(\mathcal{B})$ isomorphism from $\mathcal{A}$ to $\mathcal{B}$. The Δ_α^0 dimension of the structure $\mathcal{A}$ is the number of computable presentations of $\mathcal{A}$ up to Δ_α^0 isomorphisms.

A *Scott family* for $\mathcal{A}$ is the set Φ of formulas, with a fixed tuple of $\bar{c}$ in $\mathcal{A}$, such that 1) each tuple of parameters in $\mathcal{A}$ satisfies some formula $\varphi \in \Phi$, and 2) if both $\bar{a}, \bar{b}$ satisfy the same formula $\varphi \in \Phi$, then there is an automorphism of $\mathcal{A}$ mapping $\bar{a}$ to $\bar{b}$.

A *formally* Σ_α^0 *Scott family* is a Σ_α^0 Scott family that is made up of "computable Σ_α" formulas.

---

*supported by the grant Scientific School — 4413.2006.1 "The theory of computability and algorithmic problems"

Theorem 1 (Ash-Knight-Manasse-Slaman, Chisholm). *A computable structure is relatively Δ_α^0 categorical iff it has a formally Σ_α^0 Scott family.*

Let $\mathcal{A}$ be a computable structure and R be a relation on $\mathcal{A}$. We say that R is *intrinsically* Σ_α^0 if in all computable $\mathcal{B} \cong \mathcal{A}$ the image of R in $\mathcal{B}$ is Σ_α^0. We say that R is *relatively intrinsically* Σ_α^0 if in all computable $\mathcal{B} \cong \mathcal{A}$, the image of R is $\Sigma_\alpha^0(\mathcal{B})$. We say that R is *intrinsically* if for each automorphism f of the structure $\mathcal{A}$ the image $f(R) \subseteq R$.

Theorem 2 (Ash-Knight-Manasse-Slaman, Chisholm). *Let $\mathcal{A}$ be a computable structure. Then a relation R on $\mathcal{A}$ is relatively intrinsically Σ_α^0 iff it is defined by a computable Σ_α formula with a finite tuple of parameters.*

In the paper [1] S.S. Goncharov described the results of the author of this paper:

Theorem 3 [1]. *For each computable successor ordinal α there is a computable directed graph (symmetric, irreflexive graph) that is Δ_α^0 categorical but not relatively Δ_α^0 categorical (and without formally Σ_α^0 Scott family).*

Theorem 4 [1]. *For each computable successor ordinal α there is a computable directed graph (symmetric, irreflexive graph) with a relation that is intrinsically Σ_α^0 but not relatively intrinsically Σ_α^0.*

Theorem 5 [1]. *For each computable successor ordinal α and each finite n there is a computable directed graph (symmetric, irreflexive graph) with Δ_α^0 dimension n.*

A degree $\mathbf{d} \leq \mathbf{0}'$ is said to be a low_n degree if $\mathbf{d}^{(n)} = \mathbf{0}^{(n)}$.

Theorem 5 [1]. *For each computable successor ordinal α there is a directed graph (symmetric, irreflexive graph) with copies in just the degrees of sets X such that $\Delta_\alpha^0(X)$ is not Δ_α^0. In particular, for each finite n there is a directed graph (symmetric, irreflexive graph) with copies in just the non low_n degrees.*

1. The lattice

Let $\mathcal{A}$ be a countable infinite symmetric, irreflexive graph with a relation of adjacency E and $|\mathcal{A}| = \omega$. We used the construction from [3].

Let $\sigma = \langle E^2 \rangle$ be a signature of symmetric, irreflexive graph and $\sigma_0 = \langle \cup, \cap \rangle$ be a signature of the lattice.

Proposition 1. *For every infinite $\mathbf{d}$-computable structure $\mathcal{A}$ of signature σ there exists a structure $\mathcal{A}_0$ of signature σ_0 such that there exists an effective* *algorithm to construct for any **d**-computable constructivisation μ of graph $\mathcal{A}$ the **d**-computable constructivisation ν_μ of the lattice $\mathcal{A}_0$, and next conditions hold:*

1) ν_μ is not autoequivalent to $\nu_{\mu'}$ $\Leftrightarrow$ μ and μ' are not equivalent enumerations.

2) for any constructivisation ν of $\mathcal{A}_0$ there exists a constructivisation μ of $\mathcal{A}$ such that constructivisations ν and ν_μ are autoequivalent.

3) if $\mathcal{A}$ has no formally Σ^0_α Scott family then $\mathcal{A}_0$ has no formally Σ^0_α Scott family.

Proof. Let $\mathcal{A}$ be countable infinite symmetric, irreflexive graph and $A \subseteq \omega$. We consider a construction for $\mathbf{d} = \mathbf{0}$. For other degrees it can be done in the same manner. If A is the basic set of $\mathcal{A}$, then as the basic set of the lattice $\mathcal{A}_0$ we consider the set A_0, equal to

$$\{a,\, b,\, k\} \cup \{c_i,\, e_i \,:\, i \in \omega\,(= A)\} \cup \{d_{i,j} \,:\, i < j \wedge E(i,\, j)\}.$$

Now we define the predicates $\cup$, $\cap$ in the following manner:

(a) If $E(i,\, j)$ and $i < j$ then $c_i \cup c_j = d_{i,j}$;

(b) $c_i \cup k = e_i$ for all $i \in \omega$;

(c) For other elements x, y we define $x \cup y = a$, $x \cap y = b$.

Lemma 1. *The graph $\mathcal{A}$ is first-order definable in the $\mathcal{A}_0$.*

Proof. We define relations:

$$x \le y \rightleftharpoons \exists z\,(x \cup z = y);$$
$$x < y \rightleftharpoons \exists z\,(x \cup z \ne x \wedge x \cup z = y).$$

All elements a, b, k are definable. The element $x = a$ is an unique element satisfying the formula

$$\phi_a(x) \rightleftharpoons \exists\, z_0\, z_1\, z_2\, (\wedge_{i \ne j}(z_i \ne z_j) \wedge x > z_0 > z_1 > z_2).$$

The element $x = b$ is an unique element satisfying the formula

$$\phi_b(x) \rightleftharpoons \exists\, z_0\, z_1\, z_2\, (\wedge_{i \ne j}(z_i \ne z_j) \wedge x < z_0 < z_1 < z_2).$$

The element $x = k$ is an unique element satisfying the formula

$$\phi_k(x) \rightleftharpoons \forall y\,(\exists z\,(b < y < z < a) \wedge x \cup y = z \wedge x < z \wedge x \ne b).$$

We define the first-order definable set

$$D(x) \rightleftharpoons \{x \in A_0 \,:\, (k \cup x \ne a) \wedge (h \cup x \ne x)\} = \{c_i \,:\, i \in \omega\}$$

and the relation

$$R(x, y) \rightleftharpoons \{(x, y) \,:\, (x \ne y) \wedge D(x) \wedge D(y)\,(x \cup y \ne a)\}$$

The set D and the relation R are relatively intrinsically computable.

We define the isomorphism between the structure $\mathcal{A}$ and the structure $\mathcal{D} \rightleftharpoons \langle D, R^2 \rangle$. Let f be a one to one map from A on D such that $f(i) = c_i$ and

$$\mathcal{A} \models E(i, j) \Leftrightarrow \mathcal{D} \models R(f(i), f(j)),$$

then f is an isomorphism. Lemma is proved.

Let $\mathcal{A}$ and $\mathcal{B}$ be isomorphic computable structures and φ be an isomorphism. We define the isomorphism φ^* from $\mathcal{A}_0$ to $\mathcal{B}_0$, corresponding to the isomorphism φ in the following manner: $\varphi^*(c_i) = c_{\varphi(i)}$, $\varphi^*(e_i) = e_{\varphi(i)}$, $\varphi^*(d_{i,j}) = d_{\varphi(i), \varphi(j)}$, $\varphi^*(a) = a$, $\varphi^*(b) = b$, $\varphi^*(k) = k$.

Lemma 2. *The map from $Iz(\mathcal{A}, \mathcal{B})$ to $Iz(\mathcal{A}_0, \mathcal{B}_0)$ is bijection.*

Proof. Let $\varphi^* : \mathcal{A}_0 \to \mathcal{B}_0$ then the map $\psi = \varphi^* \upharpoonright D(\mathcal{A}_0)$ is an isomorphism from the structure $\mathcal{D}(A_0)$ to the structure $\mathcal{D}(B_0)$. Let f_0 be an isomorphism from $\mathcal{A}$ to $\mathcal{D}(A_0)$ and f_1 be an isomorphism from $\mathcal{B}$ to $\mathcal{D}(B_0)$ then the map φ such that $\varphi = f_1^{-1} \cdot \varphi^* \cdot f_0$ is an isomorphism between the structures $\mathcal{A}$ and $\mathcal{B}$. Lemma is proved.

Lemma 3. *Let μ_0, ν_0 be constructivisations of $\mathcal{A}$ and $\mathcal{A}_0$ respectively. Then for each computable structure $\mathcal{A}'_0$ of the signature σ_0 which is* **d**-*isomorphic to the computable structure $\mathcal{A}_0$ there exists a constructivisation μ of $\mathcal{A}$ of signature σ such that the computable structure $(\mathcal{A}_0, \nu_\mu)$ corresponding to computable structure $(\mathcal{A}, \mu)$ is a structure of signature σ_0 and it is* **d**-*isomorphic to the computable structure $(\mathcal{A}_0, \nu_0)$*

Proof. Let $\mathcal{A}'_0$ be a structure **d** isomorphic to the computable structure $\mathcal{A}_0$ of signature σ_0 and φ^* be this isomorphism. Let h be a computable bijection from ω to $D(\mathcal{A}'_0)$. We define the computable predicate E^* on $|\mathcal{A}|$ such that for all $i, j \in \omega : \mathcal{A}'_0 \models R(h(i), h(j)) \Rightarrow \mathcal{A} \models E^*(i, j)$, then there exists a constructivisation μ of structure $\mathcal{A}$. Let $\mathcal{A}^* = (A, \mu)$ be a structure $\mathcal{A}$ with a constructivisation μ. Now we define computable automorphism $\zeta = \mu \cdot h$ from $\mathcal{D}(\mathcal{A}'_0)$ to $\mathcal{D}(\mathcal{A}^*_0)$ such that

$$R^{\mathcal{A}'}(x, y) \Leftrightarrow E^*(h(x), h(y))$$

$$\Leftrightarrow E(\mu(h(x)), \mu(h(y))) \Leftrightarrow R^{\mathcal{A}^*}(c_{\mu(h(x))} c_{\mu(h(y))})$$

Then there exists the automorphism φ_0^* from $\mathcal{A}_0$ to $\mathcal{A}_0^*$ which extended the automorphism ζ and it is a **d** automorphism. We denote $\mathcal{A}_0^*$ as $(\mathcal{A}_0, \nu_\mu)$ then $(\mathcal{A}_0, \nu_0)$ and $(\mathcal{A}_0, \nu_\mu)$ are **d** autoequivalent. Lemma is proved.

Lemma 4. *If there exists formally Σ_α^0 Scott family for structure $\mathcal{A}_0$ then there exists formally Σ_α^0 Scott family for structure $\mathcal{A}$.*

Proof. Suppose that there exists formally Σ_α^0 Scott family Φ for structure $\mathcal{A}_0$. Let $\phi_a(x)$, $\phi_b(x)$, $\phi_k(x)$ be formulae defined above for elements a, b, k, then $\phi_a(x)$, $\phi_b(x)$, $\phi_k(x) \in \Phi$. Let

$$D(x) \rightleftharpoons \exists z_0 \exists z_1 \, (\phi_h(z_0) \wedge \phi_a(z_1) \wedge z_0 \cup x \neq z_1 \wedge z_0 \cup x \neq x),$$
$$R(x, y) \rightleftharpoons \exists z(\phi_a(z) \, D(x) \wedge D(y) \wedge x \neq y \wedge x \cup y \neq z)$$

then $D(x)$, $R(x, y)$ be a computable Σ_α formula.

We consider c.e set Φ of computable Σ_α formulae and define the set Φ_0 of computable Σ_α formulae by the choice of formulae $\phi_{n_i}(\bar{x})$, $i \in \omega$ in enumeration of Φ which satisfies in $\mathcal{A}_0$ by tuples from $D(\mathcal{A}_0)$.

The set Φ_0 is a formally Σ_α^0 Scott family of structure $\mathcal{D} \rightleftharpoons \langle D, R \rangle$. Since the structure $\mathcal{D}$ is isomorphic to the structure $\mathcal{A}$, there exists a formally Σ_α^0 Scott family of structure $\mathcal{A}$. Lemma is proved.

2. The ring

Let $\mathcal{A}$ be a countable infinite symmetric, irreflexive graph with a relation of adjacency E and $|\mathcal{A}| = \omega$. We use the construction of the ring from [3].

Let $\sigma = \langle E^2 \rangle$ be a signature of symmetric, irreflexive graph and σ_1 be a signature of the ring.

Proposition 2. *For every infinite $\mathbf{d}$-computable structure $\mathcal{A}$ of signature σ there exists a ring $\mathcal{A}_0$ of signature σ_1 such that there exists an effective algorithm to construct for any $\mathbf{d}$-computable constructivisation μ of graph $\mathcal{A}$ a $\mathbf{d}$-computable constructivisation ν_μ of the ring $\mathcal{A}_0$, and next conditions hold:*

1) ν_μ is not autoequivalent to $\nu_{\mu'} \Leftrightarrow \mu$ and μ' are not equivalent enumerations.

2) for any constructivisation ν of $\mathcal{A}_0$ there exists a constructivisation μ of $\mathcal{A}$ that constructivisations ν and ν_μ are autoequivalent.

3) if $\mathcal{A}$ has no formally Σ_α^0 Scott family then $\mathcal{A}_0$ has no formally Σ_α^0 Scott family.

Proof. Let $\mathcal{A}$ be a countable infinite symmetric, irreflexive graph which contains circles of length 3 and $A \subseteq \omega$. We consider a construction for $\mathbf{d} = \mathbf{0}$. For other degrees it can be done in the same manner. If A is a basic set of $\mathcal{A}$, then

1. The ring $\mathcal{A}_0$ is generated by the set $\{a, b, d, e, c_i : i \in \omega(= A)\}$.
2. Multiplication is commutative.
3. The ring $\mathcal{A}_0$ has characteristic 0.
4. $a^2 = b^2 = ab = ad = bd = ae = be = 0, e^2 = a, de = d^3 = b$.

5. For all $i \in \omega : c_i^2 = a, ac_i = bc_i = dc_i = 0, ec_i = b$.

6. For all $i \in \omega$: if $E(i,j)$ then $c_i c_j = b$ (notice that if $E(i,j)$ then $i \neq j$).

The structure $\mathcal{A}_0$ satisfies the ring axioms.

Elements of the ring have a form:

$$n_0 + n_1 a + n_2 b + n_3 d + n_4 d^2 + n_2 e + \sum_{i=0}^{k} n_{i+6} c_i, \tag{1}$$

where $k \in \omega$, $n_0, \ldots, n_{k+6} \in \mathbb{Z}$. We define the sets D, P, P^2 and the relation $R(x,y)$:

$$D(x) = \{x \in A_0 : x^2 = a \wedge dx = 0, \wedge ex = b\},$$
$$P = \{x \in A_0 : x^4 = 0\},$$
$$P^2 = \{y \in A_0 : \exists x \in P(y = x^2)\}$$
$$R(x,y) = \{(x,y) : D(x) \wedge D(y) \wedge xy = b\}.$$

We also define the relation of equivalence

$$Q(x,y) = \{(x,y) : D(x) \wedge D(y) \wedge xy = a\}.$$

The sets D, P, P^2 and the relations $R(x,y), Q(x,y)$ are relatively intrinsically computable.

In the paper [3] the next properties of elements and automorphisms of the ring $\mathcal{A}_0$ are proved:

(a) Let x be of form (1) then $x \in P \Leftrightarrow n_0 = 0$,

(b) For each element x from P we have

$$x^2 = (n_5^2 + \sum_{i=0}^{k} n_{i+6}^2)a + 2(n_3 n_4 + n_3 n_5 + n_5 \sum_{i=0}^{k} n_{i+6})$$
$$+ \sum_{i=0}^{k} (\sum_{j \leq k : E'(i,j)} n_{i+6} n_{j+6})b + n_3 n^2 d^2, \tag{2}$$

where E' is image of the relation E under an automorphism of structure $\mathcal{A}$, or i.e. every element of P^2 is of form $ka + lb + md^2$, where $k, m \in \omega, l \in \mathbb{Z}$.

(c) Let f be an automorphism of $\mathcal{A}_0$. Then $f(a) = a$, $f(b) = b$, $f(d)x = dx$, $f(e)x = ex$ for all $x \in A_0$ such that $x^2 = a$.

(d) The sets D, P, P^2 and the relations $R(x,y), Q(x,y)$ are invariant.

(e) Each element $x \in D$ has a form $x = ma + nb + c_i$ for some $m, n \in \mathbb{Z}$, $i \in \omega$.

Lemma 1. *The graph $\mathcal{A}$ is first-order definable in the ring $\mathcal{A}_0$.*

Proof. We consider the first-order definable set D, relations $R(x,y)$, $Q(x,y)$ and Q^* where $Q^*(x,y) \rightleftharpoons D(x) \wedge D(y) \wedge x \neq y \wedge xy \neq b$. Let $x,y \in D(A_0)$ then we have $x = ma + nb + c_i$ and $y = m'a + n'b + c_j$. We define the class of equivalence by the relation $Q(x,y)$ on elements of $D(A_0)$. Let $\bar{x} \rightleftharpoons \{y : Q(x,y)\}$. We define the structure $\mathcal{S}^*$ with the basic set $S^* \rightleftharpoons \{\bar{x} : x \in D(A_0)\}$ and the relation $\bar{R}(\bar{x}, \bar{y})$ on the set S^* such that $\mathcal{S}^* \models \bar{R}(\bar{x}, \bar{y}) \Leftrightarrow A_0 \models R(x, y)$, where $x, y \in D(A_0)$. It is clear that $\mathcal{A}$ and $\mathcal{S}^*$ are isomorphic. Lemma is proved.

Let $S \rightleftharpoons \{s_i : s_i \in \bar{c}_i\}$ be a set of representatives of elements from S^*. We define the $\mathcal{S}(A_0) \rightleftharpoons \langle S, R \rangle$. We notice that if the structure $\mathcal{A}_0$ is computable then the structure $\mathcal{S}(A_0)$ is computable as well.

Lemma 2. *The map from $Iz(\mathcal{A}, \mathcal{B})$ to $Iz(\mathcal{A}_0, \mathcal{B}_0)$ is a bijection.*

Proof. Let φ be an isomorphism from $\mathcal{A}$ to $\mathcal{B}$, $f_A(i) = s_i$ be a bijection from $|\mathcal{A}|$ to $|\mathcal{S}_A|$, and $f_B(i) = s'_i$ be a bijection from $|\mathcal{B}|$ to $|\mathcal{S}_B|$. We define an isomorphism ψ from $\mathcal{S}_A$ to $\mathcal{S}_B$ such that $\psi = f_A^{-1} \cdot \varphi \cdot f_B$. We define isomorphism φ^* from $\mathcal{A}_0$ to $\mathcal{B}_0$ (we suppose that $a, b, d, e \in A_0, B_0$) in the following manner: Let x be of the form

$$n_0 + n_1a + n_2b + n_3d + n_4d^2 + n_5e + \sum_{i=0}^{k} N_{i+6}s_i$$

then
$$\varphi^*(x) \rightleftharpoons n_0 + n_1a + n_2b + n_3d + n_4d^2 + n_5e + \sum_{i=0}^{k} n_{i+6}\psi(s_i),$$

where $n_0, \ldots, n_{k+6} \in \mathbb{Z}, k \in \omega$.

Let $\mathcal{A}_0$ and $\mathcal{B}_0$ are isomorphic structures and φ^* is this isomorphism. We consider the structures $\mathcal{S}(A_0)$ and $\mathcal{S}(B_0)$. These structures are isomorphic and $\psi \rightleftharpoons \varphi^* \upharpoonright S(A_0)$ is this isomorphism. Let $f_0(f_1)$ be a bijections from $\mathcal{S}_A$ to ω ($\mathcal{S}_B$ to ω), then the map $\varphi = f_1 \cdot \varphi^* \cdot f_0^{-1}$ is an isomorphism from $\mathcal{A}$ to $\mathcal{B}$. Lemma is proved.

Lemma 3. *Let μ_0, ν_0 be constructivisations of $\mathcal{A}$ and $\mathcal{A}_0$ respectively. Then for each computable structure $\mathcal{A}'_0$ of the signature σ_1 which is **d**-isomorphic to the computable structure $\mathcal{A}_0$ there exists a constructivisation μ of $\mathcal{A}$ of signature σ such that the computable structure $(\mathcal{A}_0, \nu_\mu)$ corresponding to computable structure $(\mathcal{A}, \mu)$ is a structure of signature σ_1 and it is **d**-isomorphic to the computable structure $(\mathcal{A}_0, \nu_0)$*

Proof. Let $\mathcal{A}'_0$ be a structure **d** isomorphic to the computable structure $\mathcal{A}_0$ of signature σ_0 and φ^* be this isomorphism. Let h be a computable bijection from ω to $S(\mathcal{A}'_0)$, where $S(\mathcal{A}'_0)$ is a set of representatives of elements from S^* which is a class of equivalence on the structure $\mathcal{A}'_0$. We define the computable predicate E^* on $|\mathcal{A}|$ such that for all $i, j \in \omega$: $\mathcal{A}'_0 \models R(h(i), h(j)) \Rightarrow \mathcal{A} \models E^*(i, j)$, then there exist a constructivisation μ of structure $\mathcal{A}$ and $\mathcal{A}^* \rightleftharpoons (\mathcal{A}, \mu)$.

There exists a construction of a structure $\mathcal{A}_0^*$ corresponding to the $\mathcal{A}^*$. Now we define computable automorphism φ_0^* from $\mathcal{A}'_0$ to $\mathcal{A}_0^*$ by follow manner: Let $\psi = \mu \cdot h^{-1}$ is a computable permutation then element $x \in |\mathcal{A}'_0|$ of form $n_0 + n_1 a + n_2 b + n_3 d + n_4 d^2 + n_2 e + \sum_{i=0}^{k} n_{i+6} s_i$ have image $\varphi_0^*(x)$ of form:
$$n_0 + n_1 a + n_2 b + n_3 d + n_4 d^2 + n_2 e + \sum_{i=0}^{k} n_{i+6} s_{\psi(i)}.$$

Then the automorphism φ_1^* from $\mathcal{A}_0$ to $\mathcal{A}_0^*$ is a $\mathbf{d}$ automorphism. We denote $\mathcal{A}_0^*$ as $(\mathcal{A}_0, \nu_\mu)$ then $(\mathcal{A}_0, \nu_0)$ and $(\mathcal{A}_0, \nu_\mu)$ are $\mathbf{d}$ autoequivalent. Lemma is proved.

Lemma 4. *There exists formally Σ_α^0 Scott family for the structure $\mathcal{A}_0$ if and only if there exists formally Σ_α^0 Scott family for the structure $\mathcal{A}$.*

Proof. We can easily construct a formally Σ_α^0 Scott family for the structure $\mathcal{A}$ from the formally Σ_α^0 Scott family for the structure $\mathcal{A}_0$. We can prove the back condition in the same way. Lemma is proved.

An ordinal α is a *ordinalsuccesor* if $\alpha = \beta + 1$ for some ordinal β. Proposition 1 and Proposition 2 imply the next results.

Theorem. *Suppose $\mathcal{A}$ is a symmetric, irreflexive graph, and the lattice (ring) $\mathcal{A}_0$ is constructed from $\mathcal{A}$ in the way that was described in Proposition 1 and Proposition 2.*

Then $\mathcal{A}$ has a computable copy iff $\mathcal{A}_0$ has a computable copy. More generally, for any X such that $\Delta_\alpha^0(X)$ is not Δ_α^0, $\mathcal{A}$ has a X computable copy iff $\mathcal{A}_0$ has a X computable copy.

In addition,

(a) if $\mathcal{A}$ is Δ_α^0 categorical then $\mathcal{A}_0$ is Δ_α^0 categorical;

(b) if $\mathcal{A}$ has Δ_α^0 dimension n then $\mathcal{A}_0$ has Δ_α^0 dimension n;

(c) if $\mathcal{A}$ has no formally Σ_α^0 Scott family then $\mathcal{A}$ has no formally Σ_α^0 Scott family.

3. The basic results

We give the following lifting of the result of Goncharov on structures that are computably categorical but not relatively computably categorical.

Corollary 1. *For each computable successor ordinal α there is a computable lattice (ring) $\mathcal{A}_0$ that is Δ_α^0 categorical but not relatively Δ_α^0 categorical (and without formally Σ_α^0 Scott family).*

We give the following lifting of the result of Manasse on relation that are intrinsically *c.e.* but not relatively intrinsically *c.e.*

Corollary 2. *For each computable successor ordinal α there is a computable lattice (ring) $\mathcal{A}_0$ with a relation that is intrinsically Σ_α^0 but not relatively intrinsically Σ_α^0.*

Here is our lifting of the result of Goncharov on structures with finite computable dimension.

Corollary 3. *For each computable successor ordinal α and each finite n there is a computable lattice (ring) $\mathcal{A}_0$ with Δ_α^0 dimension n.*

Here is our lifting of the result of Slaman and Wehner.

Corollary 4. *For each computable successor ordinal α there is a lattice (ring) $\mathcal{A}_0$ with copies in just the degrees of sets X such that $\Delta_\alpha^0(X)$ is not Δ_α^0. In particular, for each finite n there is a lattice (ring) with copies in just the non low_n degrees.*

References

1. *Goncharov S.S.* Isomorphisms and Definable Relations on Computable Models, Proceeding of the Logic Colloquim 2005, Athens.
2. *Goncharov S.S.* The quantity of non-autoequivalent constructivisations, Algebra and Logic, vol. 16 (1977), pp. 257–282.
3. *Hirschfeldt D.R., Khoussainov B., Shore R.A., Slinko A.M.* Degree spectra and computable dimension in algebraic structures. Annals of Pure and Applied Logic. 115, 2002, pp. 71–113.
4. *S.S. Goncharov, V.S. Harizanov J.F. Knight, C. McCoy, R. Miller, R.Solomon.* Enumerations in computable structure theory, Annals of Pure and Applied Logic, v. 136 (2005), N 3, pp. 219–246.
5. *Manasse M.S.* Techniques and Counterexamples in Almost Categorical Recursive Model Theory, Ph.D. Thesis, University of Wisconsin-Madison, 1982.
6. *Wehner S.* Enumeration, countable structures and Turing degrees, Proc. of the Amer. Math. Soc.,vol.126(1998), pp 2131-2139.
7. *Slaman T.* Relative to any non-recursive set. Proc. of the Amer. Math. Soc., vol. 126 (1998), pp. 2117–2122.
8. *Ershov Yu.L., Goncharov S.S.* Constructive Models, Siberian school of Algebra and Logic, v. 6, Kluver Academic/Plenum Press, Consultant bureau, New York, 2000.
9. *Chisholm J.* Effective model theory versus recursive model theory, J. Symbolic Logic 55, 1990, pp. 1168–1191.
10. *Ash C.J., Knight J.F.* Computable Structures and the Hyperarithmetical Hierarchy, Elsevier, 2000.

THE LOGIC OF PREDICTION

EVGENII VITYAEV

*Sobolev Institute of Mathematics, Russian Academy of Science,
4 Acad. Koptyug avenue, Novosibirsk, 630090, Russia,
E-mail: vityaev@math.nsc.ru*

We consider the predictions provided by the inductive theories. For these theories predictions are performed by the Inductive Statistical (I-S) inferences. It was noted by Hempel that the I-S inference is statistically ambiguous. To avoid this ambiguity we need to use the rules that satisfy the Requirement of Maximum Specificity (RMS). The formal definition of the RMS wasn't given by Hempel. We define the notions of law and probabilistic law, and also the sets of all laws $\mathcal{L}$, and probabilistic laws $\mathcal{LP}$. We prove that the set SPL of Strongest Probabilistic Laws (with the maximum values of conditional probability) contains the set $\mathcal{L}$, so we have $\mathcal{L} \subset \mathrm{SPL} \subset \mathcal{LP}$. We prove that the maximum specific rules — the strongest SPL rules for prediction of atoms — satisfy the RMS condition. The maximum specific rules may be used in I-S inference. We prove that the set MSR of all Maximum Specific Rules is consistent and the I-S inferences based on MSR rules avoid the problem of statistical ambiguity. We define Semantic Probabilistic Inferences (SP-inference) that infer the sets $\mathcal{L}, \mathcal{LP}$, SPL, MSR. Finally, we mention the program system 'Discovery', which realize the SP-inference and discovers the sets $\mathcal{L}, \mathcal{LP}$, SPL, MSR. This system was applied for solution of many practical tasks (see website www.math.nsc.ru/AP/ScientificDiscovery).

1. Induction

1.1. *The statistical ambiguity problem*

One of the major results of the Philosophy of Science is so-called *Covering Law Model* that was introduced by Hempel in the early sixties in his famous article 'Aspects of Scientific Explanation' (see Hempel [1, 2], and Salmon [3] for a historical overview). The basic idea of this covering law model is that a fact is explained by subsumption under so-called *covering law*, i.e. the task of an explanation is to show that a fact can be considered as an instantiation of a law. In the covering law model two types of explanation are distinguished: *Deductive-Nomological* explanations (D-N explanations) and *Inductive-Statistical* explanations (I-S explanations). In D-N explanations the law is *deterministic*, whereas in I-S explanations the law is *statistical*. Right from the beginning it was clear to Hempel that two I-S explanations can yield contradictory conclusions. He called this phenomenon the *sta-*

tistical ambiguity of I-S explanations [1, 2]. Let us consider the following example of the statistical ambiguity.

Suppose that we have the following statements about Jane Jones. 'Almost all cases of streptococcus infection clear up quickly after the administration of penicillin' (L1). 'Almost no cases of penicillin resistant streptococcus infection clear up quickly after the administration of penicillin' (L2). 'Jane Jones had streptococcus infection' (C1). 'Jane Jones received treatment with penicillin' (C2). 'Jane Jones had a penicillin resistant streptococcus infection' (C3). From these statements it is possible to construct two contradictory arguments, one explaining why Jane Jones recovered quickly (E), and the other one, explaining its negation why Jane Jones did not recover quickly ($\neg$E).

$$\begin{array}{cc} Argument1 & Argument2 \\ L1 & L2 \\ \hline C1, C2 & C2, C3 \\ \hline E \; [r] & \neg E \; [r] \end{array}$$

The premises of both arguments are consistent with each other, they could all be true. However, their conclusions contradict each other, making these arguments rival ones.

Hempel hoped to solve this problem by forcing all statistical laws in an argument to be maximally specific. That is, they should contain all relevant information with respect to the domain in question. In our example, then, premise C3 of the second argument invalidates the first argument, since the law L1 is not maximally specific with respect to all information about Jane Jones. So, we can only explain $\neg$E, but not E.

1.2. *Inductive-statistical inference*

Hempel proposed the formalization of the statistical inference as Inductive-Statistical Inference (I-S inference) and the property of the maximal specific statistical laws as the Requirement of Maximal Specificity (RMS). The Inductive-Statistical Inference has the form:

$$\frac{\dfrac{L_1, \ldots, L_m}{C_1, \ldots, C_n}}{G} [r]$$

It satisfies the following conditions:

- $L_1, \ldots, L_m, C_1, \ldots, C_n \vdash G$;
- $L_1, \ldots, L_m, C_1, \ldots, C_n$ are consistent;
- $L_1, \ldots, L_m \nvdash G, C_1, \ldots, C_n \nvdash G$;

- $L_1,\ldots,L_m$ are composed of statistical quantified formulas.
- $C_1,\ldots,C_n$ are quantifier-free;
- RMS: All laws $L_1,\ldots,L_m$ are maximal specific.

In Hempel's [1, 2] the RMS is defined as follows. An I-S argument of the form:

$$\frac{\dfrac{p(G;F)}{F(a)}}{G(a)}\,[r]$$

is an acceptable I-S explanation with respect to a "knowledge state" K, if the following Requirement of Maximal Specificity is satisfied. For any class H for which the following two sentences are contained in K

$$\forall x(H(x) \Rightarrow F(x)), \tag{1}$$
$$H(a),$$

there exists a statistical law p(G;H) = r' in K such that r = r'. The basic idea of RMS is that if F and H both contain the object a, and H is a subset of F, then H provides more specific information about the object a than F, and therefore the law p(G;H) should be preferred over the law p(G; F).

1.3. *The requirement of maximal specificity in default logic*

Nowadays the same problems arise in non-monotonic logic and especially in default logic. Hempel's RMS produces also non-monotonic effects in inductive statistical reasoning. The streptococcus infection example is non-monotonic in the following sense. It was observed that the conflict between argument 1 and the argument 2 depends on the knowledge state K. If K contains only the information that John is infected, then RMS determines that argument 1 is the best explanation. In that case K implies the conclusion that John will recover quickly. However, if K is expanded with the premise C3, i.e. the information that John had a penicillin resistant streptococcus infection, then RMS determines that argument 2 is the best explanation and John will not recover quickly. Hence, the conclusion that John will recover quickly is not preserved under expansion of K.

Yao-Hua Tan [4] showed that there is a remarkable resemblance between two research traditions: default logic and inductive-statistical explanations. Both research traditions have the same research objective; to develop formalisms for reasoning with incomplete information. In both research traditions the crucial problem that had to be dealt with is the problem of *Speci-*

ficity, i.e. when two arguments conflict with each other the most specific argument has to be preferred to the less specific argument. This criterion of specificity that was proposed in AI research is very similar to the criterion of maximal specificity suggested by Hempel in the early sixties.

Let us formulate the Requirement of Maximal Specificity (RMS*) in default logic. Essentially, default logic is an ordinary first-order predicate logic extended with extra inference rules that are called default rules. The logical form of a *default rule* follows:

$$(\alpha(x) : \beta_1(x), \ldots, \beta_n(x)/\omega(x))$$

The subformulas $\alpha(x)$, $\beta_i(x)$, and $\omega(x)$ are predicate logical formulas with free variable x. The subformula $\alpha(x)$ is called the *prerequisite*, $\beta_i(x)$ are the *justifications* and $\omega(x)$ is the *consequent* of the default rule. The intuitive interpretation of a default rule follows: if the prerequisite $\alpha(x)$ is valid, and all justifications $\beta_i(x)$ are consistent with the available information (i.e. $\neg\beta_i(x)$ is not derivable from the available information), then one can assume that the consequent $\omega(x)$ is valid.

A set of formulas E is an *extension* of the default theory $\Delta = \langle W;D\rangle$, D — the set of default rules, W — a set of predicate logical formulas, if E is the smallest set such as: $W \subset E$; $E = \mathrm{Th}(E)$; for each default rule $(\alpha(x):\beta_1(x), \ldots ,\beta_n(x)/\omega(x)) \subset D$, and each term t: if $\alpha(t) \in E$, and $\neg\beta_1(t), \ldots ,\neg\beta_n(t) \notin E$, then $\omega(t) \in E$.

RMS*: If a default theory has multiple conflicting extensions, then the extension is preferred which is generated by the most specific defaults [4].

The default rule with the 'most specific' prerequisite is preferred in case of conflicts. Let A(x) and B(x) be the prerequisites of the default rules D1 and D2. The prerequisite A(x) is *more specific* than B(x) if the set that the predicate A refers to is a subset of the set that B refers to, i.e. if the sentence $\forall x(A(x) \Rightarrow B(x))$ is valid. It is obvious that this criterion can be considered as the analogue of RMS in default logic.

1.4. *The solution of the statistical ambiguity problem*

From the previous consideration we see that the statistical ambiguity problem raises in AI in different forms, but it isn't solved hitherto. We will once again state the problem that wasn't solved by Hempel and his followers:

Statistical Ambiguity Problem. Is it possible to define the RMS in such a way that it solves the statistical ambiguity problem? Can we define the RMS in such a way that the set of sentences satisfying the RMS be consistent?

This problem is very important, because it means the consistency of predictions. The predictions nowadays are produced by different AI systems: expert systems, knowledge bases, robotics, intelligent data analysis and etc.

In this paper we present the solution of this problem. We define the set of Maximum Specific Rules (MSR) and the Requirement of Maximal Specificity (RMS) and prove that sentences from MSR satisfy RMS and the set of Maximum Specific Rules (MSR) is consistent.

2. Laws

Let $\mathcal{L}$ be the first-order logic with signature $\mathfrak{S} = \langle P_1, \ldots, P_m \rangle$, $m > 0$, where $P_1, \ldots, P_m$ are the predicate symbols of arity $n_1 \ldots, n_m$. An empirical system [5] is taken to mean a finite model $\mathfrak{M} = \langle B, W \rangle$ of the signature $\mathfrak{S}$, where B is the basic set of the empirical system, and $W = \langle P_1, \ldots, P_m \rangle$ is the tuple of predicates of the signature $\mathfrak{S}$ defined on B. Let $\mathrm{Th}(\mathfrak{M})$ be the set of all rules that are true on empirical system $\mathfrak{M}$ and has the form:

$$C = (A_1 \,\&\, \ldots \,\&\, A_k \Rightarrow A_0), \quad k \geq 0 \tag{2}$$

where $A_0, A_1, \ldots, A_k$ are literals. A literal is a predicate symbol or its negation with variables instantiated for arguments.

Proposition 2.1. *The rule $C = (A_1 \,\&\, \ldots \,\&\, A_k \Rightarrow A_0)$ logically follows from any rule of the form:*

$$(A_{i1} \,\&\, \ldots \,\&\, A_{ih} \Rightarrow A_0), \{A_{i1}, \ldots, A_{ih}\} \subset \{A_1, \ldots, A_k\}, \quad 0 \leq h < k, \tag{3}$$
$$\text{that is} \quad (A_{i1} \,\&\, \ldots \,\&\, A_{ih} \Rightarrow A_0) \vdash (A_1 \,\&\, \ldots \,\&\, A_k \Rightarrow A_0).$$

Definition 2.1. By *subrule* of the rule $C = (A_1 \,\&\, \ldots \,\&\, A_k \Rightarrow A_0)$ we mean any logically stronger rule of the form (3).

Corollary 2.1. *If a subrule of the rule C is true on $\mathfrak{M}$, then the rule C is also true on $\mathfrak{M}$.*

Definition 2.2. By the *law* on $\mathfrak{M}$, we mean any rule C of the form (2) that satisfies the following conditions [6]:
    (1) C is true on $\mathfrak{M}$;
    (2) the premise of the rule is not always false on $\mathfrak{M}$;
    (3) none of its subrules is true on $\mathfrak{M}$.

Let $\mathcal{L}$ be the set of all laws on $\mathfrak{M}$. From the logic and methodology of science it is known that those hypotheses are laws that are most refutable, simple and contain the minimal number of the parameters. In our case, all these properties, that are usually difficult to define, follow from the deductive power of the laws. The 'subrules' are (i) logically stronger than the rules and more prone to become false (falsifiable) because they contain weaker premises and, therefore, applicable to bulkier data; (ii) simpler as containing less number of atomic expressions than the rule; (iii) including a smaller number of 'parameters' (the number of atomic expressions may be regarded as parameters 'tuning' the rules to data).

Theorem 2.1. $\mathcal{L} \vdash Th(\mathfrak{M})$.

3. The Probability of Events and Sentences

Let us generalize the notion of the law into the probabilistic case. For this purpose we introduce the probability on the model $\mathfrak{M}$. For the sake of simplicity we will follow paper [7], and introduce the probability μ as a discrete function on B, μ: B $\rightarrow$ [0,1], such that

$$\sum_{a \in B} \mu(a) = 1, \quad \text{and} \quad \mu(a) \neq 0, \ a \in B; \quad \mu(D) = \sum_{b \in D} \mu(b), \ D \subseteq B \quad (4)$$

We define the probability μ on the product of B^n as a probability function $\mu^n(a_1, \ldots, a_n) = \mu(a_1) \times \ldots \times \mu(a_n)$ More general definitions of the probability function μ are considered in [7].

Let us define the interpretation of the language $\mathcal{L}$ on the empirical system $\mathfrak{M} = \langle B, W \rangle$ as mapping I: $\mathfrak{S} \rightarrow$ W, which associates with every signature symbol $P_j \in \mathfrak{S}$, j = 1,...,m, the predicate P_j from W of the same arity. Let X = $\{x_1, x_2, x_3, \ldots \}$ be the set of all variables of the language $\mathcal{L}$. By the validation ν is meant the function ν: X $\rightarrow$ B, mapping variables into the set of objects B.

Let us define the probability for the sentences of the language $\mathcal{L}$. Let $U(\mathfrak{S})$ be the set of all atomic formulas of the language $\mathcal{L}$; $\mathfrak{R}(\mathfrak{S})$ is the set of all the sentences of the language $\mathcal{L}$, obtained by the closure of the set $U(\mathfrak{S})$ with respect to standard Boolean constructs $\&, \vee, \neg$. By the $\hat{\varphi}$, $\varphi \in \mathfrak{R}(\mathfrak{S})$ we define the formula, where the predicate symbols of $\mathfrak{S}$ are substituted by the predicates of W via interpretation I and by the $\nu\hat{\varphi}$ we define the formula, where variables of the formula $\hat{\varphi}$ are substituted by the objects of A via the validation ν. In particular, $\nu\hat{P}_j(x_1^j, \ldots, x_{nj}^j)^{\varepsilon j} = P_j(a_1, \ldots, a_j)^{\varepsilon j}$, $\nu(x_1^j) = a_1, \ldots, \nu(x_{nj}^j) = a_j$. Let us define the probability η of the sentences of $\mathfrak{R}(\mathfrak{S})$. If $x_1, \ldots, x_n$ are all variables of the sentence $\varphi \in \mathfrak{R}(\mathfrak{S})$, then

$$\eta(\varphi) = \mu^n(\{(a_1, ..., a_n) \mid \nu\hat{\varphi} \text{ is true on } \mathfrak{M}, \nu(x_1) = a_1, ..., \nu(x_n) = a_n\}) \quad (5)$$

4. The Probabilistic Laws on $\mathfrak{M}$

Let us revise the concept of the law on $\mathfrak{M}$ in terms of probability. We do it in such a way that the concept of the law on $\mathfrak{M}$ would be a particular case of this definition. The law on $\mathfrak{M}$ is such a true rule, which subrules are false on $\mathfrak{M}$ or in other words the law is such a true rule, that cannot be made simpler or logically stronger without losing truth. This property of the law "not to be simplified" allows stating the law not only in terms of truth but also in terms of probability.

For any rule $C = (A_1 \& \ldots \& A_k \Rightarrow A_0)$ we will define the conditional probability of the rule $\eta(C) = \eta(A_0/A_1 \& \ldots \& A_k) = \eta(A_0 \& A_1 \& \ldots \& A_k)/\eta(A_1 \& \ldots \& A_k)$.

Theorem 4.1. *For any rule* $C = (A_1 \& \ldots \& A_k \Rightarrow A_0)$, *the following two conditions are equivalent:*

(1) *the rule C is the law on $\mathfrak{M}$ that satisfies the properties* $(1), (2),$ *and* (3) *of the definition 2.2 ;*

(2) *(a)* $\eta(C) = 1;$
    (b) $\eta(A_1 \& \ldots \& A_k) > 0;$
    (c) the conditional probability $\eta(C)$ of the rule C is strictly more than conditional probabilities of each of its subrules.

Proof. $(1)(1) \leftrightarrow (2)(a)$. The rule C is true on $\mathfrak{M}$ iff due to the property (5) of the probability η of sentences $\eta(C) = 1$.

$(1)(2) \leftrightarrow (2)(b)$. The premise of the rule C is not always false on $\mathfrak{M}$ iff there exist a validation ν such that $\nu(A_1 \& \overbrace{\ldots \& A_k})$ is true on $\mathfrak{M}$. Due to the property (4) of the probability μ and the property (5) of the probability η it means that $\eta(A_1 \& \ldots \& A_k) > 0$.

$(1)(3) \leftrightarrow (2)(c)$. The the conditional probability $\eta(C)$ of the rule C is equal to 1. We need to proof that conditional probability of each of its subrules is strictly less then 1. Let us consider one of its subrule $(A_{i1} \& \ldots \& A_{ih} \Rightarrow A_0), \{A_{i1}, \ldots, A_{ih}\} \subset \{A_1, \ldots, A_k\}, 0 \leq h < k$. This subrule is not true on $\mathfrak{M}$ iff due to the property (5) of the probability η its probability is strictly less then 1. $\qquad\square$

This theorem gives us the equivalent definition of the law on $\mathfrak{M}$.

Definition 4.1. By a *probabilistic law* on $\mathfrak{M}$ *with conditional probability 1* is meant the rule $C = (A_1 \& \ldots \& A_k \Rightarrow A_0)$ of the form (2) satisfying the following conditions:

(1) $\eta(C) = 1, \eta(A_1 \& \ldots \& A_k) > 0;$

(2) conditional probability of the rule $\eta(C)$ is strictly greater than conditional probabilities of each of its subrules.

The next corollary follows from the theorem 4.1.

Corollary 4.1. *The rule is a probabilistic law on $\mathfrak{M}$ with conditional probability 1 iff it is a law on $\mathfrak{M}$.*

Let us consider items 1 and 2 of the theorem 4.1 from the standpoint of the 'not to be simplified' law:

- A law is such a true on $\mathfrak{M}$ rule, that cannot be simplified or to become logically stronger without a loss of the truth.
- Any logically stronger subrule of the rule has strictly less conditional probability (less than 1), so the rule cannot be simplified without loosing the value 1 of the conditional probability.

A more general definition of the law follows from these formulations:

Definition 4.2. The *law* is such a rule of the form (2) based on the truth values, conditional probability or other evaluations of the sentences, which cannot be made logically stronger without reducing their values.

Therefore, we can define the probabilistic law for the more general case by omitting the condition $\eta(C) = 1$ from the point (1) of the definition 4.1.

Definition 4.3. By a *probabilistic law* on $\mathfrak{M}$, we designate such a rule $C = (A_1 \& \ldots \& A_k \Rightarrow A_0)$, of the form (2), the conditional probability of which is defined and strictly more than the conditional probabilities of each of its subrules. For a particular case of the subrule $\Rightarrow A_0$ the conditional probability $\eta(C)$ of the rule C is strictly greater than the probability $\eta(A_0)$.

Let us define by the $\mathcal{LP}$ the set of all probabilistic laws. It follows from the Theorem 4.1 and the definition 4.3 that the set $\mathcal{LP}$ includes the set $\mathcal{L}$.

Corollary 4.2. $\mathcal{L} \subset \mathcal{LP}$.

Definition 4.4. By the Strongest Probabilistic Law (SPL-rule) on $\mathfrak{M}$, we designate such a probabilistic law $C = (A_1 \& \ldots \& A_k \Rightarrow A_0)$, which is not a subrule of any other probabilistic law.

We define as SPL the set of all SPL-rules.

Proposition 4.1. $\mathcal{L} \subset SPL \subset \mathcal{LP}$.

5. Semantic Probabilistic Inference

Let us define the Semantic Probabilistic Inference of the set of laws $\mathcal{L}$ and the set of probabilistic laws $\mathcal{LP}$.

Definition 5.1. By the *Semantic Probabilistic Inference* (SP-inference) of the some SPL rule C we mean such a sequence of probabilistic laws, which we denote as the sequence $C_1 \sqsubset C_2 \sqsubset \cdots \sqsubset C_n$, that:

$$C_1, C_2, \ldots, C_n \in \mathcal{LP},$$
$$C_i = (A_1^i \ \& \ \ldots \ \& \ A_{ki}^i \Rightarrow G), \quad i = 1, 2, \ldots n, \quad n > 0, \tag{6}$$
$$\text{the rules } C_i \text{ are subrules of the rules } C_{i+1},$$
$$\eta(C_{i+1}) > \eta(C_i), \quad i = 1, 2, \ldots, n-1,$$
$$C \text{ is } C_n$$

Proposition 5.1. *Any probabilistic law from $\mathcal{LP}$ belongs to some SP-inference. For any SPL-rule there is some SP-inference of that rule.*

Corollary 5.1. *For any law from $\mathcal{L}$ there is some SP-inference of that law.*

Let us consider the set of all inferences of the sentence G. This set constitutes the Semantic Probabilistic Inference tree (SPI-tree) of this sentence.

Definition 5.2. By the maximum specific rule MS(G) for the I-S inference of the sentence G we mean the SPL rule of the SPI-tree of the sentence G, which has the maximum value of conditional probability.

We define as MSR the set of all maximum specific rules.

Proposition 5.2. $\mathcal{L} \subset MSR \subset SPL \subset \mathcal{LP}$

6. Probabilistic Maximum Specific Laws

Now we define the Requirement of Maximal Specificity (RMS). We will suppose that the class H of objects in (1) is defined by some sentence H $\in \Re(\Im)$ of the language $\mathcal{L}$. In this case the RMS says that p(G;H) = p(G;F) = r for this sentence. In terms of probability η it means that $\eta(G/H) = \eta(G/F) = r$ for any H $\in \Re(\Im)$, satisfying (1).

Definition 6.1. The *Requirement of Maximal Specificity* (RMS):
if we add any sentence H $\in \Re(\Im)$ to the premise of the rule (F $\Rightarrow$ G), $\eta(G/F) = r$, such that F(**a**) & H(**a**) for some object **a**, then for the new rule (F & H $\Rightarrow$ G) we have $\eta(G/F \ \& \ H) = \eta(G/F) = r$.

In other words the requirement RMS means that there is no other sentence H in $\Re(\Im)$ that increases (or decreases, see lemma 6.1 below) the conditional probability $\eta(G/F) = r$ by adding it to the premise.

Lemma 6.1. *If the sentence $H \in \Re(\Im)$ decreases the probability $\eta(G/F \& H)$ $< \eta(G/F)$ then the sentence $\neg H$ increases it: $\eta(G/F \& \neg H) > \eta(G/F)$.*

Proof. Let us denote $a = \eta(G \& F \& H)$, $b = \eta(F \& H)$, $c = \eta(G \& F \& \neg H)$, $d = \eta(F \& \neg H)$. Then the inequality $\eta(G/F \& H) < \eta(G/F)$ may be represented as $a/b < (a+c)/(b+d)$. From the inequality $a/b < (a+c)/(b+d)$ it follows that $(a+c)/(b+d) < c/d \Leftrightarrow \eta(G/F) < \eta(G/F \& \neg H)$ $\qquad \square$

Lemma 6.2. *For any rule $C = (B_1 \& \ldots \& B_t \Rightarrow A_0)$, $\eta(B_1 \& \ldots \& B_t) > 0$, of the form (2) there is a probabilistic law $C' = (A_1 \& \ldots \& A_k \Rightarrow A_0)$ on $\mathfrak{M}$ which is subrule of the rule C and $\eta(C') \geq \eta(C)$.*

Theorem 6.1. *Any MS(G) rule satisfies the RMS requirement.*

Proof. We need to prove that for any sentence $H \in \Re(\Im)$ the equalities $\eta(G/F \& H) = \eta(G/F) = r$ take place for any MS(G) rule $C = (F \Rightarrow G)$. From the definition 6.1 it follows that there exists an object **a** such that $F(\mathbf{a}) \& H(\mathbf{a})$. Due to the property (5) of the probability η we have that $\eta(F \& H) > 0$ and, hence, the conditional probability is defined.

Let us consider the case when the sentence H is some atom B or its negation $\neg B$ and $\eta(G/F \& H) \neq r$. Then, according to the lemma 6.1 one of the rules $(F \& B \Rightarrow G)$, $(F \& \neg B \Rightarrow G)$ has the greater value of the conditional probability $\eta(F \& B \Rightarrow G) > r$ or $\eta(F \& \neg B \Rightarrow G) > r$. According to lemma 6.2 there exists a probabilistic law C', which is a subrule of the rule C and $\eta(C') \geq \eta(C) > r$. The rule C' belongs to the SPI-tree and has the greater value of the conditional probability, that is contradict to the presupposition that C is MS(G) rule.

Let us consider the case when the sentence H is a conjunction of two atoms $B_1 \& B_2$ for which the theorem is true. If one of the inequalities $\eta(G/F \& B_1 \& B_2) > r$, $\eta(G/F \& \neg B_1 \& B_2) > r$, $\eta(G/F \& B_1 \& \neg B_2) > r$, $\eta(G/F \& \neg B_1 \& \neg B_2) > r$, takes place then according to lemma 6.2, there exists a probabilistic law $C' \in$ SPI-tree, which is a subrule of the rule C and $\eta(C') \geq \eta(C) > r$. This is impossible because C is a MS(G) rule. Hence, for all these inequalities we may have only equality = or inequality <. The last case is impossible due to the following equation

$$\frac{\eta(G \,\&\, F)}{\eta(F)} = r = \frac{GFB_1B_2}{FB_1B_2}, \quad \text{where}$$

$$GFB_1B_2 = \eta(G \,\&\, F \,\&\, B_1 \,\&\, B_2) + \eta(G \,\&\, F \,\&\, \neg B_1 \,\&\, B_2) +$$
$$\eta(G \,\&\, F \,\&\, B_1 \,\&\, \neg B_2) + \eta(G \,\&\, F \,\&\, \neg B_1 \,\&\, \neg B_2),$$
$$FB_1B_2 = \eta(F \,\&\, B_1 \,\&\, B_2) + \eta(F \,\&\, \neg B_1 \,\&\, B_2) +$$
$$\eta(F \,\&\, B_1 \,\&\, \neg B_2) + \eta(F \,\&\, \neg B_1 \,\&\, \neg B_2)$$

The case when the sentence H is a conjunction of some atoms or its negations may be proved by induction.

In general case the sentence $H \in \Re(\Im)$ may be presented as a disjunction of disjoint conjunctions of atoms and their negations. For completing the proof we need to consider the case when the sentence H is a disjunction of two disjoint sentences $D \vee E$, $\eta(D \,\&\, E) = 0$, for which the theorem is true and $\eta(G/F \,\&\, D) = \eta(G/F \,\&\, E) = \eta(G/F) = r$. It follows from the equation:

$$\eta(G/F \,\&\, (D \vee E)) = \frac{\eta(G \,\&\, F \,\&\, (D \vee E))}{\eta(F \,\&\, (D \vee E))} = \frac{\eta(G \,\&\, F \,\&\, D) + \eta(G \,\&\, F \,\&\, E)}{\eta(F \,\&\, D) + \eta(F \,\&\, E)} = r$$

The case of disjunction of more than two disjoint sentences is followed by induction from the case of two disjoint sentences. $\qquad\square$

Corollary 6.1. *Any law on $\mathfrak{M}$ satisfies the RMS requirement.*

7. The Solution of the Statistical Ambiguity Problem

Theorem 7.1. *The I-S inference is consistent for any theory $Th \subset MSR$.*

Proof. Let us prove that for the sentences from $Th \subset MSR$ it is impossible to obtain a contradiction when we have two inferences $\{A \Rightarrow G, B \Rightarrow \neg G\} \subset Th \subset MSR$, where $\eta(A \,\&\, B) > 0$. We prove that in this case one of the following rules is stronger (has a greater value of conditional probability) than the rules $A \Rightarrow G$, $B \Rightarrow \neg G$.

$$A \,\&\, B \Rightarrow G, \quad A \,\&\, B \Rightarrow \neg G, \quad A \,\&\, \neg B \Rightarrow G, \quad \neg A \,\&\, B \Rightarrow \neg G \tag{7}$$

Then, according to lemma 6.2, there exist probabilistic laws with conditional probability more then the rules $A \Rightarrow G$, $B \Rightarrow \neg G$, which contradicts the condition $Th \subset MSR$.

By contradiction the rules (7) have the conditional probability no more than the rules $A \Rightarrow G$, $B \Rightarrow \neg G$.

(1) Let us consider the first rule A & B $\Rightarrow$ G. By contradiction $\eta(G/A \,\&\, B)$ $\leq \eta(G/A)$. Let us consider two cases:

(a) η (A & $\neg$B) $\neq 0$. Since η(A & B) > 0, then

$$\eta(G/A) = \frac{\eta(A \,\&\, G)}{\eta(A)} =$$

$$\frac{\eta(A \,\&\, G \,\&\, B) + \eta(A \,\&\, G \,\&\, \neg B)}{\eta(A \,\&\, B) + \eta(A \,\&\, \neg B)} \geq \frac{\eta(A \,\&\, G \,\&\, B)}{\eta(A \,\&\, B)} \Leftrightarrow$$

$$\frac{\eta(A \,\&\, G \,\&\, \neg B)}{\eta(A \,\&\, \neg B)} \geq \mu(G/A) \geq \frac{\eta(A \,\&\, G \,\&\, B)}{\eta(A \,\&\, B)} \Leftrightarrow$$

$$\mu(G/A \,\&\, \neg B) \geq \mu(G/A) \geq \mu(G/A \,\&\, B)$$

If the first inequality is strong, then the other inequalities are also strong. Therefore from the inequality $\eta(G/A \,\&\, B) < \eta(G/A)$ it follows that $\eta(G/A \,\&\, \neg B) > \eta(G/A)$. It completes the proof for this case. The remaining case is $\eta(G/A \,\&\, B) = \eta(G/A)$.

(b) η (A & $\neg$B) $= 0$. Since η(A & B) > 0, then

$$\eta(G/A) = \frac{\eta(A \,\&\, G)}{\eta(A)} = \frac{\eta(A \,\&\, G \,\&\, B) + \eta(A \,\&\, G \,\&\, \neg B)}{\eta(A \,\&\, B) + \eta(A \,\&\, \neg B)} =$$

$$\frac{\eta(A \,\&\, G \,\&\, B)}{\eta(A \,\&\, B)} = \eta(G/A \,\&\, B)$$

The remaining case is the same $\eta(G/A \,\&\, B) = \eta(G/A)$.

(2) Let us consider the rule A & B $\Rightarrow$ $\neg$G. By contradiction we have $\eta(\neg G/A \,\&\, B) \leq \eta(\neg G/B)$. By similar argumentation we have

$$\mu(\neg G/\neg A \,\&\, B) \geq \mu(\neg G/B) \geq \mu(\neg G/A \,\&\, B)$$

If the inequality $\eta(\neg G/A \,\&\, B) < \eta(\neg G/B)$ is strong, then $\eta(\neg G/\neg A \,\&\, B) > \eta(\neg G/B)$ and the theorem is proved for this case. The remaining case is $\eta(\neg G/A \,\&\, B) = \eta(\neg G/B)$.

(3) Let us consider the cases 1,2 when we have the equality:

μ (G / A & B) $= \mu$ (G / A)

$\mu(\neg$ G / A & B) $= \mu$ ($\neg$ G / B)

Then $\mu(G/A \,\&\, B) + \mu(\neg G/A \,\&\, B) = 1 = \mu(G/A) + \mu(\neg G/B)$

Since the rules A $\Rightarrow$ G and B $\Rightarrow$ $\neg$G are probabilistic laws and satisfy the conditions $\eta(\neg G/B) > \eta(\neg G)$, $\eta(G/A) > \eta(G)$. Then $1 = \eta(G/A) + \eta(\neg G/B) > \eta(G) + \eta(\neg G) = 1$

We obtained the contradiction with the presupposition. $\qquad\square$

Let us illustrate this theorem by the example of Jane Jones. We can define the maximum specific rules MS(E), MS($\neg$E) for the sentences E, $\neg$E as follows:

$\widehat{L1}$: 'Almost all cases of streptococcus infection, that are not resistant to streptococcus infection, clear up quickly after the administration of penicillin';

L2 : 'Almost no cases of penicillin resistant streptococcus infection clear up quickly after the administration of penicillin'.

The rule $\widehat{L1}$ has the greater value of conditional probability, than the rule L1 and, hence, it is a MS(E) rule for the sentences E. These two rules can't be fulfilled on the same data.

Conclusion. We can predict without contradictions if we use the set MSR as statistical laws in I-S inference.

8. The Relational Data Mining and Program System 'Discovery'

Based on the semantic probabilistic inference the Relational Data Mining (RDM) approach to the intensive area of applications - Knowledge Discovery in Data Bases and Data Mining (KDD & DM) - was developed [8-10]. The program system 'Discovery', which utilizes this approach, has been implemented. In the frame of this approach we may discover the full (in the sense of theorem 2.1) and consistent (in the sense of theorem 7.1) set of rules. In [6] we argue that using RDM we may cognize the object domain. The system 'Discovery' realizes the Semantic Probabilistic Inference and can discover the sets of laws $\mathcal{L}$, $\mathcal{LP}$ and the sets SPL, MSR. The system 'Discovery' has been successfully applied to solving many practical tasks: cancer diagnostic systems, time series forecasting, psychophysics, bioinformatics, and many others (see www-site Scientific Discovery [11]).

Acknowledgments

The work is partially supported by the Russian Foundation for Basic Research 05-07-90185-v, Scientific Schools grant of the President of the Russian Federation 4413.2006.1.

References

1. Hempel, C. G. (1965) Aspects of Scientific Explanation, In: C. G. Hempel, *Aspects of Scientific Explanation and other Essays in the Philosophy of Sci-*

ence, The Free Press, New York.
2. Hempel, C. G.: 1968, 'Maximal Specificity and Lawlikeness in Probabilistic Explanation', *Philosophy of Science* **35**, 116–33.
3. Salmon, W. C. (1990) *Four Decades of Scientific Explanation*, University of Minnesota Press, Minneapolis.
4. Yao-Hua Tan (1997) is default logic a reinvention of inductive-statistical reasoning? *Synthese* **110:** 357–379, *Kluwer Academic Publishers.*
5. Krantz, D.H., Luce, R.D., Suppes, P., Tversky, A. (1971, 1989, 1990), Foundations of measurement, Vol. 1,2,3, NY, London: Acad. press, (1971) 577 p., (1989) 493 p., (1990) 356 p.
6. Evgenii Vityaev, Boris Kovalerchuk, Empirical Theories Discovery based on the Measurement Theory. *Mind and Machine*, v. 14, # 4, 551–573, 2004
7. Halpern, J.Y. (1990), 'An analysis of first-order logic of probability', *Artificial Intelligence* **46**, pp.311-350.
8. Kovalerchuk, B., Vityaev, E. (2000), Data Mining in finance: Advances in Relational and Hybrid Methods, Kluwer Academic Publishers, 308 p.
9. Kovalerchuk, B., Vityaev, E., Ruiz, J.F. (2001), 'Consistent and Complete Data and "Expert" Mining in Medicine'. In: Medical Data Mining and Knowledge Discovery, Springer, pp. 238–280.
10. Evgenii Vityaev, Boris Kovalerchuk. Data Mining For Financial Applications. In: O. Maimon and L. Rokach (eds.), Data Mining and Knowledge Discovery Handbook: A Complete Guide for Practitioners and Researchers, Springer 2005, pp. 1203–1224.
11. Scientific Discovery http://www.math.nsc.ru/AP/ScientificDiscovery

THE CHOICE OF STANDARDS FOR A REPORTING LANGUAGE

MICHAŁ WALICKI and UWE WOLTER

Department of Informatics,
University of Bergen
Post Box 7800,
N-5020 Bergen, Norway
E-mail: {michal, Uwe.Wolter}@ii.uib.no

JACK STECHER

Norwegian School of Economics and Business Administration,
Helleveien 30,
N-5045 Bergen, Norway
E-mail: jack.stecher@nhh.no

1. Introduction

The problem of information exchange among agents who see the world subjectively arises in many contexts, ranging from computer science [1] to linguistics [7] to financial accounting [2]. One aspect of this problem is the choice of whether to use a standardized terminology among all agents or to use different languages for different audiences; this has been of particular interest in international accounting contexts [3, 17].

We use topological structures to model reporting under subjective information. The points in a topological space represent what an agent in principle could wish to communicate — i.e., the world as the agent subjectively understands it. An agent need not have a distinct way of conveying everything he knows, and some terms in the agent's language may have multiple meanings. We thus think of an agent as not being able to communicate specific points, but only opens in a topological space.

There is some evidence that this idea is what standard-setting bodies have in mind in financial reporting contexts. For example, the Financial Accounting Standars Board, in [4], specifies that reported information should be "understandable" to those with a general familiarity with how businesses operate, and that reported information is approximate in nature, and should be changed only when such changes make a "material" difference (i.e., when such changes yield reports that are not within some neighborhood of each other). The International Accounting Standards Board adopts a similar position in [6].

Our approach follows the spirit—though not the details of the technical development—of formal topology [9, 12], which postulates separation of points and opens of a topological system into two distinct (yet related) sets, i.e., structures (O, pt, D). The separation invites some degree of independence in treatment of points, D, and opens, O, with one extreme being simply removing the points completely. Keeping both sets present, one can endow O with various algebraic structures, each leading to a corresponding requirement on the relation pt between opens and points. Thus, the frames of pointfree topology are complete lattices with finite meets distributing over infinite joins. The relation to points is then required to respect these. One can think of weaker structures on O, e.g., as only meet-semilattice or just partial orders, with the respective restrictions (of meet- or po-compatibility) on pt. On the other extreme to that of frames, one may allow O, D to be arbitrary sets and pt an arbitrary relation. This will be our setting, which is investigated under the name "basic pair" in [10].

Compatibility of the relation pt and the structure of O ensures that the algebraic properties of O reflect, as far as possible, the topological properties of D. In case of frames, the collection of preimages $pt^-(o)$ for all $o \in O$ gives the full topology (all open sets) on D. In the case of meet-semilattices, such preimages yield only a basis for a possible topology. In our case of arbitrary sets and relations, a topology on D is obtained by taking the preimages of O as a subbasis.

The reason for our choice is the context of application. We intend the points in D as distinctions identifiable by an agent in his experience (or world), while we envision O as the possible reports the agent may give to describe his world. I.e., the members of O are thought of *names* of the opens rather than the opens themselves. The spirit of formal topology allows one then to have different reports in O which are extensionally indistinguishable, i.e., which denote the same (open) sets of points. The same spirit is discernible in the framework of [11] and of Chu spaces [8].

In contrast to formal topology utilizing frames, our application does not justify putting any specific restrictions on the relation between these two sets, nor on possible structure of either. Such structure and dependencies are to be induced exclusively by the relation between the sets. For instance, one might wish to endow O with a partial ordering representing the specialization of reports (as in [13]). We find it natural to introduce such an ordering by means of the very relation between reports and points, namely, to view a report r as (weakly) more specific than s simply when its extension is included in that of s, $pt(r) \subset pt(s)$.

The proofs of the results here will be available in a technical report [16]. For this article, we restrict ourselves to shorter proofs and to sketches of the general arguments that underly our findings.

2. The category Rep

Objects in our category are multialgebras over a signature with two sort symbols O, D, and one operation symbol $pt : O \to D$. A multialgebra A over this signature is a pair of (possibly empty) sets, O^A, D^A, with a set-valued function $pt^A : O^A \to \mathcal{P}(D^A)$.[a] Given a multilagebra A, we write $\Omega(A)$ for the topology induced on D^A by the relation pt^A, i.e., by taking as the subbasis $\mathcal{S}B(A) = \{pt^A(o) \mid o \in O^A\} \cup \{\varnothing, D^A\}$. Notice that we do not require totality or surjectivity of pt^A; e.g., there may be points $d \in D^A$ such that for all $o \in O^A : d \notin pt^A(o)$. By adding the whole set D^A to the subbasis, we only ensure that a topology is always induced on the whole D^A and not only on its subset. Likewise, there may be "empty" reports $o \in O^A$ which are not related to any points, i.e., $pt^A(o) = \varnothing$. Morphisms of such structures might seem at first to present a difficulty due to all too many choices. We are able to address this issue, following the choice presented in the overview and classification of homomorphisms of multialgebras given in [14, 15]; we will justify this further in what follows.

Definition 2.1. A homomorphism between two multialgebras, $\phi : A \to B$, is a pair of functions $\phi_O : O^A \to O^B$ and $\phi_D : D^A \to D^B$, as on the left:

$$
\begin{array}{ccc}
\mathcal{P}(D^A) \xrightarrow{\ \mathcal{P}(\phi_D)\ } \mathcal{P}(D^B) & \qquad & \mathcal{P}(D^A) \xleftarrow{\ \mathcal{P}(\phi_D^-)\ } \mathcal{P}(D^B) \\
\ \uparrow{\scriptstyle pt^A} \qquad\qquad \uparrow{\scriptstyle pt^B} & & \ \uparrow{\scriptstyle pt^A} \qquad\qquad \uparrow{\scriptstyle pt^B} \\
O^A \xrightarrow{\ \phi_O\ } O^B & & \mathcal{P}(O^A) \xleftarrow{\ \phi_O^-\ } O^B
\end{array}
$$

and such that: $\forall y \in O^B : pt^A(\phi_O^-(y)) = \phi_D^-(pt^B(y))$.

Function applications are extended pointwise to sets; i.e., we operate with the *weak-image*, where $x \in f(Y)$ iff $\exists y \in Y : x \in f(y)$ iff $f^-(x) \cap Y \neq \varnothing$. Note that commutativity of the diagram goes in the direction opposite to

---

[a]Of course, set-valued functions can be viewed as relations. However, when we focus on homomorphisms, the difference between the two viewpoints becomes significant, and the more structured/algebraic character of functions turns out to be useful. All mentioned results concerning multialgebras and their categories can be found in [15].

the arrows ϕ_O, ϕ_D, as shown on the right. This opposite direction reflects the topological tradition. In the notation, we will usually confuse the two and write both simply as ϕ. The homomorphisms do compose and yield a category Rep (in [14, 15], it was called MAlg_{OT}).

One property of this definition of homomorphism is that, for every $d \in D^B$, if d is in the image of ϕ, then so are all reports of d; i.e., the image $\phi[A]$ ($= \phi_O[O^A] \cup \phi_D[D^A]$) is closed under preimages of pt^B : $d \in \phi_O[O^A] \Rightarrow (pt^B)^-(d) \subseteq \phi_D[D^A]$. Another one is that, if o is an "empty concept" in A, i.e., $pt^A(o) = \varnothing$, then $pt^B(\phi_O(o)) \neq \varnothing$ implies $pt^B(\phi_O(o)) \cap \phi_D[D^A] = \varnothing$. In other words, the ϕ_O-image of an "empty concept" need not be "empty", but none of its points can be in the image of ϕ_D. If ϕ is viewed as a standard that translates agent A's world into agent B's world, then these two properties have the following intuitive interpretations. First, if A asks B for something, and B can interpret this request as including d, then anything B could have interpreted as including d must be acceptable to A. Second, if o is a nonsensical term in A's language, then B should only be able to interpret $\phi(o)$ as something meaningless to A.

The force of this definition is that, if a standard ϕ enables A to communicate with B, then B must be able to validate what they discussed. In accounting contexts, this is called *representational faithfulness* [4]. Note that this concept has a direction: what the reporting entity states must be meaningful to the end user, and in particular must be something the end user can validate. The fact that a homomorphism can exist from $A \to B$ without one necessarily existing from $B \to A$ captures this idea.

We often take advantage of the fact that morphisms are defined using functions rather than relations. For instance, stating continuity conditions for relations offers several choices and complications. By contrast, when ϕ is a function, $\phi^-(X \cap Y) = \phi^-(X) \cap \phi^-(Y)$, giving us the following:

Fact 2.2. If $\phi : A \to B$ is a homomorphism, then $\phi_D : D^A \to D^B$ is a continuous mapping of the topologies $\Omega(A) \to \Omega(B)$.

Proof. It is straightforward to show that the preimage of any element of the subbasis $pt^B(y) \in \mathcal{SB}(B)$ is an open in $\Omega(A)$. The result follows. $\quad\square$

The homomorphism condition is, in fact, stronger than mere continuity, as the following example illustrates:

Example 2.3. Consider two algebras:

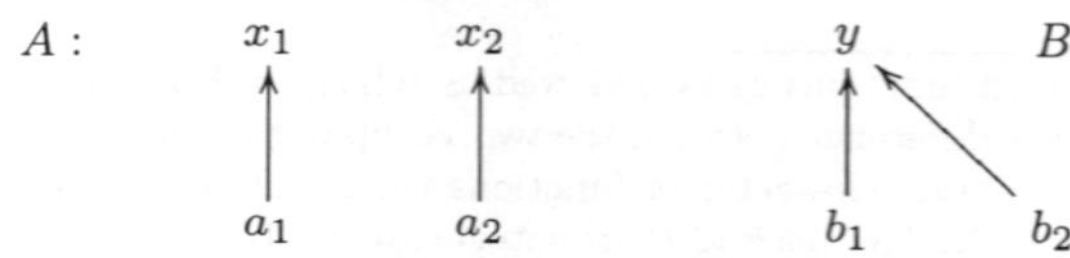

Viewed as topologies on the respective D sort, we have that $\Omega(A)$ is discrete and $\Omega(B) = \{\varnothing, \{y\}\}$. The mapping $\phi_D(x_i) = y$ is continuous. There is, however, no homomorphism $\phi : A \to B$. Its existence would require also a compatibility of reports (not only opens), namely, that preimage of B's report means for A the same as the preimage of its meaning for B, i.e., for $i \in \{1, 2\} : pt^A(\phi^-(b_i)) = \phi^-(pt^B(b_i)) = \phi^-(y) = \{x_1, x_2\}$.

The definition of continuity can vary slightly between different frameworks of formal topology, so let us only consider one simple example.

Example 2.4. The two algebras from the previous example can be viewed as basic pairs of [10] (where pt is viewed as a relation). The (continuous) morphisms proposed there are pairs of relations $r_D \subseteq D^A \times D^B$ and $r_O \subseteq O^A \times O^B$, such that $(pt^A)^-; r_O = r_D; (pt^B)^-$, where $\_;\_$ denotes the usual composition of binary relations.

The mapping $\phi(x_i) = y$ and $\phi(a_i) = b_i$ will not satisfy this condition. However, the two structures are related by the morphism with $r_D = D^A \times D^B$ and $r_O = O^A \times O^B$.

Intuitively, we would say that a homomorphism $\phi : A \to B$ describes the space of possible "adequate communication" *from A to B*. Notice that in the setting of the last example, any morphism $A \to B$ gives rise to the (inverse) morphism $A \leftarrow B$; i.e., using morphisms as "ways of communication", all communication becomes mutual and symmetric. $\phi_O(a_i) = b_i$ means that A's saying a_1 is heard by B as b_1. The adequacy is verified at A's intended distinctions: by a_i, A intends x_i. If B hears b_1, he understands by it $pt^B(b_1)$; the preimage of these points must equal what A understands by all reports which could be heard by B as b_1.

In the present case, this is impossible because B has too many reports while A has too precise language. Whatever he says will be understood by B as y, but there is no imprecision in A's language that corresponds to the imprecision of B's $pt^B(b_1) = y = pt^B(b_2)$. With such an interpretation, it seems appropriate to exlude any morphism from A to B. If B did not have the report b_2, we would obtain a possible homomorphism from A; and likewise if A had more superflous words, e.g., an a_3 with $pt^A(a_3) = x_2$.

On the other hand, an "adequate communication" from B to A is possible, since B has very little to communicate: his y can be taken as A's x_1, in which case both reports b_1, b_2 are taken by A as a_1.

2.1. *Homomorphisms, congruences and subalgebras*

The following fact gives a handy characterization of epis and monos.

Fact 2.5. A morphism is epi (mono) if and only if it is surjective (injective).

The classical congruence condition is replaced by *bireachability*, which is defined as a relation between arbitrary two algebras.

Definition 2.6. Given $A_1, A_2 \in \mathsf{Rep}$, a relation $\sim \,\subseteq A_1 \times A_2$ (i.e., a pair of relations $\sim_O \,\subseteq O^{A_1} \times O^{A_2}$ and $\sim_D \,\subseteq D^{A_1} \times D^{A_2}$) is a bireachability iff:

$$\forall a, b, a_1 : a \sim_D b \wedge a \in pt^{A_1}(a_1) \Rightarrow \exists b_1 \in A_2 : b \in pt^{A_2}(b_1) \wedge a_1 \sim_O b_1$$
$$\&\ \ \forall a, b, b_1 : a \sim_D b \wedge b \in pt^{A_2}(b_1) \Rightarrow \exists a_1 \in A_1 : a \in pt^{A_1}(a_1) \wedge a_1 \sim_O b_1$$
$$(1)$$

A bireachability C between A_1 and A_2 is given a natural algebraic structure:

$$pt^C(\langle a, b \rangle) = pt^{A_1}(a) \times pt^{A_2}(b) \cap C. \tag{2}$$

A bireachability on A is a bireachability between A and A.

Bireachability can be viewed as "bisimilarity in the opposite direction." The name refers to the following property of such relation: If two points are bireachable, $d_1 \sim_D d_2$, and $d_1 \in pt^{A_1}(o_1)$ then there exists $o_2 \sim_O o_1$ such that $d_2 \in pt^{A_2}(o_2)$, and vice versa. Since there are no operations returning elements of sort O, two arbitrary elements of this sort can be made bireachable. One verifies easily that with the algebraic structure on a bireachability C given by (2.8), the projection arrows $pr_i : C \to A_i$, $pr_i(\langle a_1, a_2 \rangle) = a_i$, are morphisms in Rep.

Fact 2.9. For every span of morphisms $\phi_i : X \to A_i$, $i \in \{1, 2\}$, the relation $R = \{\langle \phi_1(x), \phi_2(x) \rangle \mid x \in X\}$ is a bireachability between A_1 and A_2.

Bireachability between A_1 and A_2 represents some degree of "compatibility." For instance, any subset of $O^{A_1} \times O^{A_2}$ is a bireachability, which can be interpreted as saying that, as long as one does not take into account the "real distinctions" (points in D), any reports of A_1 can be related to any of A_2. However, if we also want to relate some points $d_1 \in D^{A_1}$ and $d_2 \in D^{A_2}$, "compatibility" requires that the respective reports are related: for any report $o_1 \in (pt^{A_1})^-(d_1)$, there must be a corresponding report $o_2 \in (pt^{A_2})^-(d_2)$ and vice versa. Although a span from X induces a bireachability R, the morphisms need not factor through the induced R.

Given two algebras and a collection of bireachabilities $C_i \subseteq A_1 \times A_2$, their union $\bigcup_i C_i$ satisfies condition (2.7). Thus, collecting all bireachabilities between A and B, we obtain the maximal one.

Fact 2.10. For every $A_1, A_2 \in \mathsf{Rep}$ there exists a unique maximal (wrt. set-inclusion) bireachability between A_1 and A_2.

In particular, for every algebra A, there exists a maximal bireachability on A. It will always be total on the O-part, i.e., $O^A \times O^A$. But it need not be total on D. E.g., for $O^A = \{o\}, D^A = \{d_1, d_2\}$ and $pt^A(o) = \{d_1\}$, there is no bireachability making $d_1 \sim d_2$.

Fact 2.11. The kernel of a homomorphism $\phi : A \to B$ is a bireachability equivalence on A and, given a bireachability equivalence $\sim \subseteq A \times A$, we obtain an epimorphism $e : A \to A/_\sim$, where the latter is defined as the collection of $\sim$-equivalence classes with the operation given by $pt^{A/\sim}([o]) = \{[d] \mid \exists o' \in [o]\ d' \in [d] : d' \in pt^A(o')\}$.

Henceforth, congruence will mean bireachability equivalence. The existence of a congruence on A which is the identity on O^A and non-identity on D^A implies that $\Omega(A)$ is not even T_0; i.e., there are (at least) two distinct points which belong to exactly the same opens. The above quotient by such a congruence amounts to identifying all topologically indistinct points.[b]

Define a subalgebra relation $A \sqsubseteq B$ iff the inclusion $A \subseteq B$ defines a homomorphism from $A \to B$. This relation is dual to the classical one:

Fact 2.12. $A \sqsubseteq B$ if $A \subseteq B$ and A is closed under B-preimages of operations, i.e., $\forall d \in D^A \subseteq D^B : (pt^B)^-(d) \subseteq A$.

In particular, we have the following useful facts:

Fact 2.13. For any $A \in \mathsf{Rep}$ and every $d \in D^A$, the pair $S = \langle (pt^A)^-(d), d \rangle$ with the operation $pt^S(x) = d$ for all $x \in (pt^A)^-(d)$ is a subalgebra $S \sqsubseteq A$.

Fact 2.14. For $B \in \mathsf{Rep}$ and $X \subseteq B$, there is a largest $A \sqsubseteq B$ with $A \subseteq X$.

Proof. Let $A_0 = X$. Given A_i, define A_{i+1} by removing all elements $e \in A_i$ such that for some $a_0 \in B \setminus A_i : e \in f^B(a', a_0, a)$. $A = \bigcap_{i \in N} A_i$. $\square$

3. (Co)completeness of Rep

We show that the category Rep is complete and cocomplete. The latter is a special case of the general result from [15] as are the existence of final objects and equalizers. Subsection 3.2 describes the character and construction of (binary) products, and thus shows the existence of (finite) limits in Rep.

---

[b]Vickers [12], p. 62, calls this the "localification" of the space $\Omega(A)$.

3.1. *Some earlier results*

The following are proved in [15,16].

Theorem 3.1. Rep *is cocomplete.*

Remark 3.2. See [15]. The initial object is the empty algebra $(\varnothing, \varnothing, \varnothing)$. Coproducts are disjoint unions with the natural extensions of *pt*.

StLemma 3.3. Rep has final objects and equalizers.

3.2. *Products*

We consider first the relationship between products and (maximal) bireachability between algebras. In the case of co-algebras for functors preserving mono-sources, products and maximal bisimulation coincide (theorem 8.6 in [5]). If we were to consider only the subcategory of multialgebras obtained as inverse from coalgebras (over a given polynomial functor), we could conclude the existence of products, namely, of maximal bireachabilities between the arguments. That is, if the inverse of *pt* is deterministic, the maximal bireachability becomes a product.

Example 3.4. Consider two algebras:

The maximal bireachability between them is the following:

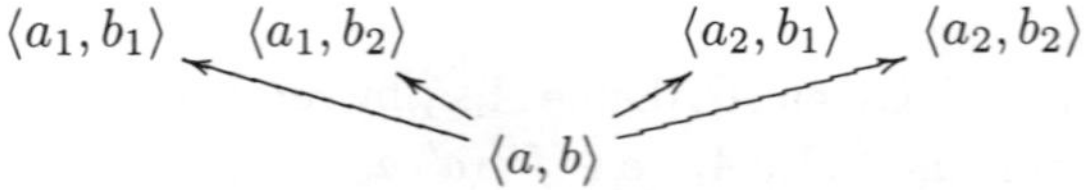

and this is the product $A \times B$.

Thus, if there are only ambiguous reports (like a or b), but no points are reported in more than one way, the product — maximal bireachability — increases the ambiguity. However, our case is both more general and more complicated. In general, maximal bireachability need not be the product: counterexamples are provided by multialgebras which are not inverses of coalgebras for polynomial functors. The problem arises when the reports are overly precise, in the sense of several reports denoting the same points.

Example 3.5. Consider two algebras:

The following are examples of bireachabilities between A_1 and A_2:

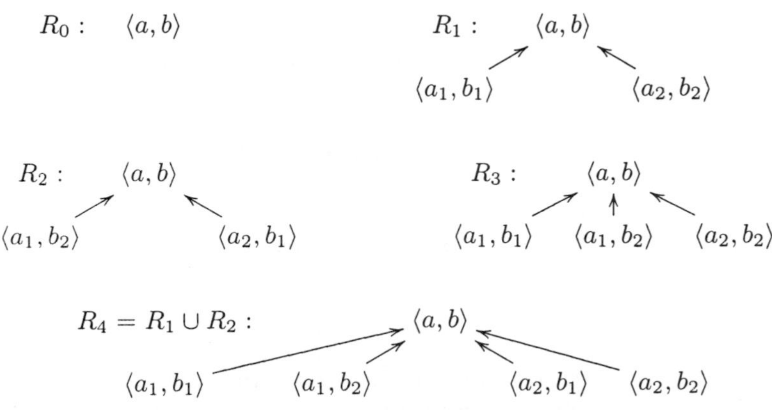

R_4 is the maximal bireachability between A and B—every other bireachability is a subset of it. However, only $R_0 \sqsubseteq R_4$; the inclusions of R_1, R_2, and R_3 are not homomorphisms. Consequently, R_4 cannot be the product $A_1 \times A_2$, as the projections from, say, R_2 would not factor through it.

Fixing the algebras A_1, A_2 and letting $\mathcal{R}_{A_1 \times A_2}$ be the collection of all bireachabilities between them, we consider the diagram $\langle \mathcal{R}_{A_1 \times A_2}, \sqsubseteq \rangle$ and its colimit P. Consider only R_1, R_3 from the above example. They have two common subobjects, $R_0 = \langle a_1, b_1 \rangle$ and $R_0' = \langle a_2, b_2 \rangle$, and these subobjects have to be identified. Writing these in bold below, the result is:

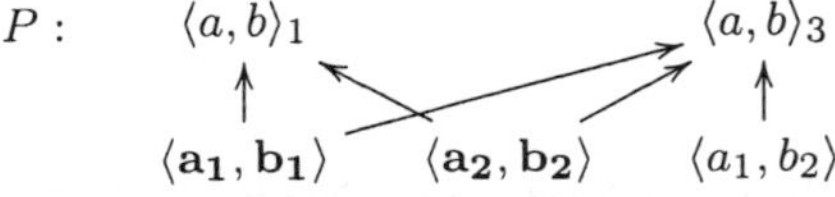

$P \not\sqsubseteq A_1 \times A_2$, so it is no longer a bireachability between A_1 and A_2. But we do have that $R_0, R_0', R_1, R_3 \sqsubseteq P$, where the inclusion of R_1 will map $i_1(\langle a, b \rangle) = \langle a, b \rangle_1$, while that of $R_3 : i_3(\langle a, b \rangle) = \langle a, b \rangle_3$. Thus, if we interpret a bireachability between A_1 and A_2 as a kind of potentially compatible communication, the colimit of $\langle \mathcal{R}_{A_1 \times A_2}, \sqsubseteq \rangle$ represents all such potentials, some of which need not be possible simultaneously. Thus, the colimit collects all possible combinations of compatible reports and assigns to them respective compatible points. Associated with the colimit P are the projections $\pi_i : P \to A_i$, $i \in \{1, 2\}$, obtained as the mediating arrows for all projections $\{p^r_A : R_r \to A_i \mid R_r \in \mathcal{R}_{A_1 \times A_2}\}$.

The following properties of colimits are proved in [16]:

Fact 3.6. Let P be the colimit of the diagram $\langle \mathcal{R}_{A_1 \times A_2}, \sqsubseteq \rangle$ for $A_i \in \mathsf{Rep}$.

(1) $O^P \simeq O^{A_1} \times O^{A_2}$,
(2) for every $d, d' \in D^P$, if $(pt^P)^-(d) = (pt^P)^-(d')$ and $\pi_i(d) = \pi_i(d')$ for $i \in \{1, 2\}$, then $d = d'$.

StLemma 3.7. Given $A_1, A_2 \in \mathsf{Rep}$, the colimit P of $\langle \mathcal{R}_{A_1 \times A_2}, \sqsubseteq \rangle$ is their product.

Theorem 3.8. *The category* Rep *is (finitely) complete.*

4. Conclusion

We present a formal structure for studying communication among agents who view the world in different, subjective ways. In particular, we develop a category Rep of reporting environments, and model the translation of one reporting environment to another as a homomorphism in this category. We show that Rep is complete and cocomplete, and observe that homomorphisms between two agents may exist in only one direction. Thus, communicating to someone is different from understanding someone.

This category provides a promising new approach for studying standardization of communication. A fulfillment of this promise will be the point of future research, for example studying settings where agents are not aware of the same things (pullbacks in Rep) and where agents need to consider outside parties when agreeing upon a standard (pushouts).

References

1. René M C Ahn. *Agents, Objects, and Events: A Computational Approach to Knowledge, Observation, and Communication*. PhD thesis, Technische Universitiet Eindhoven, 2000.
2. Harold G Avery. Accounting as a language. *The Accounting Review*, 28(1):83–7, 1953.
3. Eli Bartov, Stephen R Goldberg, and Myungsun Kim. Comparative value relevance among German, U.S., and international accounting standards: A German stock market perspective. Technical report, New York University, Grand Valley State University, and University of Missouri–Columbia, 2002.

4. FASB. Qualitative characteristics of accounting information. Statement of Financial Accounting Concepts No. 2, Financial Accounting Standards Board, 1980.

5. H. Peter Gumm and Tobias Schröder. Products of coalgebras. *Algebra Universalis*, 46:163–185, 2001.

6. IASB. Framework for the preparation and presentation of financial statements. Technical report, International Accounting Standards Board, 1989 (adopted 2001).

7. Johan Anthony Willem Kamp. A theory of truth and semantic representation. In Jeroen Groenendijk and Martin Stokhof, editors, *Formal Methods in the Study of Language*, pages 277–322. Mathematisch Centrum, 1981.

8. Dusko Pavlović and Vaughn R Pratt. The continuum as a final coalgebra. *Theoretical Computer Science*, 280(1–2):105–22, 2002.

9. Giovanni Sambin. Intuitionistic formal spaces—a first communication. In Dimiter G Skordev, editor, *Mathematical Logic and Its Applications*, pages 187–204. Plenum, 1987.

10. Giovanni Sambin and Silvia Gebellato. A preview of the basic picture: a new perspective on formal topology. In Torstein Altenkirch, Wolfgang Narashewski, and Bernhard Reus, editors, *Types for Proofs and Programs*, volume 1657 of *LNCS*, pages 194–207. Springer, 1998.

11. Silvio Valentini. Fixed points of continuous functions between formal spaces. Technical report, Department of Pure and Applied Mathematics, University of Padua, 2001.

12. Steven Vickers. *Topology via Logic*. Cambridge University Press, 1989.

13. Jan von Plato. Order in open intervals of computable reals. *Mathematical Structures in Computer Science*, 9:103–8, 1999.

14. Michał Walicki. Bireachability and final multialgebras. In J. L. Fiadeiro, N. Harman, M. Roggenbach, and J. Rutten, editors, *Algebra and Coalgebra in Computer Science*, volume 3629 of *LNCS*, pages 408–423. Springer, 2005.

15. Michał Walicki. Universal multialgebra. Technical Report 292, Department of Informatics, University of Bergen, 2005.

16. Michał Walicki, Uwe Wolter, and Jack Stecher. A category for studying the standardization of reporting languages. Technical Report, Department of Informatics, University of Bergen, 2005.

17. Theodore L Wilkinson. Can accounting be an international language? *The Accounting Review*, 39(1):133–9, 1964.

CONCEPTUAL SEMANTIC SYSTEMS
THEORY AND APPLICATIONS*

K. E. WOLFF

Darmstadt University of Applied Sciences
Mathematics and Science Faculty
Schoefferstr. 3
D-64295 Darmstadt, GERMANY
E-mail: karl.erich.wolff@t-online.de

The purpose of this paper is to present recent developments in Conceptual Knowledge Processing. That field of research is based on the mathematical theory of Formal Concept Analysis (FCA) which has its origins in the logical, algebraical, and geometrical roots that gave rise to the development of lattice theory by Garrett Birkhoff. The broad applicability of FCA stems from the notion of a *formal concept* as introduced by Rudolf Wille. Following the suggestions of Birkhoff FCA has been applied during the last twenty-five years very successfully in practice and in many branches of science. Based on these theoretical and practical experiences the author has introduced the notion of a Conceptual Semantic System which combines in a simple and general way arbitrary relational structures with the powerful tools of conceptual scaling. For applications in practice we obtain visualizations of arbitrary relational structures in form of line diagrams of concept lattices. This paper introduces the notion of a Temporal Conceptual Semantic System for a general treatment of temporal phenomena related to distributed concepts.

1. Introduction

In this paper we present a simple and general mathematical structure, called *Conceptual Semantic System* (CSS). We use the notion of a CSS for the representation of arbitrary relational knowledge in theory and practice. In this paper we introduce a general treatment of temporal phenomena in Conceptual Semantic Systems. That is based on previous investigations of more special conceptual representations of processes in Conceptual Time Systems by the author [23].

In the following we shortly describe the main steps of the development of the notion of *Conceptual Semantic Systems*.

---

*This work is supported by DFG project COMO, GZ: 436 RUS 113/829/0-1.

1.1. *Garrett Birkhoff: Lattices*

Abstracting the similarities between logical, algebraical, and geometrical hierarchies Garrett Birkhoff [4] has introduced the theory of lattices. He has also emphasized to use lattices for applications in practice. That was done first by Barbut and Monjardet [1] and later by Wille [10] who established the powerful connection to the philosophical understanding of concepts.

1.2. *Formal Contexts and their Concept Lattices*

With the purpose of restructuring lattice theory Wille [10] has introduced the notion of a *formal context* $\mathbb{K} := (G, M, I)$ where G and M are sets and $I \subseteq G \times M$. The elements of G are called *formal objects*, the elements of M are called *formal attributes*, and for $g \in G$ and $m \in M$ the statement gIm $(:\Leftrightarrow (g, m) \in I)$ is interpreted as "object g has attribute m". For any subset $X \subseteq G$ the *upper derivation* of X is defined by

$$X^{\uparrow} := \{m \in M | \forall_{g \in X} gIm\};$$

dually for $Y \subseteq M$ the *lower derivation* of Y is defined by

$$Y^{\downarrow} := \{g \in G | \forall_{m \in Y} gIm\}.$$

Then for each formal context $\mathbb{K} = (G, M, I)$ a *formal concept* of $\mathbb{K}$ is defined as a pair (A, B) where

$$A \subseteq G, \ B \subseteq M \text{ and } A^{\uparrow} = B \text{ and } B^{\downarrow} = A;$$

A is called the *extent* and B the *intent* of (A,B). The set of all formal concepts of $\mathbb{K}$ is denoted by $\mathfrak{B}(\mathbb{K})$. On the set $\mathfrak{B}(\mathbb{K})$ an order relation $\leq$ is defined by: $(A_1, B_1) \leq (A_2, B_2) :\Leftrightarrow A_1 \subseteq A_2$; that is equivalent to $B_1 \supseteq B_2$. The ordered set $\underline{\mathfrak{B}}(\mathbb{K}) := (\mathfrak{B}(\mathbb{K}), \leq)$ is a complete lattice, called the *concept lattice* of $\mathbb{K}$. It is well-known that each complete lattice is isomorphic to a concept lattice (cf. Ganter, Wille [6]).

1.3. *Using Concept Lattices in Data Analysis*

Concept lattices are very successful in data analysis for the representation of data without any loss of information. That is possible since the concept lattice of a formal context determines his formal context uniquely. Hence for a finite formal context one can draw a line diagram, i.e. a suitably labelled Hasse-diagram of its concept lattice, such that the full information of the given formal context is visualized.

If data are given by a data table with arbitrary values one can construct formal contexts that represent the given data with or without any loss of information. For that purpose a general method, called Conceptual Scaling, has been developed (cf. Ganter, Wille [5, 6]). For the interactive construction of line diagrams with the help of computer programs the reader is referred to a recent publication of P. Becker and J. Hereth Correia [3].

1.4. *Conceptual Scaling*

For the purpose of describing arbitrary data tables Wille [10] has introduced *many-valued contexts* (G, M, W, I) where G, M, W are sets and $I \subseteq G \times M \times W$ such that $(g, m, v) \in I$ and $(g, m, w) \in I$ implies $v = w$. If (G, M, W, I) is a many-valued context and $m \in M$ then a formal context $S_m := (G_m, N_m, I_m)$ is called a *conceptual scale* of m if the set $m(G) := \{w \in W | \exists_{g \in G}(g, m, w) \in I\}$ is a subset of G_m. For any many-valued context (G, M, W, I) and any family $(S_m \mid m \in M)$ of conceptual scales the pair $((G, M, W, I), (S_m \mid m \in M))$ is called a *scaled many-valued context*.

Let $((G, M, W, I), (S_m \mid m \in M))$ be a scaled many-valued context where $S_m := (G_m, N_m, I_m)$. Then the following formal context $\mathbb{K} := (G, \{(m, n) | m \in M, n \in N_m\}, J)$ where

$$gJ(m, n) :\Leftrightarrow \exists_{w \in W}(g, m, w) \in I \text{ and } wI_m n$$

is called *the derived context* of the given scaled many-valued context.

The concept lattice of the derived context (or of some of its parts) yields valuable insight into the data. Suitable choices of the conceptual scales enable the user to study the data with respect to his purposes. The conceptual scales play a double role: first, they are used as "mental frames" which describe the meaning of the values of the many-valued context by the scale attributes; second, they are used as very effective granularity tools for focussing on important details or for abstracting from irrelevant facts.

1.5. *Rough Sets, Fuzzy Sets, and Information Channels*

Conceptual scaling is related to several other theories. We mention here only Rough Set Theory of Z. Pawlak [8], Fuzzy Theory of L.A. Zadeh [27, 28], and Information Flow in the sense of Barwise and Seligman [2].

It was shown by the author [15] that the knowledge bases of Rough Set Theory can be described by nominally scaled many-valued contexts, while the ordinal structure of a concept lattice can not be described in the partition oriented Rough Set Theory.

The author [21] has introduced a conceptual interpretation of Fuzzy Theory including a mathematical definition of linguistic variables and their direct products such that the narrow role of the unit interval in Fuzzy Theory was uncovered as the reason for the difficulties around the notion of a "fuzzy implication". The main idea for that interpretation was the formal representation of objects and the usage of a simple ordinal scaling of the membership functions of a linguistic variable.

The basic notion of an *information channel* in Information Flow Theory was shown by the author [14] to be essentially the same as a scaled many-valued context.

1.6. *Classical Interpretation of Many-Valued Contexts*

A many-valued context (G, M, W, I) is usually interpreted in practical applications as a data table where "$(g, m, w) \in I$" is read as "the attribute m has the value w for the object g" (cf. Ganter, Wille [6]). That *classical interpretation* suggests to interpret formal objects $g \in G$ as "real objects" which are "measured" by values of many-valued attributes interpreted as partial "measurement functions"; that was very successful in many applications. Before discussing some disadvantages of that classical interpretation we mention some applications in Temporal Concept Analysis which led the author to a more general interpretation of scaled many-valued contexts which will be explained in the section on Conceptual Semantic Systems.

1.6.1. *Applications in Temporal Concept Analysis*

Temporal Concept Analysis is a theory describing temporal phenomena with tools of FCA [18, 23]. The interpretation of formal objects as *time granules* led the author to the notion of a *state* of a *conceptual time system* (cf. Wolff [16]). Slightly more general is the interpretation of formal objects as *actual objects* which are defined as pairs (p, g) where p denotes a "real object" like for example a person and g a time granule; that is mathematically defined in the notion of a *Conceptual Time System with Actual Objects and a Time Relation* (CTSOT) (cf. Wolff [19]). The notion of a CTSOT allows for the conceptual description of temporal phenomena where "each object is at each time granule at exactly one place". Such objects can be described by their *trajectories* or *life tracks*. They occur for example as "runs" in automata (cf. Wolff [20]), as "input words" in Turing machines (cf. Wolff,

Yameogo [26]) or as "particles" in classical physics (cf. Wolff [22]).

1.6.2. *Problems with the Representation of Objects*

The notion of a CTSOT is very well suited for the conceptual representation of states and situations of classical objects, their transitions and life tracks. It is not suited for the representation of waves or wave packets or other "distributed objects". The reason lies in the classical interpretation of many-valued contexts whose formal objects are interpreted as "real objects": for a scaled many-valued context $((G, M, W, I), (\mathbb{S}_m \mid \mathrm{m} \in \mathrm{M}))$ with derived context $\mathbb{K} := (G, N, J)$ each formal object $g \in G$ has for each $Q \subseteq N$ in the Q–part $\mathbb{K}_Q := (G, Q, J \cap (G \times Q))$ of $\mathbb{K}$ a unique object concept $\gamma_Q(g)$. Since the state of an actual object (p, t) in a CTSOT is defined as the object concept of (p, t) in some Q–part of the derived context the "object p is at each time granule g in exactly one state", for example "... in exactly one location" if the set Q of the chosen attributes describes some spatial part of the data.

For the purpose of finding a simple common conceptual description of particles and waves in physics the author has introduced the notion of a Conceptual Semantic System (cf. Wolff [22]). Another purpose for the introduction of Conceptual Semantic Systems was the difficulty to treat relational structures with scaled many-valued contexts. In the next section we shall explain how the conceptual advantages of scaled many-valued contexts and their graphical representations in line diagrams can be preserved by generalizing their classical interpretation from a functional to a relational point of view.

2. Conceptual Semantic Systems

In this section we first define the notion of a Conceptual Semantic System.

2.1. *Basic Definitions*

Definition 2.1. "Conceptual Semantic System"
Let M be a set and, for each $m \in M$, let $\mathbb{S}_m := (G_m, N_m, I_m)$ be a formal context and $\underline{\mathfrak{B}}(\mathbb{S}_m) := (\mathfrak{B}(\mathbb{S}_m), \leq_m)$ its concept lattice; let G be a set and

$$\kappa : G \times M \to \bigcup_{m \in M} \mathfrak{B}(\mathbb{S}_m) \text{ be a mapping such that } \kappa(g, m) \in \mathfrak{B}(\mathbb{S}_m),$$
$$m(g) := \kappa(g, m) \text{ and } I := \{(g, m, m(g)) | g \in G, m \in M\}.$$

Then the quadruple $\mathfrak{K} := (G, M, (\underline{\mathfrak{B}}(\mathbb{S}_m))_{m \in M}, I)$ is called a *Conceptual Semantic System (CSS) with semantic scales* $\mathbb{S}_m$ $(m \in M)$. The elements of G are called *instances* or *information units*.

We interpret the concepts of the semantic scales as "types" and the concept lattice of a semantic scale as a "type hierarchy", also called an "ontology". The statement $m(g) = \mathbf{c}$ is interpreted as "instance g tells something about the concept $\mathbf{c} \in \mathfrak{B}(\mathbb{S}_m)$". For any instance g the tuple $(m(g)|m \in M)$ is interpreted as a short description of a statement connecting the concepts $m(g)$ where $m \in M$. That allows for the representation of arbitrary relations among the chosen concepts of the semantic scales. A special example is the parametric representation of the unit circle by triples $(t, cos(t), sin(t))$. In that sense a CSS is a parametric representation of relational conceptual knowledge.

2.2. *The Semantically Derived Context*

Definition 2.2. "Semantically Derived Context"
Let $(G, M, (\underline{\mathfrak{B}}(\mathbb{S}_m))_{m \in M}, I)$ be a Conceptual Semantic System with semantic scales $\mathbb{S}_m = (G_m, N_m, I_m)$ $(m \in M)$ and let $int(c)$ denote the intent of a concept c. Then the formal context

$$\mathbb{K} := (G, N, J) \text{ where } N := \{(m,n)|m \in M, n \in N_m\} \text{ and}$$
$$gJ(m,n) :\Longleftrightarrow n \in int(m(g))$$

is called the semantically derived context of $(G, M, (\underline{\mathfrak{B}}(\mathbb{S}_m))_{m \in M}, I)$.

It is easy to see that the semantically derived context of a CSS can be obtained also by plain scaling as the usual derived context. Therefore we write in the following only "derived context" instead of "semantically derived context". The formal concepts of the semantic scales yield "realized concepts" in the derived context $\mathbb{K} = (G, N, J)$ and in its Q−parts $\mathbb{K}_Q := (G, Q, J \cap (G \times Q))$ where $Q \subseteq N$; that is described in the following definition.

2.3. *The Realization of Concepts*

Definition 2.3. "Realization of a concept of a semantic scale"
Let $(G, M, (\underline{\mathfrak{B}}(\mathbb{S}_m))_{m \in M}, I)$ be a Conceptual Semantic System with derived context $\mathbb{K} = (G, N, J)$. Then for $m \in M$ and $Q \subseteq N$ the following mapping

$$r_{m,Q} : \mathfrak{B}(\mathbb{S}_m) \to \mathfrak{B}(\mathbb{K}_Q)$$
$$\mathbf{c} = (A_\mathbf{c}, B_\mathbf{c}) \mapsto r_{m,Q}(\mathbf{c}) := ((\{m\} \times B_\mathbf{c})^\downarrow, (\{m\} \times B_\mathbf{c})^{\downarrow\uparrow})$$

is called the m-realization of $\mathbf{c}$ *in* $\mathfrak{B}(\mathbb{K}_Q)$. Let r_m denote the mapping $r_{m,N} : \mathfrak{B}(\mathbb{S}_m) \to \mathfrak{B}(\mathbb{K})$.

We interpret the m-realization of a concept $\mathbf{c}$ in $\mathfrak{B}(\mathbb{K}_Q)$ as a "really observed concept", as for example a "really observed whale" as opposed to the semantic concept "whale"; the "really observed whale" is connected with the observation, and that is reflected in the extent of the realization of the semantic concept. The intent of the m-realization of a concept $\mathbf{c}(A_\mathbf{c}, B_\mathbf{c})$ is constructed from the intent $B_\mathbf{c}$ in a very natural way.

2.4. *Aspects of a Concept with Respect to a View*

In colloquial speech an "aspect of a thing" contains some special information about a thing. The following definition introduces the notion of an "aspect of a concept with respect to a view" in a given formal context.

Definition 2.4. "Q-aspect of a concept"

Let $\mathbb{K} := (G, N, J)$ be a formal context, $Q \subseteq N$, and γ_Q the object concept mapping of the Q−part $\mathbb{K}_Q := (G, Q, J \cap (G \times Q))$ of $\mathbb{K}$. For any concept $\mathbf{c} := (A_\mathbf{c}, B_\mathbf{c}) \in \mathfrak{B}(\mathbb{K})$ the set

$$\alpha_Q(\mathbf{c}) := \{\gamma_Q(g) | g \in A_\mathbf{c}\}$$

is called *the aspect of the concept $\mathbf{c}$ with respect to the view Q*: in short *the Q-aspect of $\mathbf{c}$*. $\mathbf{c}$ is called *distributed in Q* if $|\alpha_Q(\mathbf{c})| \geq 2$.

In this paper we shall use the notion of a "Q-aspect of a concept" only for the concepts of the derived context $\mathbb{K} := (G, N, J)$ of a CSS. Then the extent $A_\mathbf{c}$ of such a concept $\mathbf{c}$ is a subset of the set G of information units. For any view $Q \subseteq N$ this extent $A_\mathbf{c}$ is mapped by the object concept mapping γ_Q onto the Q-aspect of $\mathbf{c}$, which is the set of those object concepts in $\mathbb{K}_Q$ which are object concepts of at least one information unit g *mentioning* the concept $\mathbf{c}(A_\mathbf{c}, B_\mathbf{c})$, that is $g^\uparrow \supseteq B_\mathbf{c}$.

3. Temporal Conceptual Semantic Systems

In this section we introduce Temporal Conceptual Semantic Systems as a general conceptual granularity tool for the representation of arbitrary changes or movements of concepts in discrete or continuous spaces without specifying a location part, but introducing time relations as in a CTSOT.

For that purpose we use our experiences with CTSOTs and generalize the notion of the state of an "actual object" by the Q-aspect of an "actual concept". To define the notion of an "actual concept" we have to introduce in a suitable way some conceptual representation of time. Similar to a CTSOT we use not only some specified time semantics but also a time relation for the representation of transitions and movements of concepts (cf. Wolff [19, 23]) as for example the movement of a "high pressure zone" on a weather map (cf. Wolff [24]).

Definition 3.1.

"Temporal Conceptual Semantic System"

Let $\mathfrak{K} := (G, M, (\underline{\mathfrak{B}}(S_m))_{m \in M}, I)$ be a Conceptual Semantic System with derived context $\mathbb{K} := (G, N, J)$. Let $T \in M$. The concepts $\mathbf{t} \in \mathfrak{B}(S_T)$ are called *abstract time granules*. Let $\mathbf{C} \subseteq \mathfrak{B}(\mathbb{K})$; the concepts $\mathbf{c} \in \mathbf{C}$ are called *temporal concepts*. For each $\mathbf{c} \in \mathbf{C}$ let $\mathbf{R_c} \subseteq \mathfrak{B}(S_T) \times \mathfrak{B}(S_T)$. The elements of $\mathbf{R_c}$ are called the *base transitions* of $\mathbf{c}$.

Then the quadruple $(\mathfrak{K}, T, \mathbf{C}, (\mathbf{R_c}|\mathbf{c} \in \mathbf{C}))$ is called a *Temporal Conceptual Semantic System (TCSS)*.

The special case that the concept lattice $\mathfrak{B}(S_T)$ is a chain yields the usual linear time scale which may be discrete or continuous.

It is easy to see that the notion of a TCSS is more general than the notion of a CTSOT. As a generalization of the notion of a state of an actual object in a CTSOT we now define the state of an actual concept with respect to a view.

Definition 3.2.

"State of an actual concept with respect to a view"

Let $(\mathfrak{K}, T, \mathbf{C}, (\mathbf{R_c}|\mathbf{c} \in \mathbf{C}))$ be a Temporal Conceptual Semantic System with derived context $\mathbb{K} := (G, N, J)$. Let $\mathbf{c} \in \mathbf{C}, \mathbf{t} \in \mathfrak{B}(S_T)$ and $Q \subseteq N$. Then *the state of the actual concept $(\mathbf{c}, \mathbf{t})$ with respect to the view Q, or the Q-state of $(\mathbf{c}, \mathbf{t})$*, is defined by

$$state_Q(\mathbf{c}, \mathbf{t}) := \alpha_Q(\mathbf{c} \wedge r_T(\mathbf{t})).$$

As a generalization of the notion of the life track of an object in a CTSOT we now define the notion of the life space of a concept with respect to some view.

Definition 3.3.

"Life Space of a concept with respect to a view"

Let $(\mathfrak{K}, T, \mathbf{C}, (\mathbf{R_c}|\mathbf{c} \in \mathbf{C}))$ be a Temporal Conceptual Semantic System with derived context $\mathbb{K} := (G, N, J)$. Let $\mathbf{c} \in \mathbf{C}$ and $Q \subseteq N$. Then *the life space of $\mathbf{c}$ with respect to the view Q* is defined by:

$$space_Q(\mathbf{c}) := \bigcup\{state_Q(\mathbf{c}, \mathbf{t}) | \mathbf{t} \in \mathfrak{B}(\mathbb{S}_T)\}.$$

In the special case of a CTSOT this definition of a life space reduces to the definition of a life track.

4. An Example of a Temporal Conceptual Semantic System

To show all relevant theoretical notions of a Temporal Conceptual Semantic System explicitly we construct the following small example using some data about Albert Einstein [7].

To describe the TCSS $\mathfrak{T}_1 := (\mathfrak{K}, T, \mathbf{C}, (\mathbf{R_c}|\mathbf{c} \in \mathbf{C}))$ we first construct its CSS $\mathfrak{K} := (\mathrm{G}, \mathrm{M}, (\underline{\mathfrak{B}}(\mathbb{S}_m))_{m \in M}, I)$ where the set of instances is $G := \{1, 2, 3, 4, 5\}$ and the set of many-valued attributes is $M := \{Person, Action, Time, Location\}$. The ternary incidence relation I is described by the mapping κ which is graphically represented in Table 1:

Table 1. A data table of a TCSS representing some events of the life of Albert Einstein

instances	*Person*	*Action*	*Time*	*Location*
1	**Einstein**	**born**	**1879**	**Ulm**
2	**Einstein**	**studied**	**1896**	**Zürich**
3	**Einstein**	**Prof.**	**1914**	**Berlin**
4	**Einstein**	**Prof.**	**1932**	**Princeton**
5	**Einstein**	**died**	**1955**	**Princeton**

For each $m \in M$ the values $\kappa(g, m)$ ($g \in G$) are interpreted as formal concepts of the following semantic scales $\mathbb{S}_m$.

The semantic scale for *Person* is the formal context $\mathbb{S}_P := (\{Einstein\}, \{Einstein\}, \{(Einstein, Einstein)\})$, hence **Einstein** $:= (\{Einstein\}, \{Einstein\})$ is the single formal concept of $\mathbb{S}_P$, representing in colloquial speech the phrase "Einstein is Einstein".

The semantic scale $\mathbb{S}_A$ for *Actions* is chosen to be the nominal scale represented in Table 2:

The object concepts of $\mathbb{S}_A$ are the values of *Action*. The semantic scale $\mathbb{S}_T$ for *Time* is chosen to be the interordinal scale in Table 3:

The attributes in Table 3 are defined by: a:=$(= 1879)$, b:=(≤ 1896), c:=(≤ 1914), d:=(≤ 1932), e:=(≤ 1955), f:=(≥ 1879), g:=(≥ 1896), h:=(≥ 1914), i:=(≥ 1932), j:=$(= 1955)$.

The semantic scale $\mathbb{S}_L$ for *Locations* is given in Table 4:

Table 2. The semantic scale S_A for the *Actions*

	born	studied	Prof.	died
born	×			
studied		×		
Prof.			×	
died				×

Table 3. The semantic scale S_T for *Time*

	a	b	c	d	e	f	g	h	i	j
1879	×	×	×	×	×	×				
1896		×	×	×	×	×	×			
1914			×	×	×	×	×	×		
1932				×	×	×	×	×	×	
1955					×	×	×	×	×	×

Table 4. The semantic scale S_L for *Locations*

	Ulm	Berlin	Germany	Zürich	Swiss	Princeton	USA
Ulm	×		×				
Berlin		×	×				
Zürich				×	×		
Princeton						×	×

Now we have described the CSS $\mathfrak{K}$ of the TCSS $\mathfrak{T}_1 := (\mathfrak{K}, T, \mathbf{C}, (\mathbf{R_c} \mid \mathbf{c} \in \mathbf{C}))$. The time attribute T is chosen to be $T := Time$, the set $\mathbf{C}$ is defined by $\mathbf{C} : \{r_P(\mathbf{Einstein})\}$, that is we decide to take the realized concept $r_P(\mathbf{Einstein}) : (G, \{(Person, Einstein), (Time, \geq 1879), (Time, \leq 1955)\})$ as the only temporal concept (cf. the top concept in Figure 1). Its time relation $\mathbf{R_c}$, intuitively described as the sequence of time concepts $\mathbf{1879} \to \mathbf{1896} \to \mathbf{1914} \to \mathbf{1932} \to \mathbf{1955}$, is formally defined as the set $\mathbf{R_c} := \{(\mathbf{1879}, \mathbf{1896}), (\mathbf{1896}, \mathbf{1914}), (\mathbf{1914}, \mathbf{1932}), (\mathbf{1932}, \mathbf{1955})\}$ where for example the concept $\mathbf{1879}$ is defined as the object concept $\gamma_T(1879)$ in the semantic scale S_T.

The derived context $\mathbb{K} = (G, N, J)$ of $\mathfrak{T}_1$ has $|G| = 5$ instances and $|N| = 1 + 4 + 10 + 7 = 22$ attributes. The concept lattice of $\mathbb{K}$ is represented by the transition diagram in Figure 1 where each base transition $(\mathbf{s}, \mathbf{t})$ is represented graphically by an arrow as defined for CTSOTs. The more complicated graphical representation of base transitions in TCSSs will be discussed in another paper.

To describe an example of a state we start intuitively with the idea of the "country location of Einstein since 1914" which should be, in our data, the set $\{Germany, USA\}$. For a formal description of that state we

298

employ the attribute concept mapping μ of $\mathbb{K}$ and μ_T of $\mathbb{S}_T$. The state of the actual concept $(r_P(\mathbf{Einstein}), \mu_T(\geq 1914))$ with respect to the view $Q := \{(Location, Swiss), (Location, Germany), (Location, USA)\}$ is $\alpha_Q(r_P(\mathbf{Einstein}) \wedge \mu((Time, \geq 1914))) = \{\gamma_Q(3), \gamma_Q(4)(\gamma_Q(5))\}$ which is $\{\mu_Q((Location, Germany)), \mu_Q((Location, USA))\}$ as shown in the transition diagram in Figure 2. Hence $r_P(\mathbf{Einstein}) \wedge \mu((Time, \geq 1914))$ is a distributed concept in Q. Figure 2 also shows the practically very relevant possibility to factorize Temporal Conceptual Semantic Systems.

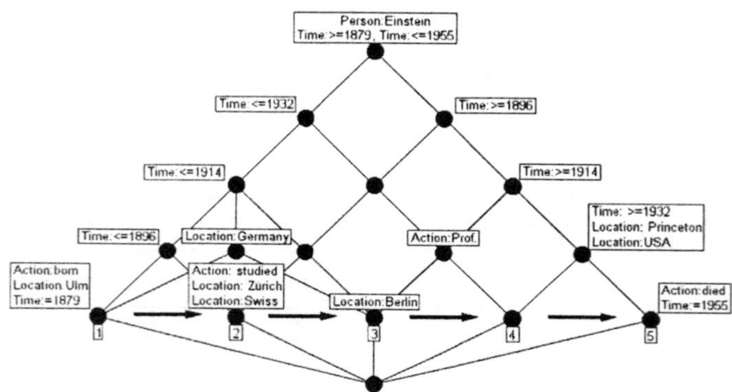

Figure 1.   The life track of Einstein in a Person-Action-Time-Location diagram

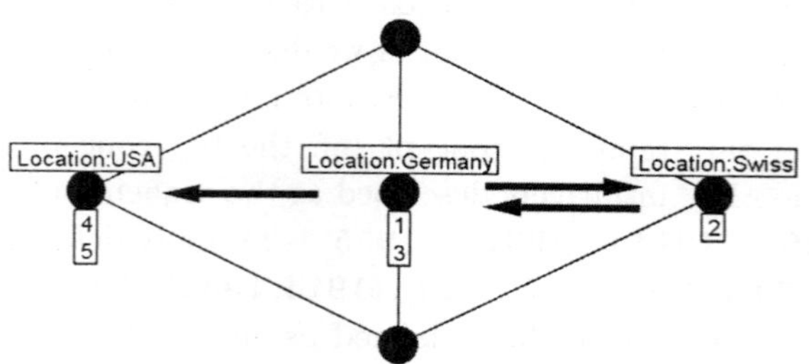

Figure 2.   Einstein's factorized life track in a location view

5. Conclusion and Future Research

We have introduced the notion of a Temporal Conceptual Semantic System as a relatively simple and general structure for the investigation of temporal phenomena. The most important feature is a mathematically clear distinction between "abstract" and "realized" concepts which is based on the introduction of "information units" coding relational statements. That leads via "Q-aspects" to the notion of distributed concepts.

The Q-aspects of concepts can be visualized with the computer program TOSCANAJ (cf. Becker, Hereth Correia [3]). That yields very useful applications in practice, for example the investigation of dynamic processes in industry (cf. Wolff [17]) and in psychosomatic process research (cf. Wolff [22]). Applications in physics yield a common conceptual understanding of particles and waves [22], as well as the conceptual understanding of the notion of the state of a quantum mechanical system, and a conceptual interpretation of Heisenberg's Uncertainty Relation [25].

Future research will focus on a conceptually based temporal logic for distributed systems and applications in the research field of Conceptual Graphs (cf. Sowa [9]), Concept Graphs and Power Context Families (cf. Wille [11–13]).

References

1. M. Barbut, B. Monjardet: Ordre et Classification, Algèbre et Combinatoire. 2 tomes. Paris, Hachette, 1970.
2. J. Barwise, J. Seligman: *Information Flow: The Logic of Distributed Systems*. Cambridge University Press, Cambridge 1997.
3. P. Becker, J. Hereth Correia: The ToscanaJ Suite for Implementing Conceptual Information Systems. In: B. Ganter, G. Stumme, R. Wille (Eds.): Formal Concept Analysis. Lecture Notes in Artificial Intelligence LNAI **3626**. Springer, Heidelberg 2005, 324–348.
4. G. Birkhoff: *Lattice theory*, 3rd ed., Amer.Math.Soc., Providence 1967.
5. B. Ganter, R. Wille: Conceptual scaling. In: F. Roberts (ed.): *Applications of combinatorics and graph theory in the biological and social sciences*. Springer-Verlag, New York 1989, 139–167.
6. B. Ganter, R. Wille: *Formal Concept Analysis: Mathematical Foundations*. Springer, Heidelberg 1999.
7. J. Neffe: *Einstein. Eine Biographie*. Rowohlt Verlag GmbH, Reinbek bei Hamburg, 2005.
8. Z. Pawlak: *Rough Sets: Theoretical Aspects of Reasoning About Data*. Kluwer Academic Publishers, 1991.
9. J. F. Sowa: *Conceptual structures: information processing in mind and machine*. Adison-Wesley, Reading 1984.
10. R. Wille: Restructuring lattice theory: an approach based on hierarchies of concepts. In: I. Rival (ed.): *Ordered Sets*. Reidel, Dordrecht-Boston 1982, 445–470.
11. R. Wille: Conceptual graphs and formal concept analysis. In: D. Lukose, H. Delugach, M. Keeler, L. Searle, J. F. Sowa (eds.): *Conceptual structures: fulfilling Peirce's dream*. LNAI **1257**. Springer, Heidelberg 1997, 290–303.
12. R. Wille: Contextual Logic summary. In: G. Stumme (ed.): *Working with conceptual structures: Contributions to ICCS 2000*. Shaker-Verlag, Aachen

2000, 265–276.

13. R. Wille: Existential concept graphs of power context families. In: U. Priss, D. Corbett, G. Angelova (eds.): *Conceptual structures: integration and interfaces*. LNAI **2393**. Springer, Heidelberg 2002, 382–395.

14. Wolff, K.E.; Information Channels and Conceptual Scaling. In: Stumme G. (ed.): Working with Conceptual Structures–Contributions to ICCS 2000, (8th International Conference on Conceptual Structures: Logical, Linguistic, and Computational Issues). Shaker Verlag, Aachen, 277–283.

15. K.E. Wolff: A Conceptual View of Knowledge Bases in Rough Set Theory. In: W. Ziarko, Y. Yao (eds.): *Rough Sets and Current Trends in Computing*. Second International Conference, RSCTC 2000, Banff, Canada, October 16–19, 2000, Revised Papers, 220–228.

16. K.E. Wolff: Concepts, States, and Systems. In: D.M. Dubois, (ed.): Computing Anticipatory Systems. CASYS-99 — Third International Conference, Liège, Belgium, 1999, American Institute of Physics, Conference Proceedings 517, 2000, 83–97.

17. K.E. Wolff: Towards a Conceptual System Theory. In: B. Sanchez, N. Nada, A. Rashid, T. Arndt, M. Sanchez (eds.): Proceedings of the World Multiconference on Systemics, Cybernetics and Informatics, SCI 2000, Vol. II: Information Systems Development, International Institute of Informatics and Systemics, 2000, ISBN 980-07-6688-X, 124–132.

18. K.E. Wolff: Temporal Concept Analysis. In: E. Mephu Nguifo & al. (eds.): *ICCS-2001 International Workshop on Concept Lattices-Based Theory, Methods and Tools for Knowledge Discovery in Databases,* Stanford University, Palo Alto (CA), 91–107.

19. K.E. Wolff: Transitions in Conceptual Time Systems. In: D.M. Dubois (ed.): *International Journal of Computing Anticipatory Systems*, vol. **11**, CHAOS 2002, p. 398–412.

20. K.E. Wolff: Interpretation of Automata in Temporal Concept Analysis. In: U. Priss, D. Corbett, G. Angelova (eds.): *Integration and Interfaces*. Tenth International Conference on Conceptual Structures. LNAI **2393**, Springer 2002, 341–353.

21. K.E. Wolff: Concepts in Fuzzy Scaling Theory: Order and Granularity. 7th European Congress on Intelligent Techniques and Soft Computing, Aachen 1999. *Fuzzy Sets and Systems* **132**, 2002, 63–75.

22. K.E. Wolff: 'Particles' and 'Waves' as Understood by Temporal Concept Analysis. In: K.E. Wolff, H.D. Pfeiffer, H.S. Delugach (eds.): *Conceptual Structures at Work*. 12th International Conference on Conceptual Structures, ICCS 2004. Huntsville, AL, USA, July 2004. Proceedings. Springer Lecture Notes in Artificial Intelligence, LNAI 3127, Springer-Verlag, Berlin Heidelberg 2004, 126–141.

23. K.E. Wolff: States, Transitions, and Life Tracks in Temporal Concept Analysis. In: B. Ganter, G. Stumme, R. Wille (eds.): *Formal Concept Analysis*, LNAI **3626**, Springer-Verlag, Heidelberg 2005, 127–148.

24. K.E. Wolff: States of Distributed Objects in Conceptual Semantic Systems. In: F. Dau, M.-L. Mugnier, G. Stumme (eds.): *Conceptual Structures: Com-*

mon Semantics for Sharing Knowledge, LNAI **3596**, Springer-Verlag, Heidelberg 2005, 250–266.

25. K.E. Wolff: A Conceptual Analogue of Heisenberg's Uncertainty Relation. In: B. Ganter, L. Kwuida (eds.): Contributions to ICFCA 2006. Verlag Allgemeine Wissenschaft, 2006, 19–30.

26. K.E. Wolff, W. Yameogo: Turing Machine Representation in Temporal Concept Analysis. In: B. Ganter, R. Godin (eds.): Formal Concept Analysis. Third International Conference ICFCA 2005. Springer Lecture Notes in Artificial Intelligence, LNAI **3403**, Springer-Verlag, Heidelberg 2005, 360–374.

27. L. A. Zadeh: Fuzzy sets. Information and Control 8, 1965, 338–353.

28. L. A. Zadeh: The concept of a linguistic variable and its application to approximate reasoning. Part I: Inf. Science 8, 199–249; Part II: Inf. Science 8, 301–357; Part III: Inf. Science 9, 43–80, 1975.

COMPLEXITY RESULTS ON MINIMAL UNSATISFIABLE FORMULAS*

XISHUN ZHAO

Institute of Logic and cognition
Sun Yat-Sen University
510275 Guangzhou
P. R. China
Email: hsdp08@zsu.edu.cn

In this paper we will give a review of complexity results concerning minimal unsatisfiable (MU) formulas. At first we recall the complexity of classes of maximal MU formulas, marginal MU formulas, ect., and some classes which are closed under splitting. Then the complexity of detecting the existence of some simple MU subformulas, existence of homomorphisms between MU formulas is reported. Finally, we mention some generalizations to the minimal unsatisfiability.

1. Introduction

A propositional formula F in conjunctive normal form (CNF) is called minimal unsatisfiable if and only if F is unsatisfiable and any proper subformula of F is satisfiable. The class of minimal unsatisfiable formulas is denoted as MU. Please note that any unsatisfiable formula in CNF contains a minimal unsatisfiable sub-formula. A deeper understanding of MU formulas might be helpful for (1) developing new algorithms solving the satisfiability problem, (2) looking for more tractable classes of formulas, and (3) developing new hard formulas for some proof calculi.

Furthermore, MU formulas also have some application interest in practice. Take belief revision as example, let K be a Knowledge Base represented as a set of clauses, φ a clause which representing our new knowledge and should be added to K. However, $K \cup \{\varphi\}$ is inconsistent due to our former wrong knowledge, and we have to revise K to preserve the consistency. One method is Safe Revision [2]: First compute $S := \{F \subseteq K : F \cup \{\varphi\}$ is minimal unsatisfiable $\}$, then from each $F \in S$ select one clause $\gamma(F)$ according to some principle, delete all the selected clauses.

---

*This research was partially supported by the NSFC projects under grant numbers: 60573011, 10410638 and a MOE project under grant number: 05JJD72040122

The revised knowledge base consists of all the remaining clauses and φ. In addition, minimal unsatisfiable formulas have some applications in formal verification, model checking, diagnose, etc. In formal verification, an abstract model should be refined because some property fails. Refinement is achieved by identifying the cause of infeasibility. Although a unsatisfiable formula itself is an explanation of the infeasibility, we are interested in a "minimal" explanation since it excludes irrelevant information. Thus minimal unsatisfiable subformulas provide useful insight on the cause of infeasibility.

In the past decade, many breakthroughs has been made in order to have a deeper understanding of MU–formulas. In this paper, we shall report main results on the complexity concerning minimal unsatisfiability.

In 1988, MU has been shown to be D^P–complete [30]. D^P is the class of problems which can be described as the difference of two NP–problems. It is strongly conjectured that D^P is different from NP and from $coNP$.

There are several approaches for defining natural subclasses of MU–formulas. For example, the deficiency, the difference between the number of clauses and the number of variables, can be restricted. It is known that any minimal unsatisfiable formula over n variables consists of at least $n + 1$ clauses [1, 6, 9, 27].

There exist some minimal unsatisfiable formulas such that removing or adding some literal to some clause will not destroy the minimal unsatisfiability. Please see the following example. Let

$$F = (\neg a \vee c) \wedge (b \vee a \vee c) \wedge (\neg b \vee a) \wedge \neg c.$$

It is easy to see that the resulting formula by removing c from the first clause or by adding c to the third clause is still minimal unsatisfiable. This motivates us to investigate subclasses of minimal unsatisfiable formulas to which (resp. from which) we can not add (resp. delete) any occurrence of a literal with minimal unsatisfiability still preserved.

Another class of MU–formulas which is closely related to unique satisfiability is the class of MU formulas which after removing any clause will have exactly one satisfying truth assignment.

A powerful tool for investigating the structure of minimal unsatisfiable formulas is splitting. We take a variable x, set $x = 1$ and $x = 0$, and consider the resulting formulas. For a minimal unsatisfiable formula the resulting formulas contain again minimal unsatisfiable formulas.

A disjunctive splitting means that the formula can be divided into two separate minimal unsatisfiable formulas by setting the variable true resp.

304

false. The minimal unsatisfiable formulas with disjunctive splitting on every variable is also interesting because removing from it any clause results in a uniquely satisfiable formula.

Unfortunately, the above mentioned classes of minimal unsatisfiable formulas are not closed under splitting, that is, a formula in the class may have splitting formulas not in the class. Therefore, we investigate classes closed under splitting.

Since some classes of simple minimal unsatisfiable formulas are polynomial-time solvable, it should be interesting to decide the unsatisfiability by testing the existence of simple minimal unsatisfiable subformulas.

We also review some results on homomorphisms between minimal unsatisfiable formulas and some generalizations of minimal unsatisfiability.

In this paper, clauses (disjunctions of literals) are considered as sets of literals, while CNF formulas (conjunctions of clauses) are multi-sets of clauses. The symbol "+" stands for the union operation of multi-sets.

2. MU-Formulas with Fixed Deficiency

Given a *CNF* formula F, the deficiency, denoted as $d(F)$, is the difference between the number of clauses of F and the number of variables occurring in F.

For any fixed natural number k, we denote by $MU(k)$ the class of all minimal formulas with deficiency k.

Please note that the satisfiability problem for formulas with fixed deficiency is still *NP*–complete.

In 1996, H. Kleine Büning propose the **question:** for fixed k, whether $MU(k)$ can be solved in polynomial time.

Lemma 2.1. (G. Davydov, I. Davydova, and H. Kleine Büning [6]) *The Problem of determining if a CNF formula belongs to $MU(1)$ can be solved in linear time.*

The proof of Lemma 2.1 is based the following two nice properties of MU formulas.

Proposition 2.1. [6, 14] $MU(k)$ *is closed under* $(1, *)$*–resolution. That means, if* $F = \{L \vee f, \neg L \vee g_1, \neg L \vee g_2, \cdots, \neg L \vee g_s\} + F' \in MU(k)$ *and* L *and* $\neg L$ *don't occur in* F' *then* $\{f \vee g_1, f \vee g_2, \cdots, f \vee g_s\} + F' \in MU(k)$.

Proposition 2.2. [6, 14] *Any formula* $F \in MU(1)$ *always contains a literal which occurs in* F *exactly once.*

Then Kleine Büning in [14] proved that after iteratively applying $(1,*)$-resolution each formula in $MU(2)$ can be transformed to the following formula with respect to renaming (here each column represents a clause):

$$\begin{pmatrix} x_1 & \neg x_1 & \neg x_2 & \cdots & \neg x_{n-1} & \neg x_n & \neg x_1 \\ x_2 & x_2 & x_3 & \cdots & x_n & x_1 & \neg x_2 \\ x_3 & & & & & & \neg x_3 \\ \vdots & & & & & & \vdots \\ x_n & & & & & & \neg x_n \end{pmatrix}$$

Consequently, $MU(2)$ can be solved in polynomial time.

In 1998, Xishun Zhao and Decheng Ding [34, 35] proved the following.

Suppose $F \in MU(k)$. If F contains a **complete** clause f, i.e., every variable of F occurs in f (either positively or negatively), then F can be renamed to a formula with has at most k non-Horn clauses, and the renaming can be defined efficiently. And consequently, for formulas with a complete clause, whether it is in $MU(k)$ can be computed in polynomial time.

But finally the question was completely solved by H. Fleischner, O. Kullmann, S.Szeider in 2001.

Theorem 2.1. (H. Fleischner, O. Kullmann, and S. Szeider [10, 27]) *For each fixed k, $MU(k)$ can be solved in polynomial time.*

The following assertions and notations play the key role in their proofs.

(1) Maximum deficiency of F: $d^*(F) := \max\{d(F) \mid F' \subseteq F\}$
(2) If $d^*(F) = 0$ then F is satisfiable.
(3) If $F \in MU$ then F is stable, i.e., $d(F') < d(F)$ for any $F \subset F$.
(4) For any F we can find in polynomial time a stable subformula G of F such that $d(G) = d^*(F)$ and that G and F have the same satisfiability.
(5) Suppose F with $d(F) = k$ is stable and satisfiable, then there is a partial truth assignment v defining on k variables such that $v(F)$ has maximum deficiency 0.

3. Minimal Formulas with Simple Structures

Clearly, the class $2\,CNF\text{-}MU$, i.e. $2\,CNF \cap MU$, is solvable in quadratic time since the satisfiability of $2\,CNF$ formulas can be decided in linear time. Please note for a $2\,CNF\text{-}MU$ formula that each literal occurs in it at most twice. Then each $2\,CNF\text{-}MU$ formula can be reduced by iterative $(1,*)$–resolution in linear time to a $2\,CNF\text{-}MU$ formula in which each literal occurs exactly twice. The linear time solvability of $2\,CNF$ -MU follows from the following nice structural property of $2\,CNF$ -MU formulas [26].

Every $2\,CNF$ -MU formula in which each literal occurs at least twice has the following form up to renaming:

$$\begin{pmatrix} x_1 & \neg x_2 & \cdots & \neg x_{n-1} & \neg x_n & \neg x_1 & x_2 & \cdots & x_{n-1} & x_n \\ x_2 & x_3 & \cdots & x_n & x_1 & \neg x_2 & \neg x_3 & \cdots & \neg x_n & \neg x_1 \end{pmatrix}$$

where each column represents a clause.

However, $3\,CNF$ -MU is still D^p–complete. More generally, we have

Theorem 3.1. (Hans Kleine Büning, Xishun Zhao [26]) *For any fixed $k \geq 4$, and $p \geq 2$, and for a $k\,CNF$ formula F in which each literal occurs at least p times, the problem of determining whether F is minimal unsatisfiable is still D^p–complete.*

Open Question 1. Construct a $3\,CNF$ -MU formula in which each literal occurs at least 5 times.

Open Question 2. For a $3\,CNF$ formula in which each literal occurs at least 4 times, what is the complexity to decide whether it is in MU?

For any Horn formula F, if F is minimal unsatisfiable then F must be in $MU(1)$, and it have at least one positive unit clause. More Precisely, F is of the following form.

$$\begin{pmatrix} - & + & & & & \\ * & * & + & & & \\ * & * & * & + & & \\ \vdots & \vdots & \vdots & \vdots & \ddots & \\ * & * & * & * & \cdots & + \end{pmatrix}$$

where each row represents a variable, each column represents a clause, and entries "*" are wildcards for "$-$" (for negative occurrence) or "0" (for no occurrence).

Consequently, *Horn-MU* can be solved in linear time.

4. Maximal MU Formulas

A formula F in MU is called *maximal*, if for any clause $f \in F$ and any literal L which is not in f, adding L to f yields a satisfiable formula. In a certain sense maximal formulas are maximal extensions of MU–formulas. In this section we will show the D^P–completeness of MAX–MU, the class of all so-called maximal minimal unsatisfiable formulas.

Definition 4.1. For a formula $F \in MU$ and a clause $f \in F$ we say f is **maximal** in F if for any literal L occurring neither positively nor negatively in f the formula obtained from F by adding L to f is satisfiable.

Clearly, $F \in MU$ is maximal minimal unsatisfiable, i.e. $F \in MAX\text{-}MU$, if and only if every clause in F is maximal.

That $MAX\text{-}MU$ is in D^P is not hard to see. For the D^P-hardness we establish a reduction from the D^P-complete problem MU. At first we introduce an auxiliary function by associating to a formula F, a clause $f \in F$, and a new variable z a formula $\xi(F, f, z)$ preserving the minimal unsatisfiability. Later on the formula $\xi(F, f, z)$ will be used in order to associate in polynomial time to each formula in MU a maximal formula.

Definition 4.2. For a clause $f = L_1 \vee \cdots \vee L_k$, we use $\rho(f)$ to denote the formula consisting of the following clauses:

$$\neg L_1 \vee L_2 \vee L_3 \vee \cdots \vee L_k,$$
$$\neg L_2 \vee L_3 \vee \cdots \vee L_k,$$
$$\neg L_3 \vee \cdots \vee L_k,$$
$$\cdots$$
$$\neg L_k.$$

The formula $\rho(f) + \{f\}$ is a maximal minimal unsatisfiable formula. That means we have $\rho(f) + \{f\} \in MAX\text{-}MU$.

For a formula $F = \{f\} + H$ let z be a new variable. Then we define

$$\xi(F, f, z) = z \vee_{cl} H + \{f\} + \neg z \vee_{cl} \rho(f),$$

where $L \vee_{cl} \{g_1, \cdots, g_m\}$ denotes the formula $\{L \vee g_1, \cdots, L \vee g_m\}$

Lemma 4.1. (Hans Kleine Büning, Xishun Zhao [20])

(1) $F \in MU$ if and only if $\xi(F, f, z) \in MU$, and f is maximal in $\xi(F, f, z)$.
(2) If $F \in MU$, then any clause of $\neg z \vee_{cl} \rho(f)$ is maximal in $\xi(F, f, z)$.
(3) If $F \in MU$, then $g \in H$ is maximal in F if and only if $z \vee g$ is maximal in $\xi(F, f, z)$.

Now run the following procedure.
    Procedure MU–MAX
Input: A formula F in CNF
Output: A formula $\delta(F)$ in CNF
    begin
    $C :=$ the set of clauses in F
    while C is non-empty

for a clause f in C; for a new variable z

$F := \xi(F, f, z)$

$C := z \vee_{cl} (C - \{f\})$

end while

$\delta(F) := F$

end

The procedure requires not more than $O((mn)^3)$ steps, where m is the number of clauses in F, n the number of variables of F.

Now from the above lemma we can see that $F \in MU$ if and only if $\delta(F) \in MAX\text{-}MU$.

Theorem 4.1. (Hans Kleine Büning, Xishun Zhao [20]) *MAX–MU is D^P–complete.*

5. Marginal MU Formulas

A *MU–formula* F is called *marginal* if, and only if removing an arbitrary occurrence of a literal from F leads to a unsatisfiable formula which is not in *MU*. The class of all marginal formulas is denoted as *MARG–MU*.

Obviously, the class *MARG–MU* is in D^P. We will show the D^P–hardness by a reduction from the D^P–complete problem *MU* [30]. We establish a procedure running in polynomial time generating a formula $\sigma(F)$ from a formula F, such that $F \in MU$ if and only if $\sigma(F) \in MARG\text{-}MU$. The procedure is based on an iterative application of the following function ζ.

Let $F = \{L \vee f, L \vee g\} + H$ be a formula with at least two occurrences of the literal L. For new variables y and z we define

$$\zeta(F, L \vee f, L \vee g, y, z) = \{y \vee f, z \vee g, \neg y \vee z, y \vee \neg z, \neg y \vee \neg z \vee L\} + H.$$

The formula describes the equivalence of y and z, the two occurrences of L are replaced by one occurrence, and $\zeta(F, L \vee f, L \vee g, y, z) \models F$. For short we write $\zeta(F)$.

For a formula $F \in MU$ and a literal L we say F is **marginal w.r.t. the literal** L if removing any occurrence of L from F results in a unsatisfiable formula which is not in *MU*. Clearly, F is marginal if and only if F is marginal w.r.t. all literals.

Lemma 5.1. (Hans Kleine Büning, Xishun Zhao [20])

(1) *$F \in MU$ if and only if $\zeta(F) \in MU$*

(2) *For $F \in MU$, $\zeta(F)$ is marginal w.r.t. the new literals $y, \neg y, z, \neg z$.*

(3) *For $F \in MU$, if F is marginal w.r.t. a literal K different from L, then $\zeta(F)$ is marginal w.r.t the literal K. That is ζ preserves the marginality.*

Now we introduce the above mentioned procedure.

Procedure MU–MARG

Input: A formula F in CNF

Output: A formula $\sigma(F)$ in CNF.

> **begin**
>> $\mathcal{L}$:=the set of literals occurring at least twice in F
>> **while** $\mathcal{L}$ is non-empty
>>> for some $L \in \mathcal{L}$
>>> for two clauses $L \vee f, L \vee g \in F$; for new variables y, z
>>> $F := \zeta(F, L \vee f, L \vee g, y, z)$
>>> remove from $\mathcal{L}$ literals occurring in F exactly once
>> **end while**
>> $\sigma(F) := F$
> **end**

The running time of the procedure MU–MARG is bound by a polynomial depending on the length of F, because within the **while**-loop a double occurrence of a literal L is replaced by one occurrence. Please note, that any literal of the input formula occurs exactly once in $\sigma(F)$.

By an iterative application of the above lemma, we see that $F \in MU$ if and only if $\sigma(F) \in MU$.

Now it remains to show that for a formula $F \in MU$ the formula $\sigma(F)$ is marginal, that means marginal w.r.t any literal. Since a literal $L \in lit(F)$ occurs in $\sigma(F)$ exactly once, $\sigma(F)$ is marginal w.r.t L. By means of the above lemma and, we see that $\sigma(F)$ is marginal w.r.t. the introduced literals. Thus, $\sigma(F)$ is marginal w.r.t. any literal and therefore marginal. Therefore we have the following theorem.

Theorem 5.1. (Hans Kleine Büning, Xishun Zhao [20]) *MARG–MU is D^p-complete.*

6. Unique MU Formulas

Another class of restrictions is based on a limited number of satisfying truth assignments. Beside the unsatisfiability, minimal unsatisfiable means that for any clause f the formula $F - \{f\}$ is satisfiable. If for any clause f, $F - \{f\}$ has exactly one satisfying truth assignment, that means $F - \{f\}$ is in $Unique\text{-}SAT$, then F is called uniquely minimal unsatisfiable. The class of these formulas is denoted as $Unique\text{-}MU$. At the first glance, to demand that for all clauses there is exactly one satisfying truth assignment seems to be very strong.

310

It has been proved that the problem *Unique–MU* is as hard as the *Unique–SAT*-problem and therefore probably not D^P–complete, because it is not known whether *Unique–SAT* is D^P–complete. It is strongly conjectured that *Unique–SAT* is neither D^P–complete nor in *NP* or *coNP*. A slight modification of *Unique–MU* is the class *Almost–Unique–MU* of almost unique minimal unsatisfiable formulas. A formula $F \in MU$ is in *Almost–Unique–MU* if for at most one clause f, $F-\{f\}$ may have more than one satisfying truth assignment. Under the assumption that *Unique–SAT* is not D^P–complete, *Almost–Unique–MU* is harder than *Unique–MU*, because we have shown the D^P–completeness of *Almost–Unique–MU*.

Theorem 6.1. (Hans Kleine Büning, Xishun Zhao [20])

(1) *Unique–MU $\equiv_p$ Unique–SAT, that is, the unique minimal unsatisfiability problem is as hard as the unique satisfiability problem with respect to the polynomial reduction.*

(2) *Almost–Unique–MU is D^p-complete.*

Proof. We just present the reductions we need.

1. We first define a polynomial time reduction θ from *Unique–MU* to *Unique–SAT* such that $F \in$ *Unique–MU* if and only if $\theta(F)$ is in *Unique–SAT*. In order to simplify the construction we demand that any literal occurs negatively and positively in the formula. If this is not the case then obviously the formula F is not in *MU* and therefore not in *Unique–MU*. For $F = \{f_1, \cdots, f_m\}$ we define

$$\theta(F) := ((F - \{f_1\}) + \{\overline{f_1}\}) \wedge \bigwedge_{1 \leq i \leq m} (F - \{f_i\})^{i+1}.$$

$(F - \{f_i\})^{i+1}$ is the formula we obtain by renaming the variables of the formulas $(F - \{f_i\})$, such that the formulas $(F - \{f_j\})^{j+1}$ $(1 \leq j \leq m)$ and $((F - \{f_1\}) + \{\overline{f_1}\})$ have pairwise different variables. $\overline{f_1}$ is the conjunction of the negated literals of f_1.

The reduction from *Unique–SAT* to *Unique–MU* will be very complicated. At first we introduce the transformation $\omega(F)$, which will be used later on as a basis for our desired reduction. Let $F = \{f_1, f_2, \cdots, f_m\}$ be a $3CNF$ formula over variables $\{x_1, x_2, \cdots, x_n\}$ with clauses $f_i = L_{i1} \vee L_{i2} \vee L_{i3}$. We introduce new variables $\{y_1, y_2, \cdots, y_m\}$. π_i $(1 \leq i \leq m)$ denotes the clause

$$y_1 \vee \cdots \vee y_{i-1} \vee y_{i+1} \vee \cdots \vee y_m.$$

$\omega(F)$ is the conjunction of the following groups of clauses:

(A) The clauses $\quad f_1 \vee \pi_1, \; f_2 \vee \pi_2, \; \cdots, \; f_m \vee \pi_m$

(B) The clauses

$$\neg L_{11} \vee \pi_1 \vee \neg y_1, \; \neg L_{21} \vee \pi_2 \vee \neg y_2, \; \cdots, \; \neg L_{m1} \vee \pi_m \vee \neg y_m$$
$$\neg L_{12} \vee \pi_1 \vee \neg y_1, \; \neg L_{22} \vee \pi_2 \vee \neg y_2, \; \cdots, \; \neg L_{m2} \vee \pi_m \vee \neg y_m$$
$$\neg L_{13} \vee \pi_1 \vee \neg y_1, \; \neg L_{23} \vee \pi_2 \vee \neg y_2, \; \cdots, \; \neg L_{m3} \vee \pi_m \vee \neg y_m$$

(C) The clauses $\quad \neg y_i \vee \neg y_j \quad (1 \leq i < j \leq m)$

(D) The clause $\quad y_1 \vee y_2 \vee \cdots \vee y_m$

It is not hard to see that F is satisfiable if and only if $\omega(F)$ is minimal unsatisfiable. However, $\omega(F)$ is not necessarily in *Unique–MU* even if F is uniquely satisfiable. This is because the resulting formula after deleting a clause in group (B) may have multiple satisfying truth assignments.

For each $f_i \in F$, χ_i is the disjunction of all literals $\neg x$, where $x \in var(F) - var(f_i)$, and χ denotes the disjunction of all literals $\neg x$, where $x \in var(F)$. $\Omega(F)$ is the formula consisting of the following groups of clauses:

(A') For each clause $(f_i \vee \pi_i) \in \omega(F)$:

$$f_i \vee \pi_i \vee \chi_i, \quad f_i \vee \pi_i \vee x \quad \text{for all } x \in var(F) - var(f_i)$$

(B') For each clause $(\neg L_{ik} \vee \pi_i \vee \neg y_i) \in \omega(F)$:

$$\neg L_{ik} \vee \pi_i \vee \neg y_i \vee \chi_i, \quad \neg L_{ik} \vee \pi_i \vee \neg y_i \vee x \quad \text{for all } x \in var(F) - var(f_i)$$

(C') For each clause $(\neg y_i \vee \neg y_j) \in \omega(F)$:

$$\neg y_i \vee \neg y_j \vee \chi, \quad \neg y_i \vee \neg y_j \vee x \quad \text{for all } x \in var(F)$$

(D') The clause $y_1 \vee y_2 \vee \cdots \vee y_m$

Then we can show that F is uniquely satisfiable if and only if $\Omega(F) \in$ *Unique–MU*.

2. The membership in D^P is easy. For the hardness we recall the D^P-complete problem *SAT-UNSAT* of determining for a given pair of formulas one is satisfiable and the other is not [30]. Next we define a reduction from *SAT-UNSAT* to *Almost–Unique–MU*. For a pair of formulas F_1, F_2, We can also assume that $\Omega(F_1)$ and $\Lambda(F_2) := \Omega(F_2) - \{y_1 \vee \cdots \vee y_m\}$ have different variables. Let h_1 be a clause in $\Omega(F_1)$ such that $\Omega(F_1) - \{h_1\} \in$ *Unique–SAT* (from our construction we can easily find such a clause). For a fixed clause $h_2 \in \Lambda(F_2)$ we define

$$G := (\Omega(F_1) - \{h_1\}) + \{h_1 \vee h_2\} + (\Lambda(F_2) - \{h_2\}).$$

We can show that $F_1 \in SAT$ and $F_2 \in UNSAT$ if and only if $G \in$ *Almost–Unique–MU*.

For a technical reason, in the above construction we assume F_2 contains at least six negative clauses whose variables are distinct. Otherwise, we extend F_2 for new variables $x_1, \cdots, x_{18}$ to the formula

$$F_2 + \{\neg x_1 \vee \neg x_2 \vee \neg x_3, \cdots, \neg x_{16} \vee \neg x_{17} \vee \neg x_{18}\},$$

which has the same satisfiability with F_2. $\qquad\qquad\square$

7. MU Formulas with Disjunctive Splitting

In order to characterize and to analyze minimal unsatisfiable formulas, we can split formulas in MU into two minimal unsatisfiable formulas. For a variable x we remove the clauses with literal $\neg x$ (set $\neg x = 1$) resp. x (set $x = 1$). In the remaining clauses we delete the occurrences of the literal x resp. $\neg x$. The formulas are unsatisfiable and contain therefore minimal unsatisfiable subformulas, say F_x and $F_{\neg x}$.

More precisely, given a minimal unsatisfiable formula F and a variable $x \in var(F)$, F can be represented as the following form.

$$F = \{(x \vee g_1), \cdots, (x \vee g_r)\} + B_x + C + B_{\neg x} + \{(\neg x \vee f_1), \cdots, (\neg x \vee f_q)\},$$

such that formulas $\{g_1, \cdots, g_r\} + B_x + C$, denoted as F_x, and $C + B_{\neg x} + \{f_1, \cdots, f_q\}$, denoted as $F_{\neg x}$, are minimal unsatisfiable. Where B_x, C, $B_{\neg x}$ are pairwise disjoint and contains no occurrence of x or $\neg x$. We call $(F_x, F_{\neg x})$ a splitting of F on x, and accordingly, F_x, $F_{\neg x}$ are called splitting formulas.

Generally speaking, splitting formulas F_x and $F_{\neg x}$ have common clauses, that is, C is non-empty. Whenever C is empty we call $(F_x, F_{\neg x})$ a disjunctive splitting of F on x.

A more detailed analysis of the class *Unique–SAT* leads to class *Dis–MU*. A minimal unsatisfiable formula F is in *Dis–MU* if and only if F has a disjunctive splitting on any variable. That means, for any variable x of F, F can be split into two disjoint subformulas in *MU*. *Dis–MU* is of interest, because *Dis–MU* is a proper subclass of *Unique–MU* and its close relation to tree–like decision procedures [19]. For the polynomial-time reduction Ω we also can prove that $F \in$ *Unique-SAT* if and and only if $\Omega(F) \in$ *Dis–MU*. Therefore, *Dis–MU* is at least as hard as *Unique–SAT*. We did not succeed in finding a reduction from a D^P–complete problem. But we conjecture that the problem *Dis–MU* is not D^P–complete.

Theorem 7.1. *(Hans Kleine Büning, Xishun Zhao [20]) Dis–MU is at least as hard as the unique satisfiability problem with respect to polynomial reduction.*

Open Question 3 Is *Dis–MU* D^P-complete?

8. MU Formulas Closed under Splitting

From the above results we see that the restrictions of maximality, marginality, and disjunctive splitting, etc. can not reduce the complexity greatly. One reasoning is probably that these features are not closed under splitting. Take maximality as example, suppose F is a maximal *MU* formula and $(F_x, F_{\neg x})$ a splitting of F on x, then the splitting formulas are not necessarily maximal.

Example 1: The following example is a maximal formula with a non-maximal splitting formula: (split on x)

$$F = \begin{pmatrix} x & x & x & \neg x & \neg x & \neg x & \neg x \\ \neg a & a & \neg a & a & \neg a & \neg a & a \\ \neg b & \neg b & b & b & \neg b & b & \neg b \\ & & & \neg c & \neg c & \neg c & c \end{pmatrix}, \quad F_{\neg x} = \begin{pmatrix} a & \neg a & \neg a & a \\ b & \neg b & b & \neg b \\ & \neg c & \neg c & \neg c & c \end{pmatrix}$$

The formula $F_{\neg x}$ is not maximal, because we can add the literal $\neg c$ to the first clause of the formula.

Example 2: Let

$$F := \begin{pmatrix} a & a & a & \neg a & \neg a & \neg a \\ b & c & \neg b & \neg b & \neg c & b \\ & & \neg c & & & c \end{pmatrix}$$

Splitting on a leads to $F_a = b \wedge c \wedge (\neg b \vee \neg c)$ and $F_{\neg a} = \neg b \wedge \neg c \wedge (b \vee c)$.
Splitting on b leads to $F_b = a \wedge (\neg a \vee c) \wedge (\neg a \vee \neg c)$ and $F_{\neg b} = \neg a \wedge (a \vee \neg c) \wedge (a \vee c)$.
Splitting on c leads to $F_c = a \wedge (\neg a \vee b) \wedge (\neg a \vee \neg b)$ and $F_{\neg c} = \neg a \wedge (a \vee \neg b) \wedge (a \vee b)$.
Thus, F is in *Dis–MU*, but a splitting of F_b on c leads to a non–disjunctive splitting.

Definition 8.1. Let $K \subseteq MU$ be any non-empty class of *MU* formulas. Then we define

$$K^* = \{F \in K \mid \forall x \in var(F) : \forall \text{ splitting } (F_x, F_{\neg x}) : F_x, F_{\neg x} \in K^* \cup \{\{\sqcup\}\}\}$$

The classes K^* is the largest subclass of K closed under splitting.

Let $F = \{f_1, \cdots, f_r\}$ be a formula not necessarily minimal unsatisfiable. Two clauses f_i and f_j $(i \neq j)$ hit each other, if there is some literal L with $L \in f_i$ and $\neg L \in f_j$. That means f_i and f_j can be resolved over L. The literal L is called a hitting literal. We say F is a hitting formula if any two different clauses of F hit each other. We use HIT to denote the class of all hitting formulas. Please notice that any unsatisfiable hitting formula must be minimal unsatisfiable. $HIT\text{-}MU$ is the class of all unsatisfiable hitting formulas.

By a result of Iwama [12], we know that for a hitting formula F,

$$F \text{ is unsatisfiable if and only if } \Sigma_{f \in F} 2^{-|f|} = 1,$$

here $|f|$ is the number of literals in F. Therefore, $HIT\text{-}MU$ can be solved in polynomial time.

Theorem 8.1. (Hans Kleine Büning, Xishun Zhao [19]) $MAX\text{-}MU^* = HIT\text{-}MU$, and hence, $MAX\text{-}MU^*$ can be solved in polynomial time.

We have some examples which shown that the three classes $MARG\text{-}MU$, $Unique\text{-}MU$ and $Dis\text{-}MU$ are pairwise different.

The formula in Example 2 is in $Dis\text{-}MU$ but not marginal. That can be seen by removing the literal a in the third clause and by removing the literal $\neg a$ in the last clause. The resulting formula remains minimal unsatisfiable. Thus, $Dis\text{-}MU \not\subseteq MARG\text{-}MU$.

Example 3: $MARG\text{-}MU \not\subseteq D$ and $MARG\text{-}MU \not\subseteq Unique\text{-}MU$.
The following formula is in $MARG\text{-}MU$, but neither in $Dis\text{-}MU$, nor in $Unique\text{-}MU$.

$$\begin{pmatrix} x_1 & \neg x_3 & \neg x_2 & \neg x_1 & \neg x_1 \\ x_2 & x_1 & x_3 & x_2 & \neg x_2 \end{pmatrix}$$

Example 4: $Unique\text{-}MU \not\subseteq MARG\text{-}MU$ and $Unique\text{-}MU \not\subseteq Dis\text{-}MU$.
Let

$$F = \begin{pmatrix} \neg z & \neg z & \neg z & \neg z & \neg z & & & & & & z & z & z & z \\ x & x & \neg y & \neg y & \neg x & x & \neg x & y & \neg y & \neg y & y & \neg a & \neg a \\ \neg a & \neg a & \neg b & \neg b & y & a & b & a & b & \neg x & x & \neg b & \neg b \\ \neg b & b & \neg a & a & & & & & & & & \neg x & x \end{pmatrix}$$

F is not marginal (remove $\neg a$ from the first clause). It is not difficult to check that F is in $Unique\text{-}MU$ and F has two different splittings on z, hence $F \notin D$. Take for $F_{\neg z}$ the first 7 clauses and remove $\neg z$, and for $F'_{\neg z}$ take the first 5 clauses and the clauses $(y \vee a)$ and $(\neg y \vee b)$ and remove $\neg z$.

Proposition 8.1. (Hans Kleine Büning, Xishun Zhao [19]) $Dis\text{-}MU \subseteq Unique\text{-}MU$.

However, we have proved the following theorem.

Theorem 8.2. (Hans Kleine Büning, Xishun Zhao [19])
$MARG\text{-}MU^* = Unique\text{-}MU^*=Dis\text{-}MU^*$.

Some nice properties of $Dis\text{--}MU^*$ have been proved in [19], however, the following question is still open.

Open Question 4. Is the problem of determining whether $Dis\text{-}MU^*$ is solvable in polynomial time.

9. Formulas with Simple MU–subformulas

Clearly, a *CNF* formula is unsatisfiable if and only if it has a minimal unsatisfiable subformula. Thus, the problem of determining whether a formula has *MU* subformula is coNP-complete. However, we are interested in the problem of determining whether a formula has a simple *MU* subformula.

To decide whether a formula F has a *Horn-MU* subformula, we just consider the subformula F' which consists of all Horn clauses of F. If F' is unsatisfiable then F must contains a Horn subformula in *MU*. Therefore, the problem can be solved in linear time.

The most interesting problem is to determine whether a formula contains a subformula in $MU(1)$, since $MU(1)$ formulas also have nice structure [6].

However, Hans Kleine Büning and Xishun Zhao defined a reduction which transform a formula F to a formula $\mathcal{M}(F)$ such that a satisfying truth assignment of F corresponds to a $MU(1)$ subformula of $\mathcal{M}(F)$. Therefore,

Theorem 9.1. (Hans Kleine Büning, Xishun Zhao [21])
The problem of determining whether a CNF formula has a subformula in $MU(1)$ is NP-complete.

In fact the above theorem is true when replace $MU(1)$ by $MU(k)$ for any fixed $k \geq 1$ (see also [21]).

10. Homomorphisms Between MU Formulas

Let H, F be formulas in *CNF* and $\phi : Lit(H) \to Lit(F)$ a map. We call ϕ a homomorphism from H to F if

(1) $\phi(\neg L) = \neg\phi(L)$ for every literal $L \in Lit(H)$, and

(2) $\phi(C) \in F$ for every clause $C \in H$

where $\phi(C) := \{\phi(L) \mid L \in C\}$. We simply write $\phi : H \to F$ if ϕ is a homomorphism from H to F.

The notion of homomorphism is of interest because homomorphisms preserve unsatisfiability. That is, if $\phi : H \to F$ is a homomorphism, and if H is unsatisfiable, then F is unsatisfiable.

A interesting problem is whether a tractable class M of unsatisfiable formulas is **homomorphically complete**, i.e., for any unsatisfiable formula F there is a formula H in M such that H is homomorphic to F. If M is homomorphically complete, then one can prove the unsatisfiability by establishing a homomorphism from a formula in M.

Theorem 10.1.
(1) (Stefen Szeider [31]) $MU(1)$ *is homomorphically complete.*
(2) (Hans Kleine Büning, Xishun Zhao [22]) *For any fixed k, $MU(k)$ is homomorphically complete.*

However, to decide whether a formula H is homomorphic to a formula F is a hard problem even H and F are very simple.

Theorem 10.2. (Hans Kleine Büning, Daoyun Xu [18])
For formulas $H, F \in MU(1)$, the problem of determining whether there is homomorphism from H to F is NP-complete.

However, if $H, F \in HIT\text{-}MU(1)$ then the problem becomes tractable.

11. Generalizations

11.1. *Clause-Minimal Formulas*

Inspired by minimal unsatisfiability, Hans Kleine Büning and Xishun Zhao [24] proposed the notion of clause-minimal formulas. A formula F in *CNF* is said to be **clause-minimal** if for any clause f in F, $F - \{f\}$ is not equivalent to F, that is, F has no equivalent proper subformula. *CL-MIN* is the class of all clause-minimal formulas.

Please notice that a unsatisfiable formula is clause-minimal if and only if it is minimal unsatisfiable. Thus, the notion of clause-minimality is a generalization to minimal unsatisfiability.

CL-MIN is known to be NP-complete [30]. Unlike $MU(k)$ which is tractable, *CL-MIN(k)*, the class of *CL-MIN* formulas with deficiency k, is still NP-complete [24]. The main reason is that clause-minimal formulas may be not stable.

11.2. *Lean Formulas*

Generally, a unsatisfiable formula may have several minimal unsatisfiable subformulas, some of which may be very simple and some of which may be complex. Then there probably exists a subformula $F' \subseteq F$ such that the unsatisfiability decision is harder for F' than for F. Oliver Kullmann introduced the notion of lean formulas [28].

A **lean formula** F is characterized by the condition that every clause of F can be used in some (tree) resolution refutation of F [28]. For every clause-set F there is a largest lean sub-clause-set $Na(F) \subseteq F$. By reducing F to the satisfiability equivalent formula $Na(F)$ (instead of some minimally unsatisfiable formula) we have overcome the above problem by eliminating only absolutely superfluous clauses).

Please notice that every minimal unsatisfiable formula is lean.

An equivalent characterization of lean clause-set is that they do not have any nontrivial autarky, where an autarky for a formula F is a partial assignment satisfying each clause of F it touches (a clause such that to at least one of its literals a truth value is assigned) [28]. The reader might observe here, that application of an autarky cannot render a satisfiable formula unsatisfiable, since application of any partial assignment is only dangerous (can destroy satisfiability) at clauses which are not satisfied but shortened (by setting some literals in them to false).

Oliver also studied some special autarkies, e.g., matching autarkies. A formula is said to be matching lean if it has no nontrivial matching autarky. Matching lean formulas are nothing but stable formulas, that is, a formula is matching lean if and only if the deficiency of any proper subformula $F' \subset F$ is less than the deficiency of F.

Theorem 11.1. (Oliver Kullmann [28])

(1) *The problem of deciding whether a formula is lean is coNP-complete.*
(2) *The problem of deciding whether a formula is matching lean is can be solved in polynomial time.*

11.3. *Minimal False Quantified Boolean Formulas*

Any $QCNF$ -formula Φ has the form $\Phi = Q_1 x_1 \cdots Q_n x_n \phi$, where $Q \in \{\exists, \forall\}$ and ϕ is a *CNF* –formula. Sometimes we use an abbreviation and write $\Phi = Q\phi$.

A quantified boolean formula $\Phi \in QCNF$ is termed **minimal false**, if Φ is false and after removing an arbitrary clause the resulting formula is true [7]. The set of minimal false formulas is denoted as MF .

Given a formula $\Phi \in QCNF$, the deficiency of Φ, denoted as $d(\Phi)$, is not the difference between the number of clauses and the number of variables, but the difference between the number of clauses and the number of existential variables. $MF(k)$ is the class of minimal false formulas with deficiency k.

Lemma 11.1. (Decheng Ding, Hans Kleine Büning, and Xishun Zhao [7])
If $\Phi \in MF$, then for any proper subformula Φ' of Φ, $d(\Phi') < d(\Phi)$.

Open Question 5. Is $MF(k)$ solvable in polynomial time for fixed $k \geq 1$?

References

1. R. Aharoni, N. Linial: Minimal Non–Two–Colorable Hypergraphs and Minimal Unsatisfiable Formulas. Journal of Combinatorial Theory **43** (1986), 196–204.
2. C.E. Alchourrón, D. Makinson, On the Logic of Theory Change: Safe Contraction, it Studia Logic, **44** (1985), 405–422.
3. R. Bruni: On exact selection of minimally unsatisfiable subformulae, *Annals of Mathematics and Artificial Intelligence*, **43** (2005), 35–50.
4. V. Chvatal, T. Szemeredi: Many hard Examples for Resolution. Journal of the Association for Computing Machinery **35** (1988) 759–768.
5. M. Davis, H. Putnam: A Computing Procedure for Quantification Theory. Journal of the Association for Computing Machinery **7** (1960) 201–215.
6. G. Davydov, I. Davydova, and H. Kleine Büning: An Efficient Algorithm for the Minimal Unsatisfiability Problem for a Subclass of CNF. *Annals of Mathematics and Artificial Intelligence*, **23** (1998) 229–245.
7. Decheng Ding, H. Kleine Buening, and Xishun Zhao: Minimal Falsity for QBF with Fixed Deficiency, In: *Proc. of International Conference on Quantified Boolean Formulas*, 21–28, Siena, Italy, 2001.
8. T. Eiter, G. Gottlob: Identifying the Minimal Traversals of A Hypergraph and Reslated Problems. *SIAM Journal of Computing*, **24** (6), p. 1278–1304,1995.
9. J. Franco, A. V. Gelder: A Perspective on Certain Polynomial Time Solvable Classes of Satisfiability, To appear in *Discrete Applied Mathematics,* 2000.
10. H. Fleischner, O. Kullmann, and S. Szeider: Polynomial-Time Recognition of Minimal Unsatisfiable Formulas with Fiexed Clause-Variable Difference, Electronic Colloquium on Computational Complexity, Report 49, 2000.
11. A. Haken: The intractability of Resolution. *Theoretical Computer Science*, **39** (1985), 297–308.
12. K. Iwama: CNF Satisfiability Test by Counting and Polynomial Average Time. *SIAM J. Comput.*, **18** (1989), 385–391
13. K. Iwama, E. Miyano: Intractability of Read–Once Resolution. In: *Proceedings Structure in Complexity Theory,* 10th Annual Conference, p. 29–36, IEEE, 1995.
14. H. Kleine Büning: On subclasses of minimal unsatisfiable formulas, *Discrete Applied Mathematics*, **107** (2000), 83–98.
15. H. Kleine Büning: An Upper Bound for Minimal Resolution Refutation, In: *Lecture Notes in Computer Science* (Eds: G. Gottlob, E. Grandjean and K. Seyr) Volume 1584, p. 171–178, Springer Verlag, 1999.
16. H. Kleine Büning, T. Lettmann: *Propositional Logic: Deduction and Algorithms,* Cambridge University Press, 1999.

17. H. Kleine Büning, K. Subramani, and Xishun Zhao: On Boolean Models for Quantified Boolean Horn Formulas, In *Lecture Notes in Computer Scence* **2919**, 93–104, Springer-Verlag, 2004.

18. H. Kleine Büning, Daoyun Xu: The Complexity of Homomorphisms and Renamings for Minimal Unsatisfiable Formulas, *Annals of Mathematics and Artifficial Intelligence*, **43** (2005), 113–127.

19. H. Kleine Büning, Xishun Zhao: On the Structure of Some Classes of Minimal Unsatisfiable formulas, *Discrete Applied Mathematics*, **130** (2003), No. 2, 185–207.

20. H. Kleine Büning, Xishun Zhao: The Complexity of Some Subclasses of Minimal Unsatisfiable Formulas, submitted for publication

21. H. Kleine Büning, Xishun Zhao: The Complexity of Read–Once Resolution, *Annals of Mathematics and Artificial Intelligence*, **36** (2002), 419–435.

22. H. Kleine Büning, Xishun Zhao: Polynomial Time Algorithms for Computing a Representation for Minimal Unsatisfiable Formulas with Fixed Deficiency, *Information Processing Letters*, **84** (2002), 147–151.

23. H. Kleine Büning, Xishun Zhao: Read–Once Unit Resolution, In *Lecture Notes in Computer Science* **2919**, 356–369, Springer-Verlag, 2004.

24. H. Kleine Buening, Xishun Zhao: Extension and Equivalence Problems for Clause Minimal Formulas, *Annals of Mathematics and Artifficial Intelligence*, **43** (2005), 295–306.

25. H. Kleine Buening, Xishun Zhao: Satisfiable Formulas Closed under Replacement, *Electronic Notes in Discrete Mathematics*, vol. 9, Elsever Sceience, 2002.

26. H. Kleine Buening, Xishun Zhao: Minimal Unsatisfiability — Results and Open Question. Technique Report, 2001.

27. O. Kullmann: An Application of Matroid Theory to the SAT Problem, Electronic Colloquium on Computational Complexity, Report 18, 2000

28. O. Kullmann: Lean-sets: Generalizations of minimal unsatisfiable clause-sets, *Discrete Applied Mathematics*, **130** (2003), 209–249.

29. S. H. Ma, D. M. Liang: A Polynomial–Time Algorithm for Reducing the Number of Variables in MAX–SAT Problem, *Science in China*, (Series, E), **40** (3), 301–311, 1997.

30. C. H. Papadimitriou, D. Wolfe: The Complexity of Facets Resolved. Journal of Computer and System Science, **37** (1988), 2–13.

31. S. Szeider: Homomorphisms of conjunctive normal form, *Discrete Appiled Mathematics*, **130** (2003), 351–365.

32. A. Urquhart: Hard Examples for Resolution. Journal of ACM, **34** (1987), 209–219.

33. Xishun Zhao: On Kullmann's Conjectures, manusccript. Also see http://cs-svr1.swan.ac.uk/ csoliver/Artikel/OpenProblemsConflicts.html

34. Xishun Zhao, Decheng Ding: Some Classes of Minimal Unsatisfiable Formulas, *SAT'1998*, Paderborn, 1998.

35. Xishun Zhao, Decheng Ding: Two tractable subclasses of Minimal Unsatisfiable Formulas, *Science in China*, (Series, A), **42** (7), 720–731, 1999.